Method Performance Studies for Speciation Analysis

Method Performance Studies for Speciation Analysis

Philippe Quevauviller

European Commission, Brussels, Belgium

Publication no. EUR 18348 EN of the European Commission
Dissemination of Scientific and Technical Knowledge Unit,
Directorate General Telecommunications, Information Market and Exploitation of Research, Luxembourg.

ISBN 0-85404-467-1

A catalogue record for this book is available from the British Library

Published by The Royal Society of Chemistry,
Thomas Graham House, Science Park, Milton Road,
Cambridge CB4 4WF, UK

For further information see our web site at www.rsc.org

Typeset by Computape (Pickering) Ltd, Pickering, North Yorkshire, UK

Printed by MPG Books Ltd, Bodmin, Cornwall, UK

Foreword

Recent years have shown an explosive growth in methods for the quantitation of the different chemical forms of trace elements in a variety of samples. Collectively these methods are termed speciation analysis. The reasons for this growth of interest are also clear. While analytical laboratories have available a wide range of techniques for determining total elemental content, such as atomic absorption spectrometry and inductively coupled plasma-atomic emission spectrometry or inductively coupled plasma-mass spectrometry, such total elemental information yields only a partial story. For example the toxicity of an element depends dramatically upon its chemical form as does its mobility in the environment or availability to living organisms. Speciation also gives vital clues as to the origins of samples and their history.

Accordingly many procedures have been proposed for speciation. Amongst the most popular are the so-called 'hybrid techniques' which couple separatory methods such as chromatography with detection methods such as optical or mass spectrometry. This has led to a proliferation of abbreviations, such as GC-MS, GC-AAS, GC-ICP-AES, GC-ICP-MS, HPLC-AAS, HPLC-ICP-MS, HPLC-ICP-AES, SFC-ICP-MS and CZE-ICP-MS, which read more like alphabet soup than science.

Before such procedures will be widely accepted in laboratories throughout the world, or the results from them given the credibility necessary to encourage legislators to include speciation requirements in regulations, confidence has to be built in these methods. Researchers in speciation have therefore to establish networks and conference infrastructures to exchange experiences and samples. Collaborative inter-laboratory trials have been held which have done much to increase awareness of the strengths and weaknesses of many of these techniques and to significantly improve the quality of analysis. Now we are beginning to see the availability of several new certified reference materials for use in speciation analysis resulting from these endeavours. Only the ready availability of such reference materials will facilitate the ready acceptance of speciation measurements as truly valid.

The Standards, Measurement and Testing Programme of the European Commission has contributed enormously to this development of awareness and programmes to improve quality of speciation measurements. One outstanding young scientist in the Commission has made a particular contribution – Dr Philippe Quevauviller. It is therefore a special privilege to write the Foreword for this highly important book.

Using several years experience, as a laboratory scientist pioneering speciation analyses and latterly as a scientist in DGXII of the European Union leading the campaign to improve the quality and credibility of speciation measurements, Dr Quevauviller has written a masterly text. No-one is better qualified to write a text critically evaluating the performance of different methods for speciation analysis. Carefully organised and authoritatively written, I am sure this book will be an essential reference for all those who perform or use speciation analysis.

Les Ebdon
Plymouth
July 1998

Acknowledgements

This book is the result of extensive collaborative research which would not have been possible without the dedicated work of many actors. I would firstly like to thank my "Scientific Godfather", Ben Griepink, who has taught me the importance of Reference Materials and their impact on the European Measurement Infrastructure. I also thank Herbert Muntau who has always been helpful in the design and management of new challenging projects (and speciation ones belong to this category!). This book results as well from numerous writings and discussions that I shared with Eddie Maier in the past nine years.

Each chapter of this book represents a sum of collaborations and the list would be too long if I should list all the names of the experts who have participated in the projects described and in workshop discussions. All the laboratories which have participated in the projects are listed in the respective chapters and I take this occasion to acknowledge again their collaboration; without their interest, competence and open mind, we would never have achieved the improvements which have placed European laboratories at the forefront of speciation analytical science. A special thank goes to the project coordinators and partners who are listed below:

Iver Drabæk, Umberto Fortunati, Marco Filippelli, Angelo Bortoli, Herbert Muntau for the project on methylmercury
Michel Astruc and his team, Les Ebdon, Wim Cofino, Freek Ariese, Roberto Morabito, Salvatore Chiavarini, Herbert Muntau for the project on tin speciation
Fred Adams, Wilfried Dirkx, Roy Harrison, Les Ebdon, Les Pitts, Steve Hill, Cristina Rivas for the project on trimethyllead
Maurice Leroy, Alain Lamotte, Micheline Ollé, Florence Lagarde for the project on arsenic speciation
Carmen Cámara, Gloria Cobo, Maria-Antonia Palacios, Riansares Muñoz, Olivier Donard for the project on selenium speciation
Rita Cornelis, Kristien Vercoutere, Louis Mees, Steen Dyg, Jytte Molin-Christensen, Kirsten Byrialsen for the project on chromium speciation
Ben Fairman, Alfredo Sanz Medel, Les Ebdon, Miguel Válcarcel for the feasibility study on aluminium speciation
Allan Ure, Gemma Rauret, Fermin López-Sánchez, Roser Rubio, Haidi

Fiedler, Enrique Barahona, Manuel Lachica, Alain Gomez, Geffrey Bacon, Catherine Davidson, Herbert Muntau for the project on single and sequential extraction

Contents

Preface v
by Professor Leslie Ebdon

Chapter 1 Introduction **1**
1.1 Need for Method Performance Evaluation 1
1.2 EC Initiatives Related to Measurement Quality 2
1.3 Improving the Quality of Speciation Analysis 4

Chapter 2 Speciation Analysis of Environmental Samples **6**
2.1 Definitions 7
2.2 Objectives 8
Research 8
Regulations 8
2.3 Sources of Errors in Speciation Analysis 11
Extraction 11
Derivatization 13
Separation 14
Final Detection 16
Calibration 17
Primary Calibrants and Internal Standards 17
Stability of Chemical Species in Solutions 18

Chapter 3 Method Performance Studies—Aims and Principles **20**
3.1 Overview of Quality Assurance (QA) Principles 20
General 20
Statistical Control 21
Comparison with Results of Other Methods 21
Use of Certified Reference Materials 22
Interlaboratory Studies 23
3.2 Improvement Schemes 24
Definitions 24
Organization 24
General Principles 24
Participants 26
Organizer 26

3.3 Reference Materials for Method Performance Studies 27
Requirements for the Preparation of Reference Materials 28
Preparation 29
Collection 29
Sample Treatment 30
Stabilization 30
Homogenization 30
Control of the Homogeneity 31
Control of the Stability 32
Storage and Transport 34
Procedures to Certify and Assign Values 35
Certification of Reference Materials 35
Assigned Values 37
Evaluation of Results 38
Collection of the Data 38
Description of the Methods 38
Technical Evaluation 39
Statistical Evaluation 39

Chapter 4 Mercury Speciation **41**
4.1 Aim of the Project and Coordination 41
Justification 41
The Programme and Timetable 41
Coordination 42
Interlaboratory Studies 42
Certification of Fish Reference Materials 42
Certification of Sediment Reference Material 42
Participating Laboratories 42
Interlaboratory Studies 42
Certification Campaigns 43
4.2 Techniques Used in Mercury Speciation 44
Methylmercury Determination in Fish 44
Ion Exchange/AAS 44
Westöö Extraction/GC-ECD 45
Toluene Extraction/GC-ECD 45
Acid Extraction/GC-AFS 46
Toluene Extraction/ETAAS 46
Toluene Extraction/GC-FTIR 46
Toluene Extraction/GC-MIP 46
Distillation/GLC-CVAFS 47
Methylmercury Determination in Sediment 47
Toluene Extraction/GC-ECD 47
Toluene Extraction/CGC-CVAAS 48
Microwave Digestion/GC-QFAAS 48
Distillation/GC-CVAFS 48
Distillation/GC-CVAAS 48

Distillation/HPLC-CVAAS 49
Distillation/HPLC-ICP-MS 49
Toluene Extraction/HPLC-CVAFS 50
Supercritical Fluid Extraction/GC-MIP 50
4.3 Interlaboratory Studies 51
First Interlaboratory Study 51
Samples 51
Results 51
Second Interlaboratory Study 52
Samples 52
Results 53
Third Interlaboratory Study 54
Samples 54
Results 54
4.4 Certification of Fish Reference Materials, CRMs 463-464 56
Preparation of the Candidate CRMs 56
Collection 56
Homogenization and Bottling 56
Homogeneity Control 57
Stability Control 57
Technical Evaluation 59
Certified Values 61
4.5 Interlaboratory Study on Sediment 61
Preparation of the Material 61
Characterization of the Bacterial Flora 61
Sample Collection and Treatment 62
Homogeneity and Stability Tests 62
Effects of γ-Irradiation on MeHg Content 62
Results 63
4.6 Certification of Sediment Reference Material, CRM 580 64
Preparation of the Candidate CRM 64
Collection 64
Homogenization and Bottling 65
Homogeneity Control 65
Stability Control 66
Technical Evaluation 66
Certified Values 68

Chapter 5 Tin Speciation **69**
5.1 Aim of the Project and Coordination 69
Justification 69
The Programme and Timetable 69
Coordination 70
Interlaboratory Studies 70
Certification of Sediment Reference Materials 70
Certification of Mussel Reference Material 70

Participating Laboratories 70
5.2 Techniques Used in Tin Speciation 71
Butyltin Determination in Sediment 71
Hydride Generation/GC-QFAAS 72
Hydride Generation/CGC-FPD 73
Ethylation/CGC-FPD 73
SFE/Ethylation/CGC-FPD 73
Pentylation/CGC-FPD 73
Pentylation/CGC-QFAAS 74
Pentylation/CGC-MS 74
HPLC/ICP-MS 75
Polarography 75
Butyl- and Phenyl-tin Determinations in Mussel 75
Hydride Generation/GC-QFAAS 77
Ethylation/CGC-AAS 77
Hydride Generation/GC-ICP-MS 77
Hydride Generation/CGC-FPD 78
Ethylation/CGC-FPD 78
SFE/Ethylation/CGC-FPD 79
Pentylation/CGC-FPD 80
Ethylation/CGC-MIP-AES 80
Ethylation/CGC-MS 81
Pentylation/CGC-MS 81
HPLC-ICP-AES 82
HPLC-ICP-MS 82
HPLC/Fluorimetry 82
Parallel Research Developments 83
5.3 Interlaboratory Studies 83
Interlaboratory Study on Organotin in Solutions 83
Interlaboratory Study on TBT in a TBT-spiked Sediment 84
5.4 Certification of Harbour Sediment, RM 424 85
Preparation 85
Collection and Preparation 85
Homogeneity Control 85
Stability Control 85
Technical Discussion 86
5.5 Certification of Coastal Sediment, CRM 462 87
Feasibility Study on Material Preparation 87
Preparation of the Candidate CRM 89
Collection 89
Homogenization and Bottling 89
Homogeneity Control 90
Stability Control 91
Technical Evaluation 91
Tributyltin 91
Dibutyltin 93

Monobutyltin 94
Certified Values 94
Re-certification 95
5.6 Certification of Mussel Tissue, CRM 477 95
Preparation of the Material 95
Collection 95
Freeze-drying Procedure 96
Homogeneity Control 96
Stability Control 98
Technical Discussion 99
Tributyltin 99
Dibutyltin 99
Monobutyltin 99
Triphenyltin 99
Monophenyltin and Diphenyltin 103
Certified Values 103

Chapter 6 Lead Speciation **104**
6.1 Aim of the Project and Coordination 104
Justification 104
The Programme and Timetable 105
Coordination 105
Feasibility Study 105
Interlaboratory Studies 105
Certification 105
Participating Laboratories 105
6.2 Techniques Used in Lead Speciation 106
Trimethyllead Determination in Artificial Rainwater 106
Hydride Generation/ETAAS 106
Ethylation/GC-QFAAS 106
Ethylation/CGC-QFAAS 106
Propylation/GC-QFAAS 107
Ethylation/CGC-MIP-AES 107
Propylation/CGC-ICP-MS 108
Pentylation/CGC-MS 108
Butylation/CGC-MS 108
Trimethyllead Determination in Urban Dust 109
Ethylation/GC-QFAAS 109
Propylation/GC-QFAAS 109
Ethylation/CGC-MIP-AES 110
SFE/Propylation/CGC-MS 110
Pentylation/CGC-MS 111
Butylation/CGC-MS 111
Development of HPLC-ID-ICP-MS 111
6.3 Feasibility Study 112
6.4 First Interlaboratory Study 112

Preparation and Verification of Calibrants 112
Preparation of Solutions 114
Results 114
6.5 Second Interlaboratory Study 115
Feasibility Study on Reference Material Preparation 115
Samples for the Interlaboratory Study 116
Results 116
Solution A: 10 Times Dilution of 50 μg L^{-1} Solution 117
Solution B: 10 Times Dilution of 5 μg L^{-1} Solution 118
Solution C: 100 Times Dilution of 5 μg L^{-1} Solution 119
Urban Dust 119
6.6 Trimethyllead in Rainwater 120
Preparation of the Candidate Reference Material 120
Homogeneity Control 121
Stability Control 122
Preparation of Calibrant 123
Technical Evaluation 124
Conclusions 124
6.7 Certification of Trimethyllead in Urban Dust, CRM 605 125
Preparation of the Material 125
Homogeneity Control 126
Stability Control 127
Technical Evaluation 127
Certified Value 128

Chapter 7 Arsenic Speciation 130
7.1 Aim of the Project and Coordination 130
Justification 130
The Programme and Timetable 130
Coordination 130
Participating Laboratories 131
7.2 Techniques Used in Arsenic Speciation 131
Determination of As species in Fish Tissue 131
UV Irradiation/Hydride Generation/ICP-AES 131
UV Irradiation/Hydride Generation/QFAAS 132
Gas Chromatography/Hydride Generation/QFAAS 132
Liquid Chromatography/Hydride Generation/ICP-AES 132
Liquid Chromatography/QFAAS 132
Liquid Chromatography/ICP-MS 132
7.3 Preparation of Pure Calibrants 133
7.4 Interlaboratory Studies 134
Aqueous Solutions 134
Solutions Mimicking Fish and Soil Extracts 134
Fish and Mussel Cleaned Extracts 135
Mussel Raw Extract and Biological (Shark and Mussel) Tissues 135
Conclusions 136

7.5 Certification of CRMs 626 and 627 136
Preparation of the Candidate CRMs 136
Tuna Fish Material 136
Arsenobetaine Solution 137
Evaluation of the Results 138
Certified Values 139

Chapter 8 Selenium Speciation **140**
8.1 Aim of the Project and Coordination 140
Justification 140
The Programme and Timetable 140
Coordination 140
Participating Laboratories 141
8.2 Techniques Used in Selenium Speciation 141
Determination of Selenite in Solution 141
Hydride Generation/AAS 141
Hydride Generation/AFS 142
Hydride Generation/ICP-MS 142
Ethylation/GC-MIP-AES 142
Microcolumn Preconcentration/ETAAS 142
Microcolumn Preconcentration/ICP-MS 142
HPLC-ICP-MS 142
Determination of Selenate in Solution 142
Hydride Generation/HPLC-AAS 143
HPLC-ICP-MS 143
8.3 Feasibility Study 143
8.4 Interlaboratory Study 145
Results of Selenite Determination 145
Results of Selenate Determination 146
Additional Remarks on the Interlaboratory Study 146
8.5 Tentative Certification 146
Preparation of the Candidate CRMs 146
Equipment and Cleaning Procedures 146
Homogenization and Bottling 147
Homogeneity Control 147
Stability Control 148
Technical Evaluation 148
Results for Selenite 148
Results for Total Inorganic Selenium 149
Results for Selenate 149
Additional Stability Checks 149
Conclusions 154

Chapter 9 Chromium Speciation **155**
9.1 Aim of the Project and Coordination 155
Justification 155

The Programme and Timetable 156
Coordination 156
Participating Laboratories 156
9.2 Feasibility Studies 156
9.3 Techniques Used in Chromium Speciation 157
Determination of Cr Species in Lyophilized Solution 157
Ion Chromatography with Spectrometric Detection 157
Ion Chromatography with Chemiluminescence 157
UV Digestion/DPCSV 157
Anion Exchange/ETAAS 158
Anion Exchange/IDMS 158
Microcolumn Preconcentration/ICP-MS 158
Determination of Cr(VI) in Welding Dust 158
Ion Chromatography with Spectrometric Detection 158
Anion Exchange/ETAAS 159
FAAS 159
ICPMS 159
Anion Exchange/IDMS 159
9.4 First Interlaboratory Study 160
Technical Evaluation 160
Solution A, Cr(VI) 160
Solution A, Cr(III) 161
Solution B, Cr(VI) 161
Filter 161
9.5 Certification of the Lyophilized Solution, CRM 544 162
Preparation of the Candidate Reference Material 162
Homogeneity Control 162
Stability Control 163
Technical Evaluation 164
Certified Values 165
9.6 Certification of Welding Dust, CRM 545 166
Preparation of the Candidate CRM 166
Homogeneity Control 168
Stability Control 169
Technical Discussion 169
Certified Values 169

Chapter 10 Aluminium Speciation 172
10.1 Aim of the Project and Coordination 172
Justification 172
Programme and Coordination 173
10.2 Techniques Used in Aluminium Speciation 173
8-Hydroxyquinoline Extraction Procedures 173
Driscoll Methods 173
Chelex-100 Based Methods 175
Fluoride Electrode Methods 175

HPLC Methods 176
^{27}Al NMR Studies 176
10.3 Method Validation 177
10.4 Future Perspectives 178

Chapter 11 Single and Sequential Extraction **181**
11.1 Aim of the Project and Coordination 181
Introduction 181
The Programme and Timetable 182
Coordination 182
Feasibility Study 182
Certification of Sewage Sludge-amended Soils 182
Certification of Calcareous Soil 182
Certification of Sediment 182
Participating Laboratories 183
11.2 Feasibility Study 183
Temporal Stability of Extractable Contents 184
Preliminary Trial 184
11.3 Sequential Extraction for Sediment Analysis 185
Design of the Sequential Extraction Procedure 185
Interlaboratory Trials 185
Sediment Samples Used in the Interlaboratory Studies 185
Techniques Used in the Intercomparisons 185
Technical Evaluation 186
Conclusions 190
11.4 Certification of Sediment, CRM 601 190
Preparation of the Candidate Reference Material 190
Homogeneity Control 191
Stability Control 193
Technical Evaluation 193
Certified Values 195
11.5 Single Extractions for Sludge-amended Soils, CRMs 483, 484 195
Interlaboratory Study 195
Preparation of the Candidate Reference Materials 197
Sewage Sludge-amended Soil 197
Terra Rossa Soil 199
Homogeneity Control 200
Stability Control 201
Technical Discussion 201
Certified Values 204
Indicative Values 205
11.6 Single Extractions for Calcareous Soil, CRM 600 206
Interlaboratory Study 206
Selected Reference Material 206
Selection of Extractants 207
Results and Discussion 210

Preparation of the Candidate Reference Material 212
Soil Characterization 213
Homogeneity Control 213
Stability Control 214
Technical Evaluation 216
Certified Values 217
11.7 Annexes: Sequential and Single Extraction Schemes 218
Sequential Extraction Procedure 218
Apparatus 218
Reagents 219
Method 219
EDTA Extraction Protocol 221
Acetic Acid Extraction Protocol 221
NH_4NO_3, $CaCl_2$ and $NaNO_3$ Extraction Protocols Used in the Certification (Indicative Values) 222
Extraction with 0.1 mol L^{-1} NH_4NO_3 222
Extraction with 0.01 mol L^{-1} $CaCl_2$ 221
Extraction with 0.1 mol L^{-1} $NaNO_3$ 224
DTPA Extraction Protocol 224

Chapter 12 European Network on Speciation **226**
12.1 Introduction 226
12.2 The Paradox of Speciation 226
12.3 The Trend 227
12.4 The Network "Speciation '21" 228
Aim of the Network 228
Participants in the Network 228
Work Programme 229
Programme and Dissemination Strategy 229
Expert Meetings on Environmental Issues 229
Expert Meetings on Food Issues 229
Expert Meetings on Occupational Health/Hygiene Issues 230
Mid-term Workshop 230
Review of the State of the Art and Transfer of Knowledge 230
Final Workshop and Publication 231
Additional Information 231

References **232**

Appendix Assessment Forms **240**

Subject Index **265**

Abbreviations

AAS	Atomic absorption spectrometry
ACSV	Adsorptive cathodic stripping voltammetry
AES	Atomic emission spectrometry
AFS	Atomic fluorescence spectrometry
APDC	Ammonium pyrrolidinethiocarbamate
ASTM	American Society for Testing & Materials
BCR	(European) Community Bureau of Reference
CEC	Commission of the European Community
CEN	Comité Européen de Normalisation
CENELEC	Comité Européen de Normalisation Electrotechnique
CGC	Capillary gas chromatography
CRM	Certified reference material
CV	Coefficient of variation
CVAAS	Cold vapour atomic absorption spectrometry
CVAFS	Cold vapour atomic fluorescence spectrometry
CZE	Capillary zone electrophoresis
DBT	Dibutyltin
DDTC	Diethyldithiocarbamate
DDW	Deionized distilled water
DL	Detection limit
DMA	Dimethylarsinic acid
DPASV	Differential pulse anodic stripping voltammetry
DPC	Diphenylcarbazide
DPhT	Diphenyltin
DPCSV	Differential pulse cathodic stripping voltammetry
DTPA	Diethylenetriaminepentaacetic acid
EC	European Commission/European Community
ECD	Electron capture detection
EDXRF	Energy dispersive X-ray fluorescence spectrometry
EN	European Norm
ETAAS	Electrothermal atomic absorption spectrometry

ETSI	European Telecommunications Standards Institute
EU	European Union
FAAS	Flame atomic absorption spectrometry
FIA	Flow injection analysis
FID	Flame ionization detection
FPD	Flame photometric detection
FTIR	Fourier transform infrared spectroscopy
GC	Gas chromatography
GLC	Gas-liquid chromatography
HG	Hydride generation
HGAAS	Hydride generation atomic absorption spectrometry
HICP	Hydride generation inductively coupled atomic emission spectrometry
HPLC	High performance liquid chromatography
HQS	8-Hydroxyquinoline-5-sulfonic acid
HRM	In-house reference material
IC	Ion chromatography
ICP	Inductively coupled plasma atomic emission spectrometry
ID	Isotope dilution
IDMS	Isotope dilution mass spectrometry
INAA	Instrumental neutron activation analysis
IRMM	Institute for Reference Materials & Measurements
ISO	International Standards Organization
LAES	Liquid anion exchange solution
LC	Liquid Chromatography
LRM	Laboratory reference material
LMW	Low molecular weight
MBT	Monobutyltin
MIBK	Methyl isobutyl ketone
MIP	Microwave-induced plasma atomic emission spectrometry
MMA	Monomethylarsonic acid
MPhT	Monophenyltin
MPN	Most probable number
MS	Mass spectrometry
NAA	Neutron activation analysis
NIOSH	National Institute of Occupational Health & Safety
NIST	National Institute of Standards & Technology
NRCC	National Research Council of Canada

OEL	Occupational exposure limits
OES	Optical emission spectrometry
PCV	Pyrocatechol violet
QA	Quality Assurance
QC	Quality Control
QFAAS	Quartz furnace atomic absorption spectrometry
RM	Reference Material
RNAA	Neutron activation analysis into radiochemical separation
RSD	Relative standard deviation
RTD	Research & Technological Development
SD	Standard deviation
SFE	Supercritical fluid extraction
SME	Small & Medium Enterprise
SM&T	Standards, Materials & Testing
SOP	Standard operating procedure
SPDC	Sodium pyrrolidinedithiocarbamate
SRM	Standard reference material
TBT	Tributyltin
TEA	Triethanolamine
TISAB	Total ionic strength buffer
TMAH	Tetramethylammonium hydroxide
TPhT	Triphenyltin
TriEL	Triethyllead
TriML	Trimethyllead
XRF	X-ray fluorescence
ZETAAS	Zeeman electrothermal atomic absorption spectrometry

CHAPTER 1

Introduction

Sound, accurate and reliable measurements, be they physical, chemical or biological in nature, are essential to the functioning of modern society. Without them, industries, particularly high technology ones, cannot operate, trade is impaired by disputes, healthcare becomes empirical and legislation, ranging from environmental and worker protection to the operation of the Common Agricultural Policy and the Single Market, cannot be successfully implemented. For these reasons, advanced nations spend up to 6% of their GNP on measurements and measurement related operations.

Measurements hence affect the daily lives of every European citizen. Often the results of measurements or chemical analyses are taken for granted, *e.g.* in our direct contact with measuring devices when we buy food or consume gas and electricity at home. The importance of accurate measurements creates particular concern in specific cases, *e.g.* when food is tested to check whether it has been contaminated by poisonous substances or when blood is analysed as part of a hospital check-up.

It is precisely because measurements affect everyone that regulations (either national or European) are established in order to ensure that the measurements and chemical analyses are performed in a reliable way and therefore that consumer interests are properly protected. The need for harmonization of measurement systems has been recognized – some of them centuries ago – *e.g.* the verification of weights and measures to ensure fair trades and the adoption of the metric system now known as "Système International" (SI).

A considerable number of measurements are, of course, not directly evident to the general public. These concern the quality of products which determine, for example, the prices of food and/or feedstuffs; this quality is not open to bargaining but must be measured with the same accuracy and reliability in every country so that arguments about a product's acceptability are avoided and hence that a proper functioning of the Single Market may be ensured.

1.1 Need for Method Performance Evaluation

The harmonization of measurements and technical specifications is a continuous process, and is achieved either by means of Community Directives or the establishment of European Norms. However, this does not solve all the problems. Indeed, the measurements and analyses required for the implementation of these Norms and Directives are sometimes so difficult that, even when applying the same method, laboratories may still find very different results. It is obvious that such a disagreement between laboratories does not allow the Norms and Directives to be respected and therefore these have no harmonization effects. As a consequence, measures to evaluate and guarantee the quality of a laboratory's performance were established involving quality assurance rules and guidelines (*e.g.* Good Laboratory Practice, ISO 9000 and EN 45000 standard series), accreditation systems and the production of certified reference materials (CRMs).

Quality issues related to analytical measurements have been described in full detail in several books [1–3] and highlighted several principles, *e.g.* validation of methods, quality assurance protocols incorporating the use of CRMs, independent assessment of method performance by participation in proficiency testing schemes, and accreditation. Method performance studies also represent a very important aspect to evaluate the state-of-the-art of a particular type of analysis at the development stage (*e.g.* for testing the applicability of a standard method) or to improve the quality of measurements, *e.g.* prior to the certification of reference materials.

1.2 EC Initiatives Related to Measurement Quality

In order to eliminate disputes arising from doubtful measurements, the Commission of the European Communities established the Community Bureau of Reference (BCR) about 25 years ago to encourage and support technical collaboration between the laboratories of EC Member States. In this way, the Community helped laboratories in the Member States to provide accurate and reliable measurements in those sectors which are vital to the Community as a whole: trade, agriculture, food, industrial products, environment, health and consumer protection [4].

This collaborative effort on measurements was substantially increased within the second Framework Programme (Applied Metrology and Chemical Analysis, 1987–1992). It was likely to expand since the Community had embarked on an ambitious programme to unify its internal market. Major efforts were indeed required to harmonize a wider range of technical standards and measurements throughout the Community so that companies could be sure they were competing on equal terms in each Member State. In this context, it became essential that the accuracy of results be proven wherever the measurements or analyses were performed. The establishment of laboratory networks was a successful tool for the improvement of the quality of a wide variety of measurements performed in Europe. In turn, these collaborative efforts facilitated European cohesion.

To pursue this action within the Third Framework Programme (1990–1994), the European Community has implemented the Measurements and Testing programme which, by addressing the issues highlighted above, aimed to contribute to the harmonization and improvement of methods of measurement and analysis when these methods were not sufficiently accurate and laboratories obtained differing results. Through this harmonization, the programme aimed to contribute to the ease of circulation of agricultural and industrial products in the Community, to improvements in the means of monitoring environment and health and to the resolution of the new challenges faced by industry. The aims of the Measurements and Testing programme were also to improve the competitive position of European industry by promoting industrial innovation, to support pre-normative research and other technical support necessary for the development and application policies (Internal Market, environment, agriculture, health, *etc.*, and support to activities of CEN, CENELEC, ETSI, *etc.*) and to support the further development of the measurement infrastructure of Europe (coordination of national activities, promotion of good measurement practices throughout Europe, *etc.*) [5].

The Measurements and Testing Programme has developed into a wider programme called Standards, Measurements and Testing (SM&T) within the Fourth Framework Programme (1994–1998). This programme aims, through research and technological developments, to improve the competitiveness of all sectors of European industry, to support the implementation of Community policy and to meet the needs of society [6]. The main targets are:

- **To improve** the competitive position of all sectors of European industry (including SMEs) by promoting better measurements at the research and development levels, better definition and control of the quality of products, more efficient in-process measurements and technical assistance to the mutual recognition of certificates in accordance with the Global Approach to Conformity Assessment
- **To promote** research and other technical support necessary for the development and implementation of other Community policies (Single Market, environment, agriculture, health, transport and protection of the Community's external frontiers)
- **To promote** research in support of the activities of CEN, CENELEC, ETSI and other European bodies which seek to maintain or establish quality standards *via* either new and existing written standards or codes of practice
- **To support** the further development of the European measurement infrastructure by facilitating the coordination of national activities, the development of measurement standards, of advanced methods and systems and the mutual recognition of results and accreditation systems
- **To promote** the dissemination and application of good measurement practice throughout Europe, particularly in the less favoured regions, for example, by the organization of training courses and by the establishment of networks.

From this description, it is obvious that the SM&T programme is mainly oriented towards support to European industry and Community legislation. The selection of projects is carried out through a system of time-limited calls for proposals which are regularly published in the Official Journal of the

European Communities. The types of projects generally funded are of four types:

- **Interlaboratory studies** carried out by consortia of European laboratories, of which the aim is improve the state-of-the-art of different types of measurements [7]. These projects are useful to detect possible sources of errors related to particular techniques, to create networks of laboratories within the European Union and to prepare groups of expert laboratories for the certification of reference materials
- **The certification of reference materials** is the traditional BCR activity [8] which is still continuing within the SM&T programme [9]. These collaborative projects enable the production of reference materials certified in a reliable manner, which are necessary for the verification of the accuracy of analytical methodologies in various sectors (*e.g.* environment, food and agriculture, biomedical)
- **Development of new methods and instrumentation** became one of the core activities of the SM&T programme. These developments concern innovative instruments necessary to improve the quality of measurements in particular sectors (*e.g.* on-line measurement techniques, field-measurement methods *etc.*) [10]
- **Pre-normative research** is undertaken to test the feasibility of standards (*e.g.* through interlaboratory studies or method development) prior to their implementation by official normalization bodies (*e.g.* CEN, CENELEC, ISO). The same type of projects may be performed to test the requirements of draft EC Directives prior to their establishment [10].

1.3 Improving the Quality of Speciation Analysis

As illustrated by Chapter 2, speciation is one of the growing features of analytical chemistry of the 1990s. It is known that the determination of total trace element contents in environmental monitoring, toxicity studies, *etc.*, is no longer sufficient for the understanding of biogeochemical pathways of trace elements which depend on specific chemical forms (*e.g.* different oxidation states, organometallic compounds, *etc.*). Owing to the number of analyses performed by a wide range of EC laboratories, the SM&T programme has recognized the need to launch collaborative projects to establish and improve the state-of-the-art of speciation analysis in Europe. This book presents an overview of these projects which were undertaken over the past 10 years to improve the quality control of speciation analysis in various environmental matrices. The different chapters illustrate the aims of the programme and the above-mentioned activities. In particular, the results of all the interlaboratory studies which were carried out prior to certification campaigns of a series of chemical species (*e.g.* tributyltin, methylmercury) are described; the chapters also include a full description of the preparation of candidate reference materials (RMs) and of the certification results, as well as the development of new analytical techniques which were necessary in the course of certification (*e.g.* supercritical fluid extraction, isotope dilution mass spectrometry).

Chapter 2 discusses several aspects of speciation analysis, *e.g.* definitions, existing regulations and sources of errors likely to occur at various steps of the analytical procedures. It is followed by Chapter 3 describing general principles of improvement schemes (organizational aspects) and basic requirements to be followed for the certification of reference materials.

Specific chapters then focus on different projects on speciation analysis. Chapter 4 deals with interlaboratory studies on methylmercury in fish and sediment; Chapter 5 describes the collaborative projects to certify organotins in sediment RMs and mussel tissues; Chapter 6 gives an overview of the certification project on trimethyllead in simulated rainwater and urban dust; Chapter 7 describes the certification project on arsenic species in fish tissues; Chapter 8 focuses on the intercomparison and tentative certification of Se(IV) and Se(VI) in simulated freshwater; Chapter 9 deals with a feasibility study to stabilize Cr species in solution followed by the certification of Cr(III) and Cr(VI) in lyophilized solutions and welding dust; Chapter 10 gives a review of methods used for Al speciation; Chapter 11 develops the overall collaborative project to standardize single and sequential extraction procedures for soil and sediment analysis, followed by interlaboratory studies and certification of soil and sediment reference materials.

CHAPTER 2

Speciation Analysis of Environmental Samples

The number of determinations of chemical species in environmental matrices carried out in routine and research laboratories has increased considerably in the last few years. However, the quality of the results has often been neglected in environmental, food and biomedical analyses. Good reproducibility of an analysis is not sufficient. It helps, of course, to make results comparable over a limited area or a limited period ("trend" monitoring), but for a full comparability of results over time and location, and thus for a solid and universal interpretation of the findings, accuracy is a must; this has been widely demonstrated in the literature (*e.g.* [8,13,14]). Too many scientists have stated that good reproducibility in time was sufficient to follow trends and demonstrate the effects of actions taken by authorities to improve the quality of the environment or food. Such statements overlook modelling applications, theory development, *etc.*, and ignore improvements in equipment and methodology.

To achieve not only good reproducibility but also good accuracy, various measures are necessary. It is clear that good Quality Control (QC) of speciation analyses has not yet been achieved. Typical examples illustrate the lack of accuracy that may occur in the determination of inorganic [14] and organic traces [15] in environmental matrices. These examples are by no means selected but occur quite commonly in many fields of analysis, including the determination of species in environmental matrices. When results differ so much, they are not trustworthy. Moreover, in the past, too many wrong but highly reproducible results have lead to misinterpretation of environmental processes.

In the past few years, the determination of chemical species of elements (*e.g.* As, Hg, Sn species) has become of increasing concern due to their high toxic impact (see a review from Craig [16]). Some of these compounds (*e.g.* methylmercury, tributyltin) are now included in the black list of compounds to be monitored in the marine environment according to an EEC Directive (amendment of the Directive 76/464/EEC). Consequently, a wide variety of analytical techniques have been developed recently and are described in the literature, *e.g.* for Sn speciation [17,18], As speciation [19], Hg speciation [20], Pb speciation [21] and Se speciation [22].

Analytical techniques used for the determination of chemical species are generally based on a succession of steps (*e.g.* extraction, separation, detection) which enhance the risks of errors. This chapter gives an overview of the different types of errors that may occur in speciation analyses.

2.1 Definitions

The term "speciation" is used for a wide variety of analyses, ranging from the determination of well defined "species", *e.g.* oxidation states of elements or organometallic compounds, to forms of elements which are operationally defined (*i.e.* related to an extraction procedure) and which are quoted as "bioavailable", "mobile", *etc.*, forms of elements. A recent definition has been given by Hetland *et al.* who described speciation as "a specific form (monoatomic or molecular) or configuration in which an element can occur, or a distinct group of atoms consistently present in different matrices" [23]; the official definition is presently in discussion within IUPAC. This definition tends to restrict the term "speciation" to well-identified chemical forms of elements. However, the use of this term is much wider, which creates some confusion among scientists. The confusion is even greater when legislation is approached since, at a certain level, nobody can explain what is the difference between "bioavailability", "bioaccumulation", "essentiality", "lethal effect", *etc.* It should now be time for the scientific community to use clearer terminology in relation to speciation to avoid any misunderstanding; this has been well understood by scientists working in the field of organic analysis who defined specific terms in relation to different compounds or families of compounds, *e.g.* in the case of chlorinated biphenyls or polyaromatic hydrocarbons. Therefore, a trend that should be adopted in speciation analysis would be to specify clearly the actual forms which are determined, *i.e.* not speaking any more about the speciation of a given element (unless referring to a particular type of analysis in general terms as used in organic analysis, *e.g.* mercury speciation and polychloroaromatic hydrocarbon or chlorinated biphenyl determinations, respectively). With reference to elements with different oxidation states, an example can be the determination of inorganic selenite and selenate. With regard to organometallic species, the compounds may be referred to as cations or with their counter-ions: an example is tributyltin which could be referred to as TBT^+ or with its respective anion (*e.g.* TBTCl, TBTAc), depending on the actual determination technique used. Concerning "extractable trace metals", the term speciation should not be applied since these operationally defined determinations (obtained from single and/or sequential extractions) define "groups" of trace elements without clear identification; these procedures represent a useful approach for environmental studies (in particular for soil and sediment) [24], but the comparability of the measurements is only possible provided that standardized protocols are used [25]. As soon as procedures have been accepted as a standardized method, the determinations should clearly refer to the actual measurements, *e.g.* EDTA-extractable or acetic acid-extractable trace elements, and not to unclear terms such as "bioavailable", "mobile", *etc.*, which is rather an interpretation of the measurements than the exact terminology of what is measured [26].

2.2 Objectives

Research

The need for the determination of individual chemical species occurs especially where these species are known to be very toxic to humans and biota, *e.g.* As(III), Cr(VI), Se(VI), methylmercury or tributyltin. Speciation is an indispensable tool for studies on the biogeochemistry of trace elements. Consequently, the increasing concern for an improved control of environmental contamination levels and of knowledge of trace element behaviour in the environment has led to the development of new analytical techniques for the determination of a wide variety of compounds, which is also reflected by a considerable increase in the number of determinations.

As already stressed, these techniques involve many analytical steps such as extraction, derivatization, separation and detection, which should be performed in such a way that decay of the unstable species does not occur. However, the control of the quality of measurements is often hampered by the lack of suitable reference materials for speciation analyses. Research is hence directed towards the development of new (if possible simple) analytical methods, the production of reference materials, and the monitoring of chemical species for various purposes (environmental risk assessment, toxicity studies, biogeochemical cycles of trace elements, *etc.*).

Regulations

The European Community actively participates in the preparation of international conventions on the environment and in their implementation. EU Member States are free to adopt national legislation in the absence of Community legislation, but where the Community has acted, Community legislation is supreme and binding on both past and future Member State actions. The European Community can adopt (i) non-binding recommendations and resolutions, (ii) regulations that are binding and directly applicable in all Member States, (iii) decisions that are directly binding on the persons to whom they are addressed, including Member States, and (iv) directives which must be implemented by the laws or regulations of the Member States. A series of principles and priorities regarding environmental protection were set out within the first five-year environmental action programme (1973–1977) adopted by the European Community; these principles are summarized below and remain valid in subsequent action programmes:

- Prevention is better than cure
- Environmental impacts should be taken into account at the earliest possible stage in decision making

- Exploitation of nature which causes significant damage to the ecological balance must be avoided
- Scientific knowledge should be improved to enable actions to be taken
- The cost of preventing and repairing environmental damage should be borne by the polluter
- Activities in one Member State should not cause deterioration of the environment in another
- Environmental policy in the Member States must take into account the interests of the developing countries
- The EC and Member States of the European Union should promote international and worldwide environmental protection through international organizations
- Environmental protection is everyone's responsibility; therefore education is necessary
- Environmental protection measures should be taken at the most "appropriate level", taking into account the type of pollution, the action needed, and the geographical zone to be protected. This is known as the "subsidiarity principle"
- National environmental programmes should be coordinated on the basis of a common long-term concept and national policies should be harmonized within the Community, not in isolation.

On the basis of these preliminary remarks, let us examine which, to date, are the regulations in which reference is made to chemical forms of elements.

The Council Decision 75/437/EEC on Marine Pollution from Land-based Sources approves on behalf of the Community the "Paris Convention", which aims at preventing pollution of the north-east Atlantic and Arctic Oceans, the North and Baltic Seas and part of the Mediterranean Sea from land-based sources. Three categories of polluting substances are listed in the Annex: a first group of substances (Annex A, Part I), which should be eliminated, includes persistent chemical groups such as organohalogen compounds, mercury, cadmium and persistent hydrocarbon compounds; a second group of substances (Annex A, Part II), of which the pollution should be limited, includes less persistent organic compounds and certain heavy metals. The chemical forms of the elements appear in this Part II and concern, in particular, organic compounds of phosphorus, silicon, and tin and substances which may form such compounds in the marine environment, excluding those which are biologically harmless, or which are rapidly converted in the sea into substances which are biologically harmless, and the following elements *and their compounds*: arsenic, chromium, copper, lead, nickel and zinc.

The Council Decision 77/585/EEC on the Mediterranean Sea concludes on behalf of the Community the "Barcelona Convention" for the protection of the Mediterranean Sea against pollution, and the Protocol for the prevention of pollution of the Mediterranean Sea by dumping from ships and aircraft. The dumping of wastes or matter listed in Annex I is prohibited; this includes organohalogen and organosilicon compounds, organophosphorus compounds, organotin compounds, mercury and cadmium and their compounds, persistent synthetic materials, crude oil and hydrocarbons. This Decision has been supplemented by a series of three other Decisions, namely 81/420/EEC, 83/101/EEC and 84/132/EEC, in which additional substances, "the dumping of which requiring special care", are found in the Annex II, namely arsenic, lead, copper, zinc, beryllium, chromium, nickel, vanadium, selenium, antimony *and their compounds.*

The Council Decision 77/586/EEC on the Rhine River concludes on behalf of the Community the "Berne Convention" for the protection of the Rhine against chemical pollution, implying that discharges of a range of toxic substances should be gradually eliminated (Annex I); these include organohalogen, organophosphorus and organotin compounds, carcinogens, mercury and cadmium compounds and persistent mineral oils and hydrocarbons. Annex II of this Decision lists other dangerous substances for which pollution should be reduced, namely a range of 20 metalloids and metal compounds, biocides and their derivatives not appearing in Annex I.

The Council Directive 80/68/EEC on groundwater pollution caused by certain dangerous substances also aims at preventing or limiting the direct or indirect introduction to the groundwater of families or groups of dangerous substances. List I substances includes organohalogen, organophosphorus and organotin compounds; carcinogenic, mutagenic or teratogenic substances; mercury and cadmium compounds; mineral oils and hydrocarbons; and cyanides. List II substances (for which general investigation and authorization procedures are requested prior to direct discharges, disposal or tipping) include 20 metalloids and metal compounds, biocides and their derivatives not appearing on List I.

Several remarks can be made on the basis of this legislation. Firstly, the mention of a range of elements *and their compounds* implies that the laboratories should, in principle, determine a wide variety of chemical forms of the elements for which knowledge is still very scarce both in terms of toxicity and analytical methodologies. This stresses the need for a great deal of effort by research organizations to develop new methodologies and perform toxicity tests of as many chemical forms as possible; the results of such investigations should enable the EC legislation to be refined according to specific compounds (an example is the case of organotin compounds: only trialkyltin compounds are recognized to be extremely toxic to shellfish, *e.g.* tributyltin or triphenyltin, mono- and dialkyltin forms being much less harmful; mentioning "organotins" in legislation implies, however, that all the compounds should theoretically be monitored, which in turn represents a heavy burden for laboratories, without sound scientific grounds). Secondly, the compliance to this legislation requires that the data produced are comparable from one laboratory to another, *i.e.* that the analytical results are of proven quality. The latter implies that CRMs should be available for a wide variety of environmental matrices and chemical forms of elements which, to date, as described below, is far from being the case.

2.3 Sources of Error in Speciation Analysis

The sources of error that are likely to occur in speciation analysis have been discussed in technical meetings resulting from various interlaboratory studies described in the following chapters. Discussions have also been conducted among experts during scientific workshops which were held in Arcachon (France) in 1990 [27] and Rome in 1994 [28,29]. The following considerations are taken from the summary of the round-tables conducted during these workshops [30] and from the literature [31].

Basic techniques for speciation analysis were developed in the early 1980s; they generally involve various analytical steps, *e.g.* extraction either with organic solvents (*e.g.* toluene, dichloromethane) or different acids (*e.g.* acetic or hydrochloric acid), derivatization procedures (*e.g.* hydride generation, Grignard reactions), separation (GC or HPLC), and detection by a wide variety of methods, *e.g.* atomic absorption spectrometry (AAS), mass spectrometry (MS), flame photometric detection (FPD), electron capture detection (ECD), *etc.* Each of these steps includes specific sources of error which have to be evaluated.

Extraction

Extraction of chemical species from samples is a very complicated matter in which two conflicting issues need to be combined: obtaining an adequate recovery on one side and preventing losses, especially destruction of the analyte, on the other. Basically, the extraction should be done in such a way that the analyte is separated from the interfering matrix without loss or contamination, without change of the speciation and together with the minimum of interferences.

A good assessment of quality assurance (QA) implies that the extraction recoveries are verified; this can be done by spiking a sample of similar composition as the sample analysed with a known content of the analyte of concern, leaving it to equilibrate and determining the analyte after extraction. The major drawback is that the spike is not always bound the same way as the naturally occurring compounds, *i.e.* good recovery of a spiked compound does not necessarily imply that a similar recovery will be obtained for incurred compounds; however, an extraction procedure which would not enable quantitative recovery of a spiked compound should be abandoned since it will certainly not be applicable to naturally bound compounds.

Recovery may be tested using the standard addition method. It is generally agreed that spiked compounds should be added in aqueous solution to previously wetted material, allowing sufficient time to reach equilibration (*e.g.* 24 h); freeze-dried materials have to be dried again before analysis [30].

Alternatively, and only if the extraction procedure does not change the matrix composition and appearance, the recovery experiment may be carried out on the previously extracted real sample by spiking, equilibration and extraction. However, the recovery assessment can often be overestimated and this risk should be faced; CRMs may again be a tool to ascertain accuracy; they are, however, only useful in cases where they contain incurred, and not spiked, species.

Spiking experiments are particularly important, although they are not without problems. In particular, standardized (validated) protocols are missing and procedures are, at present, not sufficiently tested for different compounds and matrices. However, although the present knowledge is not perfect, the use of spiking experiments helps to minimize errors. The technique could be validated with radiolabelled materials, providing that the radioactive tracer is part of the analyte in the sample (different compounds could have different behaviour and lead to wrong conclusions on the obtained recovery, *e.g.* mono-, di- and tributyltin display different behaviour [32] and, consequently, the recovery should be assessed independently for each compound as well as for the compounds together).

It is also necessary to find a compromise between a good recovery (sufficiently strong attack) and the preservation of the speciation, *e.g.* the use of too concentrated HCl for the extraction of biological material alters the methylmercury content, which was demonstrated by the use of a radiolabelled methylmercury [33].

Extraction effects are recognized to be different according to the types of species. They will be less for stable or inert compounds such as TBT, more important for stable or not inert species [*e.g.* Cr(VI)] and paramount for unstable or not inert species (*e.g.* Al species), which stresses the need to check the procedures carefully.

Clean-up of extracts, *e.g.* by using ion exchange or other chromatographic techniques, was applied to biological materials; however, this may lead to losses as observed for TBT [32]. Clean-up could become simplified or even be redundant (reducing the risk of decay) if a more specific method of final determination could be found.

For the development of good extraction methods, materials with an incurred analyte (*i.e.* bound to the matrix in the same way as the unknown) which is preferably labelled (radioactive labelling would allow verification of the recovery) would be necessary. Such materials not being available, the extraction method used should be validated by other independent methods.

Supercritical extractions could be the methods of choice. Such methods are currently being developed, *e.g.* for butyltins [34,35], and seem to offer good possibilities for extracting the species without alteration.

These various comments clearly underline that validation of extraction procedures has still not been achieved. Much work needs to be carried out to evaluate systematically the advantages and limitations of existing techniques. The collaborative studies described in this book represent a good step forward to improve present knowledge. Collaborative projects have already identified pitfalls due to some extraction techniques [36] and research is currently performed in the frame of SM&T projects, *e.g.* systematic comparison of a range of classical extraction methods with supercritical fluid extraction for butyl- and phenyl-tin determinations [35,37].

Derivatization

Derivatization procedures are used to separate trace elements from their matrices and interferences, and to concentrate the analyte species. Some derivatization methods (*e.g.* hydride generation) are used to generate volatile species which are more easily separated from each other by chromatography. Reactions employed nowadays in speciation analysis mostly involve alkylation with Grignard reagents (*e.g.* pentylation, ethylation, butylation). Recent discussions have pointed out the difficulties inherent to derivatization which should be avoided if at all possible to simplify methods and avoid potential sources of contamination, analyte loss and artifact formation [29]. Several major problems were highlighted: derivatization yields, being often matrix dependent, are difficult to determine due to a lack of appropriate, high purity calibrants; the increased number of analytical steps prior to and after derivatization (*e.g.* extraction, preconcentration, clean-up) increase the overall uncertainty; the stability of some derivatives is generally poor and may be affected by uncontrollable factors, such as the initial sample composition.

Hydride generation is carried out in acid media, generally using sodium borohydride as reductant and hydride transfer agent to yield metal and metalloid hydrides. The main advantage of this reaction is that metal–carbon cleavage does not occur and the speciation is therefore maintained. A wide range of element species can be determined using this approach, such as As, Ge, Sb, Se and Sn [38]. By selecting a proper pH in the reaction vessel, the range of compounds that yield hydrides is restricted. In the case of As, for instance, reduction at pH 1 yields arsine from both arsenite and arsenate whilst at pH 5 only arsenite is reduced. However, hydride generation presents major drawbacks, *e.g.* difficulties were noted in recovering tributyltin hydrides in the presence of a high content of organic compounds in the matrix [39] and large variabilities in recovery were observed between different matrices; in addition, sodium borohydride is not able to convert all compounds into volatile compounds, *e.g.* arsenobetaine or -choline into arsine, triphenyltin *etc.* The binding in these molecules simply is too strong. In the case of arsenic, hydride generation is a good tool to separate species which can be very useful, *e.g.* in fish analysis where only the content of toxic (*i.e.* inorganic) As is of importance; however, for the determination of all As-containing species (*i.e.* distribution of As), hydride generation cannot be applied without pre-treatment, *e.g.* by UV irradiation [40,41].

Grignard reactions, *e.g.* butylation or pentylation, are widely used for the determination of alkyl-Pb and -Sn species; the reaction yields products which can be separated relatively easily by GC. Water destroys the reagent and the species of interest has therefore to be removed from water-based matrices, which may be achieved by extraction of a diethyldithiocarbamate complex into an organic phase prior to derivatization, as in the case of alkyl-Pb species determination [17]; this back-extraction increases the risks of contamination or losses. With the increasing use of this technique followed by GC-MIP-AES, the sub-pg detection limits obtained for Pb, Sn and Hg species has necessitated a re-evaluation of the blank levels originating from the Grignard reagents; it was noted that there may be considerable differences in reagent quality and purity between different commercial sources, between alkylating reagents and on a batch-to-batch basis [9]. Finally, as for other reactions, the verification of derivatization yields is presently hampered by the lack of suitable calibrants.

The use of sodium tetraethylborate overcomes the problem of hydrolytic instability of the Grignard reagents, allowing ethylation to be carried out in an aqueous medium. This reaction has been employed for the analysis of alkyl-Pb, -Sn and -Hg species [33,42]. Aqueous phase ethylation is growing in popularity because extraction into an organic phase is avoided and one source of analytical uncertainty is therefore eliminated. Artifact formation and problems with impurities relating to the use of Grignard reagents are also greatly reduced. Problems arose, however, for the determination of methylmercury owing to the conversion of inorganic mercury to an ethyl(methyl)-mercury derivative in the presence of $NaBEt_4$ [29]. Another problem was related to the risk of incomplete ethylation of methylmercury in samples rich in humic substances, sea salts and sulfur since the matrix may consume large amounts of the reagent; in this case, extraction is often required which must be carefully optimized for every sample type. As observed for other derivatization reagents, batch-to-batch variations in the quality of commercial reagents also stresses the caution that should be exercised. Here again, the lack of suitable calibrants hampers verification of the yield of the derivatization.

When working with derivatization procedures it should be realised that the reactions are far from being well controlled and, despite many publications, the reaction mechanisms of derivatization are still not well understood. In addition, better control of the purity of derivatization reagents should be exercised; as manufacturers are unlikely to do this, the responsibility for reagent testing depends on the analytical laboratory. Moreover, systematic collaborative studies to compare derivatization methods as applied in different laboratories should be undertaken; such a project is currently being carried out for butyl- and phenyl-tin species in mussel and sediment samples [43]. As a final remark, it should be noted that, in general, the risk of producing a wrong result increases with the number of steps in a determination and with their complexity. Therefore, if derivatization can be avoided it is worthwile considering such a possibility.

Separation

Separation is performed where the determination of different species cannot be performed with sufficient selectivity, *e.g.* when an element-selective detector is used (*e.g.* AAS). The separation of chemical species of elements can only be performed by techniques which do not destroy the chemical forms, *e.g.* by heat-induced decay of the species. Except for cold trapping methods (for water analysis), the separation is performed after extraction and suitable clean-up of the extract. Three basic methods are generally used in speciation analysis: chromatography (*e.g.* anion exchange, ion pairing reversed phase liquid chromatography (LC) or capillary GC), capillary zone electrophoresis or cold trapping.

GC has become a powerful tool in the determination of traces of organic compounds. Unfortunately, such developments are not sufficiently applied in speciation analysis. Whereas, for most environmental applications, packed columns have been abandoned for the determination of traces of organic compounds because of low resolution and time-consuming procedures, they are, however, still in use for the determination of organo-tin and -mercury compounds. Indeed, although the use of packed column GC results in lower detection limits, the coupling to a highly sensitive detector such as atomic fluorescence spectroscopy will still provide a good analytical system [29]. With regard to capillary GC, the problems are generally the same as those encountered in trace organic analysis. An exception is methylmercury determination, for which a conditioning of the column with mercury chloride is necessary, which may create artifact formation. Quality control of the column condition is recommended, as well as the use of programmable temperature injectors to prevent analyte loss during venting [29].

Thermostable and volatile compounds may be separated by gas–liquid chromatography (GLC). Several stationary phases are available on fused silica capillary columns. The separation power relies on the polarity of the compounds and of the stationary phase. This separation method often requires a derivatization step.

It is generally agreed that liquid separation techniques offer more potential than gaseous separation methods [29]. A much greater range of analytes can, in principle, be separated by liquid chromatography. Problems, however, still exist with the stability of silica-based ion exchange columns. For LC (*e.g.* HPLC) systems there is no need for derivatization prior to separation. Unfortunately, stationary phases in HPLC are less available (*e.g.* ion exchangers, ion pairing) than in GC; consequently, separation problems may still exist for some species, *e.g.* As-betaine/As(III). However, LC is better suited for element specific detections such as ICP-AES or ICP-MS, AAS, X-ray fluorescence (XRF), neutron activation analysis (NAA) or electrochemical detectors.

Cold trapping has been used successfully for the determination of alkyl-Sn and -Se compounds [44], Pb and some As compounds [45]. The technique presents both advantages to concentrate the species and to separate them sequentially according to their specific volatility. One drawback of this method is that only volatile forms of elements (hydrides, ethylated or methylated forms) may be separated; other molecules of low volatility, *e.g.* TPhT or As-betaine, cannot be separated. In addition, cold trapping requires a derivatization step. Both steps are difficult to validate and it is still unclear which physical and chemical parameters may hamper, for a given matrix, the formation and separation of volatile forms. Moreover, separation based on evaporation temperatures may not be sufficient to distinguish two compounds of similar volatility, *e.g.* $BuMeSnH_2$ and MPhT [46]. Although the technique is not always applicable, its simplicity and the fact that it can operate on-line with derivatization steps makes it a recommended method for a variety of compounds. Other powerful techniques that have proven to be suited for a wide range of applications in organic analysis should be considered for speciation. The power of capillary zone electrophoresis (CZE) has been demonstrated for non-volatile, stable polar or charged compounds. Recently, this technique was successfully tried for As species where As-betaine and -choline could hardly be separated any another way [19]. However, the retention time is not always well under control, which after fractionation may cause errors in identification and quantification.

The transfer of technology currently in use for organic compounds to the field of species determination will allow a better understanding of the procedures and should lead to easier and more rugged methods in speciation. QC procedures for such methods, taking into account the many sources of error, have been successfully developed for organic compounds [47,48]. These QC procedures could also be applied to set the QC for a hyphenated method to a high standard.

Final Detection

The detectors used for speciation analysis are either element specific (*e.g.* AAS) or non-specific (*e.g.* FID, FPD, ECD). In general, the determinand should arrive alone in the detector to avoid interferences; the choice of the detector will actually depend on the chemical forms to be determined and on the mode of separation used.

Electrothermal AAS (ETAAS), although being sensitive, cannot be applied in a continuous (on-line) mode and is not generally used in speciation analysis since the necessary manipulations caused by the off-line character of the method may increase the risks of errors considerably. Whenever applied, the precautions for the measurement are the same as for inorganic analysis; the choice of the matrix modifier, the temperature programme, *etc.*, should follow the same rules as for the determination of the element content.

Flame or quartz furnace AAS are often used as an element-sensitive detector. The technique may be operated on-line with a separation device, *e.g.* cold-trapping, providing that parameters are properly optimised (gas flow rates, temperature control, *etc.*).

ICP-AES or ICP-MS has been used on-line after HPLC separation. MS can be specific in certain cases and even would allow an on-line QA in the isotope dilution mode.

Voltammetry has been used in some cases as a species-sensitive detector; this technique is applicable, provided that the species to be determined are sufficiently electroactive and that electrode reactions proceed with a sufficiently high rate [49].

Classical detectors after LC or GC separation have also been applied for the determination of some chemical forms of elements, *e.g.* FPD or FID detection for TBT (see Chapter 5).

Calibration

Primary Calibrants and Internal Standards

General principles of calibration of course apply to speciation analysis. This step is often not sufficiently considered, since it is estimated that about 25–30% of erroneous data may be attributed to calibration errors; this has been reflected, for example, in the BCR intercomparison on extractable trace elements in soil and sediment [24], where many laboratories were demonstrated to have errors when analysing a common calibrant solution.

The importance of calibration is obvious since all efforts made to obtain a good sample and perform the extraction under the proper conditions are spoiled if the calibration is wrong. Basic principles imply that the balance should be frequently calibrated as well as all volumetric glassware. The number of laboratories that do not use volumetric glassware in calibration, but instead use gravimetric dilution of reagents, is rapidly increasing as the advantages in terms of accuracy are well understood. Calibrants should be of good (verified) purity and stoichiometry. Compounds containing water of crystallization should be stored under such conditions that stoichiometry is maintained. When making a stock calibrant solution, preferably two stocks should be made independently, one serving to verify the other; when preparing the dilution (preferably gravimetrically) it is recommended to make two independent solutions, again for verification purposes. The alternative would be to verify the new calibrant solution using the previous one. Calibrant solutions should be made prior to use, even if the solutions are acidified.

The laboratory should carefully consider the calibration mode chosen: (i) standard additions, (ii) calibration curve and (iii) bracketing standards. There is no calibration mode that in all cases should be recommended. All suffer from typical sources of error, *e.g.* for standard additions: non-linearity of the calibration curve, extrapolation difficulties, chemical form of calibrant added, *etc.* In the case of external calibration (calibration curve), points should be measured often, especially in the non-linear part of the curve; sometimes even minor changes of the calibrant matrix (*e.g.* relatively weak complexing agents) may change the calibration considerably, which leads to the recommendation to reproduce the matrix in fair detail by using matrix-matched solutions. Bracketing standards are an excellent tool to correct for fluctuations in the measurement system; they can, however, be very time-consuming for many routine laboratories.

In principle, methods should be validated for each type of matrix and for the extraction agent applied. Matrix effects may affect strongly the calibration (loss of signal, interferences *etc.*). Standard addition techniques are therefore the only way to control the validity of the detection, but only if the addition is performed with the proper identical form of the compound to be determined; it should be noted, however, that by using standard addition procedures, levels of concentration may be reached which are no longer within the linear range of the detector response (*e.g.* for ECD, AAS). It is therefore of primary importance to evaluate this linear range before starting the analysis.

Techniques involving a derivatization step require special caution for their calibration. It is current practice that these techniques be calibrated with calibrants which will be determined in the same way as the compounds in the sample, *e.g.* in the case of standard additions, spiked extracts will be derivatized, undergo separation and final detection. However, the compounds which are actually separated are in derivatized forms and not in their original anionic forms anymore. Thus tributyltin chloride present in a sample may be derivatized into tributyltin hydride (when hydride generation is used) or pentylated tributyltin (when a Grignard reaction is used, in this case pentylation). As a consequence, there are no means to control the derivatization yield and, therefore, the traceability link is lost. To overcome this problem, the most suitable approach would be to use "secondary" calibrants, *i.e.* compounds to be determined in a derivatized form, that would enable the derivatization yield to be evaluated. To date, there are very few attempts to follow this principle which, however, would guarantee a full quality control of techniques using derivatization; this is mainly due to a lack of "secondary calibrants" that are not available on the market. An on-going SM&T project is currently considering this problem in the frame of a certification on butyl- and phenyl-tin compounds; a set of derivatized butyl- and phenyl-tin compounds has been prepared for distribution to the laboratories using derivatization (*e.g.* pentylated or ethylated forms) to ensure the best quality control of the determinations for the purpose of certification; this approach had already been used for the validation of techniques using pentylation followed by gas chromatography (preparation of pentylated organotin calibrants) [50]. This principle should obviously be followed for all kinds of speciation analysis and further efforts should be directed towards the preparation of "secondary calibrants" for chemical species that are the most commonly determined.

Stability of Chemical Species in Solutions

As a complementary comment on calibration, it should be noted that the stability of chemical species in solution is of the utmost importance in terms of quality assurance. Calibrant solutions should obviously be prepared and stored in such a way that they remain stable over a specified duration.

The BCR experience related to the stability of chemical species in solution during storage has been summarized in a survey [51] in which causes of instability are discussed, *e.g.* chemical reaction between species, effects of container material, microbial activity, temperature, pH and light. The various stability experiments on chemical species in solution and the resulting storage precautions are developed in the various chapters of this book. Here we will only consider general aspects.

The container material has to be carefully chosen to avoid undesirable effects such as adsorption of the analyte onto the vessel walls or contamination of the solutions. Thus adsorption of organotins onto PTFE and glass has been noted [52] and solutions may be contaminated with dibutyltin when using PVC-containing materials [53].

Microbial activity must be prevented since it may lead to possible methylation reactions, *e.g.* of inorganic antimony [16] or mercury species [54], and biodegradation, *e.g.* debutylation of tributyltin to monobutyltin [55]. Biological effects are generally dependent on the temperature. Therefore, it is necessary to control and monitor the temperature at which solutions are kept. The pH of the media is a parameter that strongly affects stability, mainly of inorganic species.

Light is one of the major causes of instability of organometallic compounds. With the exception of some compounds (*e.g.* arsenobetaine), the metal–carbon bond is generally weak enough to be photodegraded; hence solutions containing organometallic compounds are recommended to be kept in the dark in opaque containers.

Accordingly, it is crucial to establish how these variables may affect the stability of the different chemical species to ensure that calibrant solutions are fitted to the purpose, or to verify that synthetic solutions are suitable for use in interlaboratory studies (see stability requirements of reference samples described in Chapter 3).

CHAPTER 3

Method Performance Studies – Aims and Principles

The best way to make progress in a particular field is to collaborate with colleagues who work in similar areas of analytical science. By sharing expertise, many difficulties encountered in some laboratories may be solved and improvement may be obtained in a collaborative manner. The dissemination of information through the literature, workshops or international conferences, and the exchange of ideas during meetings, is important but is not sufficient to improve drastically the state of the art of a given analytical field. Therefore, beside the transfer of knowledge *via* the "classical" route, many analysts have started to exchange samples, calibrants and reference materials, in order to test their techniques. As mentioned in Chapter 1, this approach has been structured in a systematic way by the BCR, which has conducted several interlaboratory studies aimed at evaluating and possibly improving analytical methods. With respect to environmental analysis, the projects focused mainly on inorganic, organic and speciation analysis of a wide variety of matrices (water, sediment, soils, plants, *etc.*). Nearly all the projects enabled improvement in the methods to such a degree of reliability that they could be applied to the certification of reference materials (see examples in the various chapters of this book), which, besides their use for the validation of methods, represents a tool for the dissemination of the developed knowledge. The present chapter describes the principles of improvement schemes as followed by BCR.

3.1 Overview of Quality Assurance (QA) Principles

General

Two basic parameters should be considered when discussing analytical results: accuracy ("absence of systematic errors") and uncertainty (coefficient of variation or confidence interval) caused by random errors and random variations in the procedure. In this context, accuracy is of primary importance. However, if the uncertainty in a result is too high, it cannot be used for any conclusion concerning, for example, the quality of the environment or of food. An unacceptably high uncertainty renders the result useless.

In the performance of an analysis, all basic principles of calibration, of elimination of sources of contamination and losses, and of correction for interferences should be followed [1]. These principles are briefly discussed in Chapter 2 with respect to speciation analysis. The following sections give an overview of general QA principles.

Statistical Control

When a laboratory works at a constant level of high quality, fluctuations in the results become random and can be predicted statistically [56]. This implies in the first place that limits of determination and detection should be constant and well known; rules for rounding off final results should be based on the performance of the method in the laboratory. Furthermore, if such a situation of absence of systematic fluctuations exists, normal statistics (*e.g.* regression analysis, *t*- and *F*-tests, analysis of variance) can be applied to study the results wherever necessary [57]. Whenever a laboratory is in statistical control, the results are not necessarily accurate; however, they are reproducible. The ways to verify accuracy will be described in the next paragraphs.

Control charts should be used as soon as the method is under control in the laboratory using reference materials of good quality (*i.e.* stable, homogeneous and relevant with respect to matrix and interferences).

A CRM can be used to assess accuracy. It must be emphasized that reproducible or accurate results, obtained respectively with reference materials (RMs) or CRMs, are not always sufficient in QA. It is necessary that the composition of the control materials is close to the composition of the unknown sample, this closeness involving matrix composition, possibly interfering major and minor substances, *etc.*

A control chart provides a graphical way of interpreting the method output in time, so that the reproducibility of the results and the method precision over a period of time and over different technicians can be evaluated. To do so, one or several materials of good homogeneity and stability should be analyzed with each batch of unknown material. Some 5–10% (depending on the frequency of situations being out of quality control) of all analytical runs should be used for this purpose [57]. These checks can be done with a Shewhart control chart in which the results of calibrant analyses (X) or the difference between duplicate values (R) are plotted over time. The X-chart additionally presents the lines corresponding to a risk of 5% or 1% that the results are not contained in the whole population of results obtained over a period of time. These lines are for "warning" and "action", respectively.

Cusum charts are also used to allow the detection of drifts in methods and trends by plotting the sum of the difference between the result obtained and a reference value in relation to time [57]. Examples of control charts used for the quality control of speciation analysis are given elsewhere [31].

Comparison with Results of Other Methods

The use of a Shewhart chart enables the detection of whether or not a method is under quality control over time; it is not, however, able to detect a systematic error which is present from the moment of introduction of the method in a laboratory. Results should hence be verified by other methods.

As pointed out in Chapter 2, all methods have their own particular source of error, *e.g.* for some techniques, errors may occur due to an incomplete derivatization, a step which is not necessary for other techniques such as high performance liquid chromatography (HPLC); the latter technique, however, may have errors such as incomplete separation which is not encountered, or to a lesser extent, in the former technique.

An independent method should be used to verify the results of routine analysis. If the results of both methods are in good agreement, it can be concluded that the results of the routine analysis are unlikely to be affected by a contribution of a systematic nature (*e.g.* insufficient extraction). This conclusion is stronger when the two methods differ widely, such as derivatization/gas chromatography (GC)/atomic absorption spectrometry (AAS) and HPLC/inductively coupled plasma mass spectrometry (ICP-MS).

If the methods have similarities, such as an extraction step, a comparison of the results would most likely lead to conclusions concerning the accuracy of the method of final determination, and not as regards the analytical result as a whole.

Use of Certified Reference Materials

Results can only be accurate and comparable worldwide if they are traceable. By definition, traceability of a measurement is achieved by an unbroken chain of calibrations connecting the measurement process to the fundamental units. In the vast majority of chemical analyses, the chain is broken because in the treatment the sample is physically destroyed by dissolutions, calcinations, *etc.* With respect to speciation analysis, the chain is even more complex since it involves successive analytical steps. To approach full traceability it is necessary to demonstrate that no loss or contamination has occurred in the course of the sample treatment.

The only possibility for any laboratory to ensure traceability in a simple manner is to verify the analytical procedure by means of a so-called matrix RM certified in a reliable manner. The laboratory which measures such an RM by its own procedure and finds a value in disagreement with the certified value is warned that its measurement includes an error, of which the source must be identified. Thus, CRMs having well known properties should be used to:

- Verify the accuracy of results obtained in a laboratory
- Monitor the performance of the method (*e.g.* cusum control charts)
- Calibrate equipment which requires a calibrant similar to the matrix (*e.g.* optical emission spectrometry, X-ray fluorescence spectrometry)
- Demonstrate equivalence between methods
- Detect errors in the application of standardized methods (*e.g.* ISO, ASTM).

The conclusion on the accuracy obtained for the unknown sample is always a conservative one: if the laboratory finds wrong results on an CRM, it is by no means certain of a good performance on the unknown. If, however, the laboratory finds a value in agreement with the certified value (according to ISO Guide 33 [58]), it should realise that, owing to discrepancies in composition between the CRM and the unknown, there is a risk that the result on the unknown may be wrong. The use of as many as possible relevant CRMs is therefore necessary for a good QA.

Interlaboratory Studies

When all necessary measures have been taken in the laboratory to achieve accurate results, the laboratory should demonstrate its accuracy in interlaboratory studies, which are also useful in detecting systematic errors [7]. In general, besides the sampling error, three sources of error can be detected in all analyses:

- Sample pretreatment (*e.g.* extraction, separation, clean-up, preconcentration)
- Final measurements (*e.g.* calibration errors, spectral interferences, peak overlap, baseline and background corrections)
- Laboratory itself (*e.g.* training and educational level of workers, care applied to the work, awareness of pitfalls, management, clean bench facilities).

As described in Section 3.2, interlaboratory studies are organized in such a way that several laboratories analyse a common material which is distributed by a central laboratory responsible for the data collection and evaluation. When laboratories participate in an interlaboratory study, different sample pretreatment methods and techniques of separation and final determination are compared and discussed, as well as the performance of these laboratories. If the results of such an intercomparison are in good and statistical agreement, the collaboratively obtained value is likely to be the best approximation of the truth.

Before conducting an interlaboratory study the aims should be clearly defined. An intercomparison can be held [7]: (i) to detect the pitfalls of a commonly applied method and to ascertain its performance in practice, or to evaluate the performance of a newly developed method; (ii) to measure the quality of a laboratory or part of a laboratory (*e.g.* audits for accredited laboratories); (iii) to improve the quality of a laboratory in collaborative work with mutual learning processes; (iv) to certify the content of a reference material.

In the ideal situation, where the results of all laboratories are under quality control and accurate, intercomparisons of types (ii) and (iv) will be held only. For the time being, however, types (i) and (iii) play an important role.

3.2 Improvement Schemes

Definition

Improvement schemes can be defined as a succession of individual interlaboratory studies in which several laboratories analyse the same test samples for the same characteristics (usually the content of an analyte), following a similar protocol to validate each individual step of their own analytical method in order to eliminate all sources of systematic errors [47].

Besides the classical interlaboratory studies, improvement schemes enable laboratories to develop and validate all steps of new or existing analytical procedure(s) in adequately organized successive exercises [59]. Improvement schemes may be considered as preliminary studies for laboratory or method performance studies or certification of reference materials [47]. Such programmes are very valuable when the analytical procedures include several complex and critical steps, *e.g.* for the determination of trace organic compounds or chemical species. They require a long term involvement of the organizer and participants, as well as investment of resources.

Organization

The organization of improvement schemes requires good management capability from the organizer, but also a good scientific background, to design properly the exercises and evaluate their results. The first step in the organization of such a programme is to discuss its strategy with all participants involved. This preliminary meeting also enables the collection of the existing knowledge on previous studies and available techniques, and to establish clearly a strategy in relation to the determinands and matrices to be selected. In particular, the choice of the determinands and matrices should be based on the feasibility of preparation of homogeneous and stable samples. The improvement scheme should include a series of meetings between the different exercises so that the outcome of each evaluation can be discussed with all participants in order to draw conclusions and prepare the next exercises. In some cases, additional trials are necessary (*e.g.* when the results are of insufficient quality), which necessitates the extension of the duration of the programme. All participants should preferably commit themselves at the start of the project to maintain their participation over the entire study.

General Principles

In chemical analysis, the substances to be determined are rarely directly measurable and sample pre-treatment is in most cases necessary to convert or separate the analyte in a form that is compatible with the measurement system. This may imply that the initial physical or chemical composition of the sample is changed without losing control of this change so that the traceability of the final detection to a predetermined reference (*e.g.* fundamental units) is not lost.

In speciation analysis, typical pre-treatment steps are digestion, extraction and purification; these are frequently followed by intermediate steps such as derivatization or separation, calibration and final detection. Each action undertaken in one of these steps represents a possible source of error, which adds to the total uncertainty of the final determination (see Chapter 2).

The objective of improvement schemes is to study and validate each step of the analytical procedure of each laboratory in a collaborative manner. In the best case, each critical step of the procedure should be evaluated in an adapted exercise. The individual steps may be studied with a series of different materials in a stepwise manner. In principle, the strategy consists of starting from the most simple matrix, *e.g.* pure solutions and/or mixtures of compounds in solution, which are used to test the performance of the detector. The analysis of more complex matrices (*e.g.* raw extract, purified extract) enables testing the separation and/or clean-up steps, whereas solid samples are used to test the entire procedure. Spiked samples can be analysed to evaluate the extraction procedure, keeping in mind that a complete recovery of a spiked analyte does not mean that the same performance will be achieved with a naturally bound determinand (conversely, if a poor recovery is obtained with a spiked sample, one can assume that this procedure will not work with a natural sample). Such an approach is actually similar to the steps that should be followed when developing and validating a new method in a laboratory.

The difficulty of one particular step may sometimes require it to be subdivided into one or two exercises of increasing complexity. In case too many sources of error are identified, it may be necessary to repeat the study on similar (but not the same) samples. A test sample should never be distributed twice to the participants since pre-knowledge of the material may influence the analyst.

The improvement schemes, usually involving a group of 20 to 50 laboratories, may start with a "simple" exercise, *e.g.* by distributing solutions or pure substances. This enables evaluation of the methods of final detection and possibly to optimize them. More complex samples, *e.g.* complex mixtures of compounds including interferents or extracts, are therefore analysed and the pre-treatment/separation steps are assessed; at this stage, the performance of the method is re-evaluated and the procedure may be fully reconsidered if necessary. As mentioned above, an intermediate step can be the analysis of a spiked sample, which is followed by real samples.

The outcome of the different exercises is discussed among all participants in technical meetings, in particular to identify random and/or systematic errors in the procedures. Whereas random errors can be detected and minimized by intralaboratory measures, systematic errors can only be identified and eliminated by comparing results with other laboratories/techniques. When all steps have been successfully evaluated, *i.e.* all possible sources of systematic error have been removed and the random errors have been minimized, the methods can be considered as valid. This does not imply that the technique(s) can directly be used in routine analysis since further work is likely to be needed to test the robustness and ruggedness of the method before being used by technicians for daily "routine measurements".

It should be noted that standardized methods may also be developed and tested by following a similar stepwise approach. In this case, the participants are given less freedom in terms of method development since they are requested to apply a common procedure (*e.g.* leaching or an extraction procedure) [60,61].

Participants

Achieving good quality control of chemical analysis requires a high degree of motivation from laboratory staff. The participation in interlaboratory studies is a good tool to maintain a high quality standard; this participation may be mandatory, *e.g.* to meet accreditation requirements.

In the IUPAC document "Nomenclature for Interlaboratory Studies", Horwitz stresses that at least eight sets of data should be considered as a minimum for a sound statistical treatment to test the performance of a method or a laboratory [62]. The organization and interpretation of the data depend on the number of participants. This number may vary from tens to hundreds of laboratories, depending on the objective of the study.

It should be stressed that laboratories should have installed all necessary quality assurance and quality control systems prior to participation in an interlaboratory study, *i.e.* the method(s) used should be validated and performed under statistical control for the particular matrix concerned by the study. In other words, interlaboratory studies should not serve the purpose of evaluating and/or optimizing a method in the course of its development.

Organizer

The organizer of an interlaboratory study should adapt the requirements for participation to the objective of the exercise. There is a large difference in the degree of responsibility between a method performance study (*e.g.* establishment of the state-of-the-art of a type of analysis or validation of a well-defined method) and a certification campaign (in which no systematic errors should be left undetected). In all cases, the organization will involve:

- The supply of clear information to the participants
- The production and distribution of the samples (giving evidence that they are representative for the analytical problem(s) studied and that they are homogeneous and stable)
- The collection of results
- The presentation and technical evaluation of the data (discussed, if possible, with all the participants)
- The statistical treatment of the results which have been accepted on technical grounds.

For interlaboratory studies leading either to decisions which could affect the professional status of the participants (*e.g.* proficiency testing within accreditation systems) or to the certification of reference materials, the organizer should be evaluated; in such cases, the study manager should be independent from the group of participants.

The organizer should specify clearly what use will be made of the data; in many cases, the confidentiality of the data and of laboratory coding should be guaranteed. The outcome and acceptance of an interlaboratory study will depend on the degree of confidence from the participants.

3.3 Reference Materials for Method Performance Studies

This section summarizes important aspects related to the type of existing reference materials for method performance studies. When relevant, a distinction is made between "reference materials" (RMs) and "certified reference materials" (CRMs), the fundamental difference between RMs and CRMs being that some parameters in CRMs are known with great accuracy and guaranteed by the producer.

Reference materials can be:

(a) solutions or materials intended for testing part(s) of an analytical procedure, pure extracts or digests, raw extracts, spiked samples;
(b) laboratory matrix reference materials having a composition which is as close as possible to the matrix to be analysed by the laboratory in its daily practice. These materials may be produced in the laboratory itself, *e.g.* for the statistical control of methods, or may be used for the evaluation of the performance of laboratories in interlaboratory studies.

Reference materials which are prepared for in-house control by the laboratories are often referred to as "laboratory reference materials" (LRMs) or "in-house reference materials" (HRMs).

CRMs are also referred to as "standard reference materials" (SRMs) by some producers, *e.g.* the National Institute of Standards and Technology, USA. CRMs allow the user to link his results with those of internationally recognized standards. In addition, CRMs enable the user to verify his performance at any desired moment in terms of accuracy. CRMs can be:

(a) Pure substances or solutions to be used for calibration and/or identification; pure substances are usually certified by establishing the maximum amount, in mass fractions, of the impurities which remain in the purified substance. For metals, it is possible to estimate the mass fraction of other elements within the limits of sensitivity of the available analytical methods, typically some fractions of a percent. For inorganic salts or oxides, this estimate may be more difficult. The stoichiometry has also to be assessed (*e.g.* water of crystallization).
(b) Materials of a known matrix composition for the calibration of a certain type of comparative measuring instrument; certified calibration solutions have to be prepared on a mass basis within especially trained and skilled

laboratories. Certification will be based on a metrologically valid weighing procedure after a proper purity and stoichiometry assessment.

(c) Matrix reference materials which, as far as possible, represent the matrix being analysed by the user and which have a certified content. Such materials are mainly intended for the verification of a measurement process; matrix CRMs are composed of a natural unknown or only partially known matrix in which the amounts of a certain number of substances are certified. Usually, analysts prefer fully natural materials which are as similar as possible to real samples and not artificially enriched materials. In addition, spiking of solids with solutions containing the contaminants leads to unknown losses in the preparation. Therefore, gravimetric results can only rarely be used to certify the material. Certification can only be based on an analytical approach using complex procedures with limited precision compared to mass determinations.

(d) Methodologically defined reference materials; the certified value is defined by the applied method following a very strict analytical protocol, *e.g.* a standard.

Methodologically defined CRMs are materials certified following a given analytical procedure, *e.g.* the RM may be certified following a standardized method. This prescribed procedure may represent only a part or the totality of the analytical process, *e.g.* the extraction or digestion step. Many analytical concepts which attempt to associate the parameter determined with a certain property, such as mobilizable fractions of elements or compounds in soils or sediments, follow such an approach. Extensive interlaboratory studies are necessary prior to the certification of such parameters. Examples are given in Chapter 11, *e.g.* leaching of trace elements from agricultural soils and sediments.

CRMs are products of very high added value. Their production and certification is very costly. Therefore, except for calibration materials for comparative methods, they should be reserved for the selected verification of analytical procedures and not for daily checks, *e.g.* intralaboratory statistical control, nor for interlaboratory studies (round-robins).

Requirements for the Preparation of Reference Materials

A series of requirements have to be respected for the preparation of RMs and CRMs. They include three basic properties: (i) the representativeness of the sample; (ii) the homogeneity; and (iii) the long term stability. The following sections give the general rules to be followed for the preparation of matrix RMs. The procedure has to be adapted to the type of material to be prepared and to the objective it has to fulfil.

In order to arrive at sound conclusions on the analytical performance of a method or of a laboratory, the RM should be similar to samples currently analysed by the laboratory in its daily practice. This means that the RM should pose similar difficulties, *i.e.* induce the same sources of error, that can be encountered in analysing the real samples. Materials which would pose more difficulties than the daily samples may also fulfil the requested task of evaluating the method's or laboratory's performance as long as they do not require additional handling which may induce large additional uncertainties.

The requirement of representativeness of the RM means in most cases similarity of:

- The matrix composition
- The contents of the analytes
- The way of binding of these analytes
- The fingerprint pattern of possible interferences
- The physical status of the material.

In preparing a reference material, these items should be taken into full consideration. In many cases and for practical reasons, the similarity cannot always be entirely respected. The material has to be homogeneous and stable in order to ensure that samples delivered to the laboratories are the same and compromises have often to be made at the preparation stage. Certain parameters of interest which characterize the difficulty of some real natural samples may disappear, *e.g.* inhomogeneity of real samples may need a special technique or strategy of sub-sampling. Some other problems may arise or may be enhanced due to the treatment necessary to homogenize or stabilize the material, *e.g.* drying of a soil rich in clay will make the extraction of organic contaminants more difficult than from fresh material where the clay layers are not closed. It is, however, not always possible to fulfil all the requirements of similarity; often compromises are necessary, *e.g.* highly unstable compounds or matrices may not be stabilized or their stabilization may severely affect the representativeness. It is up to the producer and to the user to define the degree of acceptability of these compromises. The user should be informed of the real status of the sample, of the treatment performed and perhaps of the treatment to be applied to bring the sample back to a more representative form prior to the analysis.

Preparation

Collection

The amount of material to be collected has to be adapted to the purpose of the analysis and is a function of the analytical sample size, stability, shelf life, frequency of use and potential market. It is sometimes better to prepare a limited batch of samples so that the stock lasts a sufficient time and to prepare a new batch of material when new requirements for modern analytical techniques appear or when regulations have changed. The amount collected may vary from a few kg of solid material (*e.g.* soils, sediments) to grams (*e.g.* plant or animal tissues), from some liters (*e.g.* water) to milliliters (*e.g.* body fluids) for the preparation of a RM for the statistical control of a method, up to hundreds of kg of solid material or several cubic meters of water in case of RMs for large interlaboratory studies or CRM production. The producer needs to be equipped to treat the necessary amount of material without substantially changing the representativeness. The treatment of 3–5 kg of raw material is already the limit for usual laboratory equipment and manual processing; for larger batches and especially larger volumes of material it is necessary to scale-up to half- or even industrial-size machines.

Sample Treatment

Typical operations for the preparation of a reference material are: stabilization, crushing, grinding, sieving, filtering, mixing or homogenization of the material. They can only be performed in specialized laboratories or industries.

Stabilization

A very sensitive and difficult step of the preparation is the stabilization of the raw material, which may affect its representativeness. Stabilization is necessary to guarantee that the material remains unchanged in time. It has to be adapted to each particular case with regards to the matrix and substance to be determined and should be studied in detail before processing the batch of the RM. Pure and synthetic mixtures of solid materials are usually stable and do not need any particular treatment. Natural products or synthetic liquid or gaseous mixtures are often highly unstable. Solid materials are often dried to avoid chemical or microbiological activity. This may be achieved by heat-drying, *e.g.* for soils or sediments to be analyzed for trace elements, or by freeze-drying (*e.g.* plant or animal tissues, sediments or soil samples to be analyzed for trace organics), or by fixing the water of the material with chemical additives.

Some materials can be sterilized by γ-irradiation (^{60}Co source). In reality this treatment is mainly possible for material to be used for element determinations, as many compounds decay upon γ-irradiation, *e.g.* tin compounds [63] or organochlorine pesticides [64]. Sterilization by pasteurization or similar methods or addition of chemical preservatives, *e.g.* butylated hydroxytoluene at 0.02% [65], are other alternatives. Stabilization by simple deep-freezing is also possible but induces difficulties in transport and storage. In addition, the material can only be used once as defreezing/refreezing may not lead to a homogeneous material.

Homogenization

A material can only serve as a reference when, at each occasion of its analysis, an identical portion is available. Therefore, when stabilized, the material must be homogenized to ensure sufficient within- and between-vial homogeneity for the property value to be certified [66]. The inhomogeneity of the material should not affect significantly the total uncertainty of the measurement. In cases where the uncertainty in the measurement due to inhomogeneity is not negligible in comparison to the uncertainty of the measurement itself, each sample has to be certified individually by a non-destructive method [66].

Homogenization is not the most difficult problem for gases and liquids. Solid materials or any material composed of various phases (*e.g.* aerosols, suspensions) are obviously more difficult to homogenize or to maintain homogeneous. Experience shows that for small particle sizes, *e.g.* less than 125 μm, with a narrow particle size distribution and if the density of the particles does not differ too much, then good homogeneity is achieved and can be maintained or reobtained even after long storage periods with simple shaking or mixing with laboratory tools. For such materials, studies conducted by the BCR [67] demonstrated that sample intakes of less than 50 mg did not show additional uncertainties in measurements due to inhomogeneity. Pauwels *et al.* [68] showed that for several CRMs of solid material (produced by the BCR) the minimum sample intake goes down to 100 μg. Such results are obtained provided a proper grinding procedure and thorough homogenization prior and during the filling procedure is achieved. Unfortunately, the low particle size presents some drawbacks as it leads to materials which are usually more easy to analyse than real samples (better extractability of analytes or easier matrix digestion because of the large contact surface). In addition, static electricity may cause sub-sampling difficulties for some materials with very low particle size and low water content, *e.g.* as observed in the case of the preparation of a human hair reference material [69].

Control of the Homogeneity

In chemical analysis, a sub-sample of a material can be used only once as it is usually destroyed during the analysis. Therefore, the quantity of material in the bottle or ampoule should be sufficient to perform possibly several determinations. The more sub-samples are taken, the higher the risk that the remaining bulk is not identical anymore to the material at the beginning. In addition, the producer must guarantee that from the first to the last vial filled the material is fully identical. Therefore, verification of the homogeneity within and between vials has to be performed. During the filling procedure the producer should set aside vials at regular intervals. The homogeneity of the material should be tested within a single vial to ensure that successive test portions from the vial lead to similar results (within-vial homogeneity). The same study should be performed to verify that there is no difference from test portions taken from various vials (between-vial homogeneity). Besides subtances of interest, the study should include some matrix constituents or major and minor constituents (in case of a matrix RM). It is current practice for producers of CRMs (*e.g.* BCR) to test this homogeneity for different sample intakes. The (in)homogeneity is estimated by examining the coefficient of variation (CV) of the replicate measurements obtained on samples from different vials with those obtained from test portions from one vial. These coefficients of variations are also compared to the coefficient of variation due to the analytical method. This *CV* is usually obtained by measuring several times, *e.g.* the same extract or digest of one sample.

The method for the study of the homogeneity should allow the attainment of very good precision. A high degree of trueness is unnecessary as the only interesting parameter is the difference between samples. In addition, it is of primary interest to obtain data for several elements or compounds. Multi-elemental techniques for inorganic determinations [67] are therefore recommended, *e.g.* NAA, ICP-AES, ICP-MS (or XRF, but only for larger test portions). For organic determinations [65] this is usually obtained by all chromatographic separations associated with a proper final detection, *e.g.* ECD, FID or MS for GC, UV or MS for LC. For inorganic trace analysis, solid sampling atomic absorption spectrometry can be applied when the homogeneity needs to be tested on very small sample intakes, even down to a few mg [68]. These investigations are performed on several elements or compounds which are present in a sufficient amount in the material so that the uncertainty of the method is minimal. For a CRM, all certified substances are usually tested for homogeneity. As already stated above, good homogeneity may be obtained and demonstrated by studying the particle size and distribution. This knowledge is particularly important when materials are prepared for organic, organometallic or all other parameters which necessitate large sample intakes (*e.g.* several g). In such cases the sensitivity of the analytical method is not sufficient to estimate the degree of homogeneity on small sample sizes. In this case it also means that possible inhomogeneity is only a marginal problem and its influence on the measurements can be neglected. The minimum sample size (if possible) for which the homogeneity is sufficient should be verified and stated by the producer. Below this level of intake the uncertainty caused by inhomogeneity contributes significantly to the uncertainty of the reference (or certified) values. An additional problem is caused by the segregation of particles during transport and long term storage. This may be minimized by reducing the particle size distribution. Special care should be taken for the rehomogenization of the material before taking a test portion. The producer has to provide this information to the user.

Control of the Stability

The composition of the material and the parameters investigated should remain unchanged over the entire period of use of the material. The study of the material stability in time will mainly depend on its role. If the material is going to be used in a short term interlaboratory study, the stability has to be monitored only over the real duration of the exercise and additionally mimic situations which may be encountered during its short lifetime, *e.g.* transport under severe climatic conditions. This may vary from some hours (*e.g.* microbiological samples) to several years for a CRM. The (in)stability should be studied or known before the RM is produced and should be monitored on the batch of RM.

The stability can be estimated by evaluating the behaviour of the material under accelerated aging conditions, *e.g.* elevated temperatures over long periods of time. The BCR is currently performing studies of the stability of CRMs at room temperature and at 35–40 °C. In order to detect long term reproducibility problems with analytical methods, measurements are also made on samples stored at −20 °C (at this temperature, chemical or biological changes are negligible). The results obtained at +20 °C may lead to an assessment of the sample stability at ambient laboratory temperature whereas the results obtained at +40 °C are used to assess the worst case conditions (*e.g.* during transport) and allow evaluation of the stability of the material over longer periods of time. It is indeed assumed that a sample stable at +40 °C during one year may be stable at +20 °C for a longer period. This assumption does not hold in cases of spoilage by certain bacteria or moulds having optimum temperatures for their metabolism at 20–35 °C. However, in most cases water is removed to a level below 5% (mass fraction), or samples are sterilized (γ-irradiation, heating, *etc.*) to reduce the germ number. This gives confidence that microbial spoilage does not occur. Stability tests may be carried out at the beginning of the storage period and after various time intervals, *e.g.* 1, 3, 6 and 12 months. Samples stored at −20 °C are used as reference for the samples stored at +40 °C and at +20 °C. The ratios (R_T) of the mean values (X_T) of five measurements made at +20 °C and +40 °C, and the mean value ($X_{-20\,°C}$) from five determinations made at the same occasion of analysis on samples stored at a temperature of −20 °C are calculated: $R_T = X_T/X_{-20\,°C}$.

The overall uncertainty U_T is obtained from the coefficient of variation (CV) of five measurements obtained at each temperature:

$$U_T = (CV_T^2 + CV_{-20\,°C}{}^2)^{1/2} R_T$$

In the case of ideal stability, the ratio R_T should be 1. In practice, however, there are some random variations due to errors in the measurement. If the value 1 is contained between $R_T - U_T$ and $R_T + U_T$, one can assume that the samples are stable under these storage conditions.

For inorganic determinations, INAA is often a good choice for stability experiments. This technique allows the avoidance of a sample pretreatment step for some elements and matrices and consequently excludes the uncertainty due to this step. In addition, a fine quality control can be achieved by performing counting statistics. For organic determinations the extraction of a larger batch of material can be performed; this extract is stored at a low temperature, *e.g.* −20 °C, and analysed at each occasion of the stability study. This allows the estimation of the performance of the method at that particular period, possibly to improve it and express the results in relation to the results of the extract. The reference to samples stored at a low temperature has its limitations. This technique is not valid for water or any solutions in which certain compounds may precipitate at low temperatures and may not redissolve reproducibly upon warming up. In such circumstances, other adapted long-term reproducibility measures and reference samples have to be developed, *e.g.* detection of the appearence of products of a chemical reaction or metabolites.

Besides the preliminary stability study, regular checks should be carried out over the entire lifetime of the material. In some rare situations for the certification projects conducted under the BCR activities, examples of "instability" of substances in materials was demonstrated. A typical example was shown by the instability of tributyltin (TBT) demonstrated in CRM 462 (coastal sediment) at a temperature of +40 °C (see Chapter 5) [70]; another instability was demonstrated for organochlorine pesticides in which the increase in *p,p'*-TDE content was generated by the decompostion of *p,p'*-DDT [71].

When instability is detected at an earlier stage of study the material has to be immediately stored in adequate conditions. In addition, a short-term stability study has to be performed to monitor the behaviour of the material during transport under possibly increased temperature conditions. Short-term drastic transport conditions may be reproduced by storing the material under temperatures up to 40–50 °C for 10 days. This covers the maximum real time transport duration (normal surface transportation) under severe high temperatures.

Storage and Transport

The above mentioned homogeneity and stability parameters were implicitly related to the storage vial. Vials may be sealed glass ampoules, *e.g.* for materials containing organic contaminants or solutions, bottles for solids, *e.g.* soils, sludges, ashes, plant or animal tissues or stainless-steel or aluminium bottles for gases or any kind of protecting vial. It is advisable to avoid light or radiation interaction by using, as often as possible, amber glass, high density polymers or metals. For some special materials with a high risk of contamination by leaching from the glass surface of the vial, *e.g.* contaminants in rain water [72], quartz is the best choice. In such cases the ampoule has to be stored in a closed tube in order to avoid exposure to light and shock.

The head space above the material should be minimal to avoid evaporation and condensation in the vial. This could induce risks of inhomogeneity of the material, *e.g.* by organic contaminants with low vapour pressures in soils. The label identifying the sample should mention all particular precautions to be taken for storage by the user or for opening of the vial.

The temperature of storage should be adequate to ensure sufficient stability. Low temperatures are often desirable but are not always necessary, *e.g.* large cool rooms for stocks of CRMs. Cooling sometimes can be harmful, *e.g.* precipitation of dissolved compounds. The storage conditions together with the delivery system should be deduced from a properly conducted stability study of the material and possibly from a preliminary study of the material under various conditions and in different storage vials (especially for CRMs). The transport should be done in the shortest period possible. Rapid delivery systems are unfortunately expensive and are solely used for certain particular cases. With the test sample the user should find a reporting form to be sent back to the organizer where he indicates the status of the samples as received. Temperature indicators may be added to the samples so that too high temperatures occurring in transport are detected upon arrival of the sample.

Procedures to Certify and Assign Values

Certification of Reference Materials

The certification of reference materials follows strict rules which are described in special ISO Guide 35 [73]: "the certified value should be an accurate estimate of the true value with a reliable estimate of the uncertainty compatible with the end use requirements". Depending on the type of property value to be certified and the type of CRM, there may be differences in the approach applied. The certification of primary calibrants such as pure compounds and calibration solutions (see Section 3.3) relies on the identification, purity and stoichiometry assessment and on gravimetric methods. Matrix CRMs cannot be certified on the basis of direct gravimetric methods since samples have to be analysed after a total transformation or removal of the matrix; in this case, there are three possible approaches:

- Certification within one laboratory using a so-called definitive method, *e.g.* by two or more independent analysts
- Certification within one laboratory using two or more so-called reference methods by two or more independent analysts
- Certification by interlaboratory studies using one or several different methods, possibly including definitive methods.

In all cases, only laboratories of the highest and proven quality should be involved.

The first two approaches using definitive methods within one single laboratory do not eliminate the risk that within the laboratory a systematic bias remains. An additional confirmation through an – even limited – interlaboratory study is therefore advisable. For some chemical parameters, so-called direct methods do not require external calibration, *e.g.* gravimetry, titrimetry, volumetry, or definitive methods exist. Isotope dilution mass spectrometry (IDMS) with electrothermal ionization as developed for nuclear measurements in the early 1950s [74] can be a definitive method and has been adapted to the non-nuclear area [74,75]. Under certain conditions, such as total matrix destruction, separation of the analyte and optimal isotopic abundance ratios, only some elements, *e.g.* Li, Na, Cu, Cr, Fe, B, Pb, Cd, Rb, Se, Ca, Si, Ba or Ni, often present as traces in biological or organic matrices, can be determined by this technique, but only in a very few highly reliable laboratories. Unfortunately, the thermal ionization system which provides the most accurate and precise determinations is not available for highly volatile elements nor those elements which are difficult to thermoionize (*e.g.* Pt). ICP-MS for isotope dilution is not yet currently used; this technique has been succesfully used, however, for the certification of Hg in fish tissue [76], and the coupling of HPLC with ID-ICP-MS has allowed the determination of trimethyllead in candidate reference materials of simulated rainwater [77] (see Chapter 6). Mononuclidic elements are rarely determined by IDMS as artificial isotopes would be necessary. Certification of matrix materials through one single definitive method (*e.g.* IDMS for multi-isotopic elements) does not give the user, who does not apply this method in a daily practice, a fair estimate of the uncertainty achievable by more classical methods. In addition, definitive methods are very limited in their field of application at the level of matrix and property values to be certified. They do not exist yet for the certification of traces of organic or organometallic compounds, nor for the determination of mono-isotopic elements, nor for the determination of forms of elements with different oxidation states. For these parameters, only an interlaboratory certification study can be applied.

Interlaboratory certification studies are organized following the same basic principles as other interlaboratory studies (see Section 3.1) but involve only highly specialised analysts. All participants should have demonstrated their quality in prior exercises. The organizer has also to fulfil many requirements and should be known and recognized for the ability to organize such studies. The best way to establish the reliability of all participants involved is to ask them to demonstrate their performance in so-called step-by-step improvement schemes as described in Section 3.2. This approach has been used by the BCR for all RMs where new property values were to be certified for the first time in matrix materials. In all studies, detailed protocols and reporting forms were prepared, requesting each participant to demonstrate the quality of the measurements performed, in particular the validity of calibration (including calibration of balances, volumetric glassware and other tools of relevance, use of calibrants of adequate purity and known stoichiometry, proper solvents and reagents). The absence of contamination has also to be proven by procedure blanks and chemical reaction yields should be known accurately and demonstrated. All precautions should be taken to avoid losses (*e.g.* formation of insoluble or volatile compounds, incomplete extraction and clean-up). If the results of entirely independent methods such as IDMS, AAS and voltammetry (between-method bias) for inorganic trace determinations as applied in different laboratories working independently (between-laboratory bias) are in agreement, it can be concluded that the bias of each method is negligible and the mean value of the results is the best approximation of the true value. Examples of such certification studies applied to speciation analysis are described in the various chapters of this book.

The certification of a property value in a material leads to a certified value which is typically the mean of several determinations or the result of a metrologically valid preparation procedure, *e.g.* weighing. The confidence intervals or uncertainty limits of this mean value should also be determined. These two basic analytical parameters should be included in a certificate of analysis. The presentation and the additional information which should also be given in the certificate are listed in ISO Guide 31 and cover in particular:

- Administrative information on the producer and the material
- Brief description of the material with main properties and its preparation
- Intended use of the material
- Information for correct use and storage of the CRM
- Certified values and confidence limits
- Other non-certified values (optional)
- Analytical method(s) used for certification
- Identification of certifying institute(s)
- Legal notices and signature of certifying body

Some information which is useful for the user of the CRM cannot be given in a simple certificate. Therefore, some CRM producers, *e.g.* BCR, make the material available together with a certification report which details the information given in the certificate. In particular, such reports highlight the difficulties encountered in the certification and typical errors which may arise using the material with some analytical procedures.

The entire certification work described in the certification report should be examined by an independent (group of) expert(s) so that all possible sources of unacceptable practice are detected and eliminated. The group of experts should have strong metrological background knowledge as well as training in high analytical practice. They should give the final decision whether or not the RM can be certified. Within the BCR, this certification committee is composed of representatives from different EU and Associated Member States, covering a wide range of expertise (chemical, biological and physical measurements).

Assigned Values

For non-certified RMs the only possibility of obtaining additional information on the accuracy of the methods is to obtain a good assigned value. However, this value is often not known or is difficult to obtain. In fact the same approach and stricter rules than for certified values should be followed to obtain good assigned values. Such a value is rarely needed for a laboratory RM used for statistical control purposes but, as far as possible, it should be made available for RMs used in laboratory performance studies. Examples of reference materials have been recently shown, to which assigned values were obtained for the purpose of proficiency testing of marine monitoring data [78]. As stressed before, no direct or definitive methods are available for organic or speciation analysis, contrary to inorganic analysis. In organic or organometallic trace analysis the assigned value may be approached by an interlaboratory study with some highly experienced laboratories. These laboratories should be of a quality sufficient to certify the concerned parameter (in another exercise) in a similar matrix and should use different methods except if a method dependent parameter is studied. The main difference between a good assigned value and a certified value lies mainly in the guarantee which is given with the certified parameter and the procedure of issuing this guarantee.

Evaluation of Results

Protocols accompanying the test samples should be provided to the participants. The instructions should specify the number of samples and replicates to be analysed, and necessary information on sample preservation, sub-sampling, sample pre-treatment, *etc.*

Collection of the Data

The ways the results are evaluated and presented depend strongly upon the objective of the study. They can be delivered in the form of tables with raw results, but it is preferable to present them in an easily interpretable manner with graphical representations (for possible technical discussions among the participants).

The organizer should ensure that the collection of results is carried out using appropriate report sheets for easy evaluation. These may include tables for raw data, data corrected for dry mass, results of recovery experiments (for organic or organometallic determinations), *etc.* Examples of report sheets as applied to projects dealing with speciation analysis are given in the Appendix.

Description of the Methods

A full description of the methods is usually not required for proficiency tests or for standardization purposes; it is, however, an essential element when the performance of a laboratory or a method has to be evaluated, *e.g.* in the frame of step-by-step projects. The BCR has elaborated reporting forms for the description of various types of analytical methods for the certification of inorganic and organic parameters (see Appendix), which may be used as a basis to prepare forms for all types of interlaboratory studies. The questions are intended to remind the participants which important parameters are to be taken into consideration (possibly affecting the quality of the final result). These forms can also be used in technical discussions with other participants when differences in the results are noticed. Finally, they may constitute the basis for the method description in the certification reports.

For the maximum benefit to the participants and the technical discussion, the organizer should prepare summaries of method descriptions including information on the most critical steps in the analytical procedures, *e.g.* calibration, pre-treatment, extraction and clean-up, separation and final detection. If possible, the methods described in brief should be presented in the form of simple table showing the most important parameter grouped by item or analytical step.

Technical Evaluation

The technical evaluation of the data consists of a scientific scrutiny of the reports submitted by the participants. To facilitate the discussion in a meeting, the data should be presented in a visualized manner and the results should be ear-marked to the laboratories (laboratory codes and method abbreviations). Examples of graphical representations are given in the different chapters of this book. A classical presentation of the results is to display them in the form of bar-graphs. Another possibility is to set-up a Youden plot in which the results obtained from one sample are plotted against the results of a similar sample with higher or lower analyte contents; this type of graph is useful to detect possible systematic errors (see example in Chapter 4). Finally, robust statistics can be applied to the results, which may be evaluated on the basis of *Z*-scores [79]. In the case of more elaborated data presentations (*e.g.* Youden plot or robust statistics), sufficient explanation should be given to the participants in order to avoid misunderstanding and possible wrong conclusions.

Technical meetings are necessary in learning programmes or in the process of the development of standardized methods; meetings enable participants to extract information by comparing and possibly discussing their performance and method with other participants using similar or different procedures.

Statistical Evaluation

A statistical analysis of the data from an interlaboratory study cannot explain deviating results nor can it alone give any information on the accuracy of the results. Statistics only treat a population of the data and provide information on the statistical characteristics of this population. The results of the statistical treatment may raise discussions on particular data not belonging to the rest of the population, but outlying data can sometimes be closer to the true value than the bulk of the population [67].

In case a given protocol is to be followed, the statistical treatment should be applied only to the data which correspond strictly to the use of this protocol (*e.g.* standardized method).

If no systematic errors affect the population of data, various statistical tests may be applied to the results, which can be treated either as individual data or as means of laboratory means. When different methods are applied, the statistical treatment is usually based on the mean values of replicate determinations.

The statistical treatment involves tests, *e.g.* to assess the conformity of the distributions of individual results and of laboratory means to normal distributions (Kolmogorov–Smirnov–Lilliefors tests), to detect "outlying" values in the population of individual results and in the population of laboratory means (Nalimov test), to assess the overall consistency of the variance values obtained in the participating laboratories (Bartlett test), and to detect "outlying" values in the laboratory variances (s_i^2) (Cochran test). One-way analysis of variance (*F*-test) may be used to compare and estimate the between- and the within-laboratory components of the overall variance of all individual results.

Together with the technical evaluation of the results, the statistical evaluation forms the basis for the conclusions to be drawn and the possible actions to be taken.

CHAPTER 4

Mercury Speciation

4.1 Aim of the Project and Coordination

Justification

All forms of mercury are considered to be poisonous. Among the different mercury species, methylmercury is of particular concern due to its extreme toxicity and its ability to bioaccumulate in fish tissues. The toxic impact of methylmercury on man was observed for the first time in Minamata (Japan) in 1955 owing to an ingestion of methylmercury-contaminated fish by pregnant women which provoked severe brain damage to their new-born children [80]. The ingestion of wheat flour produced from seeds treated with organic mercury also led to large scale poisoning by methylmercury in Iraq in 1971–72. As a consequence of these risks, methylmercury is currently monitored in fish by a number of organizations to check the levels of contamination.

The technique originally developed for the determination of methylmercury in biological materials is still widely used; it involves several analytical steps such as extraction, extract clean-up, gas chromatographic separation and electron capture detection. This method, however, suffers from a lack of specificity, high detection limits and tedious column conditioning. The increasing need for methylmercury monitoring in fish has, therefore, led to the development of several new methods within the last 10 years involving, in most cases, a succession of analytical steps. These methods are far from being validated. A programme for evaluating their performance was organized in 1987–89 by the BCR [81].

The Programme and Timetable

The programme to improve the quality of MeHg determinations started in 1987 with a consultation of European experts. From the beginning it was clear that it was essential to examine critically each step of the methods used in mercury speciation analysis. As described below, a series of exercises of increasing difficulty were designed and interlaboratory studies were conducted in 1987–88 (first round-robin with simple solutions), 1988–89 (cleaned and raw extracts) and 1989–90 (extracts and real samples). The certification of

MeHg in two fish reference materials was carried out in 1991–94, whereas a recent certification on MeHg in sediment was performed in the period 1994–96.

Coordination

Interlaboratory Studies

The preparation of the fish extracts and the verification of their homogeneity and stability were carried out by the Danish Isotope Centre and the National Food Agency (Søborg, Denmark). The aqueous solutions were prepared at the Kernforschungsanlage (Jülich, Germany), whereas the mussel and tuna samples were prepared at the Environment Institute of the EC Joint Research Centre of Ispra (Italy).

Certification of Fish Reference Materials

The project was coordinated by the Danish Technological Institute (Taastrup, Denmark). The candidate reference materials were purchased from a fish market in Venice and the preparation was performed by Ecoconsult (Gavirate, Italy) and the Environment Institute of the EC Joint Research Centre of Ispra (Italy). MeHg calibrants were prepared by the Department of Chemistry of the University of Aberdeen (United Kingdom). The verification of the homogeneity and stability was performed by the Presidio Multizonale di Prevenzione (Venice, Italy).

Certification of Sediment Reference Material

The project was coordinated by the Studio di Ingegneria Ambientale (Milano, Italy). The preparation of the sediment reference material was carried out by Ecoconsult (Gavirate, Italy) and the Environment Institute of the EC Joint Research Centre of Ispra (Italy). The homogeneity and stability were verified at two laboratories from the Presidio Multizonale di Prevenzione (La Spezia and Venice, Italy). The material characterization (with regard to bacterial flora) was performed at the University of Siena (Italy).

Participating Laboratories

Interlaboratory Studies

The following laboratories participated in the interlaboratory studies on solutions and fish extracts: Bundesforschungsanstalt für Fischerei, Hamburg (Germany); Danish Isotope Centre, Copenhagen (Denmark); IFREMER, Nantes (France); Institut Joszef Stefan, Ljubljana (Slovenia); Kernforschungsanlage, Jülich (Germany); Leicester Polytechnic, School of Applied Physical Sciences (United Kingdom); MRC Toxicology Unit, Carshalton (United

Kingdom); National Food Administration, Uppsala (Sweden); National Food Agency, Søborg (Denmark); Presidio Multizonale di Prevenzione, La Spezia (Italy); Presidio Multizonale di Prevenzione, Venezia (Italy); RIKILT, Wageningen (The Netherlands); Universidad Complutense, Departamento de Química Analítica, Madrid (Spain); Universidad de Santiago de Compostela, Departamento de Química Analítica (Spain); Universita di Genoa, Laboratorio di Chimica (Italy); Vrije Universiteit Brussel, Lab. voor Analytische Scheikunde, Brussels (Belgium).

The following laboratories participated in the interlaboratory study on sediment: De Montfort University, Leicester (United Kingdom); ENEA, Divisione de Chimica Ambientale, Rome (Italy); GKSS Forschungszentrum, Geesthacht (Germany); IAEA, Marine Environment Laboratory (Monaco); KFA, Institut für Angewandte Physikalische Chemie, Jülich (Germany); Presidio Multizonale di Prevenzione, La Spezia (Italy); Presidio Multizonale di Prevenzione, Venezia (Italy); rivo-dlo, IJmuiden (The Netherlands); Sheffield Hallam University, Sheffield (United Kingdom); Swedish Environmental Research Institute, Göteborg (Sweden); Universidad de Oviedo, Departamento de Química Física y Analítica (Spain); Universidad de Santiago de Compostela, Departamento de Química Analítica (Spain); Universität Heidelberg, Inst. Sedimentforschung, Heidelberg (Germany); Université de Bordeaux, Laboratoire de Photophysique et Photochimie Moléculaire, Talence (France); University of Umeå, Department of Chemistry (Sweden); Vrije Universiteit Amsterdam, Inst. Milieuvraagstukken, Amsterdam (The Netherlands); Vrije Universiteit Brussel, Lab. voor Analytische Scheikunde, Brussels (Belgium).

Certification Campaigns

The following laboratories participated in the certification of the CRMs 463 and 464 (total mercury and methylmercury in tuna fish): Energieonderzoek Centrum Nederland, Petten (The Netherlands); FORCE Institute, Brøndby (Denmark); KFA, Kernforschungsanlage, Jülich (Germany); National Food Administration, Uppsala (Sweden); National Food Agency, Søborg (Denmark); Presidio Multizonale di Prevenzione, La Spezia (Italy); Presidio Multizonale di Prevenzione, Venezia (Italy); Universidad de Santiago de Compostela, Departmento de Química Analítica (Spain); Service Central d'Analyse, CNRS, Vernaison (France); Swedish Environmental Research Institute, Göteborg (Sweden); Università di Genoa (Italy); Universiteit Gent, I.N.W., Gent (Belgium); University of Umeå, Department of Chemistry (Sweden); Vrije Universiteit Brussel, Lab. Analytische Scheikunde (Belgium).

The following laboratories participated in the certification of the CRM 580 (total mercury and methylmercury in sediment): De Montfort University, Leicester (United Kingdom); ENEA, Divisione de Chimica Ambientale, Rome (Italy); GALAB, Geesthacht (Germany); GKSS Forschungszentrum, Geesthacht (Germany); KFA, Institut für Angewandte Physikalische Chemie, Jülich (Germany); Presidio Multizonale di Prevenzione, La Spezia (Italy);

Presidio Multizonale di Prevenzione, Venezia (Italy); rivo-dlo, IJmuiden (The Netherlands); Sheffield Hallam University, Sheffield (United Kingdom); Swedish Environmental Research Institute, Göteborg (Sweden); Universidad de Santiago de Compostella, Departamento de Química Analítica (Spain); Universität Bayreuth, Inst. für Terrestrische Ökosystemforschung (Germany); Universität Heidelberg, Inst. Sedimentforschung, Heidelberg (Germany); Université de Bordeaux, Laboratoire de Photophysique et Photochimie Moléculaire, Talence (France); Universiteit Gent, I.N.W. (Belgium); University of Umeå, Department of Chemistry (Sweden); Vrije Universiteit Amsterdam, Inst. Milieuvraagstukken, Amsterdam (The Netherlands); Vrije Universiteit Brussel, Lab. voor Analytische Scheikunde, Brussels (Belgium).

4.2 Techniques Used in Mercury Speciation

Examples of techniques which have proven to be successful in certification campaigns are briefly described below. The possible risks of errors and critical comments on techniques (some of them having not been accepted for certification) are discussed in the sections related to the different interlaboratory studies (including certifications).

Methylmercury Determination in Fish

The verification of the concentrations of the calibration solutions was considered to be a important aspect to achieve traceability. A pure methylmercury chloride compound was prepared for the certification, the purity of the calibrant of which was verified by C, H, Hg and Cl elemental analysis and was found to be higher than 99.8%. This compound was distributed to the participants in the certification both for calibration purposes and as a means of verification of the calibrant used in their laboratory.

Ion Exchange/AAS

Approximately 250 mg of sample were extracted twice with HCl. The separation of inorganic from organic mercury was carried out by ion exchange (Dowex 1X8 100–200 mesh). An aliquot was measured before and after UV irradiation, the value obtained before UV treatment being considered as the blank; the MeHg content was measured by difference between the values obtained before and after irradiation. A solution of $MeHg^+$ was used for calibration (calibration graph). The yield for the separation of inorganic from organic mercury was estimated to be more than 95%. This technique was found to be suitable for fish analysis, assuming that all the organic mercury present in the matrix was actually in the MeHg form; the method would not be applicable in cases where other organic forms of mercury would be present.

Westöö extraction/GC-ECD

Extraction was performed on a sample of 1500–2000 mg previously wetted with water by addition of HCl and toluene (Westöö extraction), shaking the mixture for 15 min and centrifuging; the toluene extract was transferred into a separating funnel and cysteine solution was added; after shaking and centrifugation, the aqueous phase was removed and back-extracted with toluene; the final extract was dried with anhydrous Na_2SO_4; the extraction recovery was in the range 70–80%. The separation was by capillary gas chromatography (column of 10 m length, internal diameter of 0.53 mm; stationary phase CP Sil 8, film thickness of 5.35 mm; injection of 2 μL; temperature of the injector of 200 °C, temperature of the ECD detector of 300 °C, column temperature of 130 °C for 3 min at 15 °C min^{-1} to 230 °C; N_2 as carrier gas at 2 mL min^{-1} and as make-up gas type at 35 mL min^{-1}). Calibration was by calibration graph and standard additions using MeHgCl in toluene as calibrant.

Toluene Extraction/GC-ECD

First method: about 800–1000 mg of sample were treated by addition of H_2SO_4 followed by toluene extraction. The extract was dried with Na_2SO_4. The separation was by capillary gas chromatography (column of 1.5 m length, internal diameter of 3.2 mm; Chromosorb W/AW/DMCS loaded with *ca.* 7% PDEAS; injection of 10 μL; temperature of the injector of 200 °C, temperature of the ECD detector of 300 °C, column temperature of 175 °C). The quantitation was performed by peak area. The calibration was carried out with a calibration graph, using a matrix-matched solution of MeHgCl in toluene (matching the concentration of the sample, taking care that the calibrant solution did not differ more than 0.03 μg mL^{-1} from the sample concentration).

Second method: about 400–500 mg of sample were treated with HCl and extracted into toluene and cysteine, and back-extracted into toluene. The extract was dried with anhydrous Na_2SO_4. The separation was by capillary gas chromatography (column of 30 m length, internal diameter of 0.53 mm; stationary phase S.P.B. 608 Supelchem, 0.50 μm film thickness; injection of 1 μL; temperature of the injector of 250 °C, temperature of the ECD detector of 350 °C, column temperature of 90 °C; N_2 as carrier gas at 10 mL min^{-1}; Ar/CH_4 as make up gas type at 60 mL min^{-1}). Calibration was by calibration graph with CH_3HgCl in toluene; C_2H_5HgCl was used as internal standard.

Third method: a mixture of HCl/water was added to 500 mg of sample. The extraction was by addition of a mixture of acetone/toluene (clean-up) and back-extraction into toluene; extraction recovery was (90 ± 4)%. The separation was by capillary gas chromatography (column of 25 m length, internal diameter of 0.53 mm; stationary phase BP-1 (methylsilicone), 1.0 μm film thickness; automatic injection of 1 μL; temperature of the injector of 200 °C, temperature of the ECD detector of 250 °C, column temperature of 90 °C; N_2 as carrier gas at 12.2 mL min^{-1} and as make-up gas type at 47 mL min^{-1}). Calibration was by calibration graph with CH_3HgCl in toluene.

Acid Extraction/GC-AFS

About 50 mg were extracted with a mixture of H_2SO_4/MeCOOH in Milli-Q water (2 mL). The extract was injected in the headspace; separation was by capillary gas chromatography (1 m length column with 3 mm internal diameter; Chromosorb WAW 80–100 mesh loaded with 10% AT-1000 stationary phase; injector temperature of 120 °C, column temperature of 180 °C; Ar as carrier gas at 100 mL min^{-1}). Final detection was by atomic fluorescence spectrometry. Calibration was by standard additions, using MeHgCl in Milli-Q water as calibrant.

Toluene Extraction/ETAAS

500 mg of sample were added to NaCl/HCl, extracted as the chloride derivative into benzene and re-extracted by a thiosulfate solution. Inorganic mercury was converted (in the same extract) into a methyl chloride derivative by methanolic tetramethyltin at 100 °C prior to extraction. An aliquot of the thiosulfate solution was injected into the graphite furnace of the AAS for the determination of total organic mercury. A verification was performed by gas liquid chromatography. Calibration was by calibration graph using MeHgCl calibrant in ethanol.

Toluene Extraction/GC-FTIR

Methylmercury was extracted from 100 mg of sample. The sample was heated at boiling point in a water bath for 5 min with HCl. The sample was then extracted three times with toluene and the extracts were transferred to a vial containing a solution of sodium thiosulfate (0.01 mol L^{-1}). The thiosulfate solution was transferred to a beaker and heated on a hot plate to evaporate the solution in order to eliminate traces of toluene. After cooling, 50 mL of $NaBH_4$ (0.01 mol L^{-1}) was added to the sample, which was purged for 5 min with nitrogen in purge and trap apparatus. The volatile Hg species were trapped in a column held at −120 °C; after 5 min the trap was heated to 250 °C and injected automatically into the GC port. The volatile Hg compounds were separated by capillary gas chromatography (column of 50 m length, internal diameter of 0.53 mm; stationary phase CP-Sil 8, 2 μm film thickness; column temperature of 100 °C; N_2 as carrier gas at 10 mL min^{-1}). Detection was by FTIR (with infrared spectra).

Toluene Extraction/GC-MIP

About 50 mg sample was wetted with 1 mL NaCl solution (standard additions were carried out at this stage) and shaken manually for 15 min; this was followed by an addition of 200 μL (sub-boiling distilled) HCl and 20 min shaking. Neutralization was performed by addition of *ca.* 2 mL NaOH (1 mol L^{-1}) to pH 6–7. The extraction was by addition of 1 mL buffer (pH 9) and 1 mL of sodium diethyldithiocarbamate (DDTC) (0.5 mol L^{-1}), followed by shaking for 7 min and centrifugation for 5 min at 5000 rpm. The toluene phase was removed and further toluene was added; the extraction recovery ranged from 72 to 85%. After shaking and centrifugation, the organic phase was removed. Derivatization was carried out by butylation with Grignard reagent (butylmagnesium chloride (2 mol L^{-1}) in tetrahydrofuran). Separation was by capillary gas chromatography (column of 15 m length, internal diameter of 0.53 mm; stationary phase DB-1 (dimethylpolysiloxane), 1.5 μm film thickness; direct injection of 2 μL; temperature of the injector of 180 °C, column temperature of 50 °C for 1 min, then programmed to 180 °C at 40 °C min^{-1}; He as carrier gas at 18 mL min^{-1}). Detection was by microwave-induced plasma (atmospheric pressure) atomic emission spectrometry at 253.7 nm. Calibration was by calibration graph and standard additions using MeHgCl in Milli-Q water and $HgCl_2$ in HCl as calibrants.

Distillation/GLC-CVAFS

Distillation was performed on a sample of 300 mg by addition of 0.5 mL H_2SO_4 (9 mol L^{-1}) in 20% KCl in H_2O at a temperature of 145 °C; distillation recovery was *ca.* 90%. Derivatization was by addition of 1% $NaBEt_4$ in acetic acid. Separation was by gas liquid chromatography (column of 0.5 m length, internal diameter of 4 mm; Chromosorb W AW-DMSC 60–80 mesh, loaded with 15% OV-3 stationary phase; temperature of the injector of 500 °C, detector temperature at 20 °C; column temperature of 100 °C; He as carrier gas at 40 mL min^{-1}). Detection was by cold vapour atomic fluorescence spectrometry. Calibration was by calibration graph and standard additions using MeHgCl calibrant in Milli-Q water.

Methylmercury Determination in Sediment

Toluene Extraction/GC-ECD

First method: about 2000 mg of sample were treated with HCl and extracted into toluene and cysteine, and back-extracted into toluene; (88 ± 5)% recovery assessed by two spikings of a CRM (PACS-1). The extract was dried with anhydrous Na_2SO_4. The separation was by capillary gas chromatography (column of 25 m length, internal diameter of 0.53 mm; stationary phase BP-1 (methylsilicone), 1.0 μm film thickness; injection of 1 μL; temperature of the injector of 200 °C, temperature of the ECD detector of 250 °C, column temperature of 90 °C; N_2 as carrier gas at 12 mL min^{-1} and makeup gas type at 50.5 mL min^{-1}). Calibration was by calibration graph with MeHgCl in toluene verified with a $MeHgCl_2$ solution.

Second method: about 250 mg sample were treated with HCl and extracted into toluene, repeating the operation twice; EtHgCl was added as internal standard; the recovery [(96 ± 2)%] was verified by spiking the RM used in the intercomparison. Clean-up was performed by equilibration of the MeHg into an aqueous cysteine solution subsequently acidified; MeHgCl was back-extracted into 1 mL toluene. The separation was capillary gas chromatography (S.P.B. 608 Supelchem, column of 30 m length, internal diameter of 0.53 mm, 0.5 μm film thickness; N_2 as carrier gas at 5 mL min^{-1}; Ar/CH_4 as make-up gas mixture at 60 mL min^{-1}) and detection by ECD (injector temperature of 250 °C; detector at 350 °C and column temperature of 100 °C). Calibration was by calibration graph, using a MeHgCl calibrant in toluene verified with similar solution (different producer).

Toluene Extraction/CGC-CVAAS

First method: about 200 mg were mixed with H_2SO_4/NaCl and extracted with toluene and thiosulfate solution; the recovery (85%) was verified by standard additions. Derivatization was carried out with $NaBH_4$ as reducing agent. Separation was by capillary gas chromatography (CP-Sil 8) and the final detection was by cold vapour atomic absorption spectrometry. Calibration was by calibration graph, using a MeHgCl calibrant in toluene.

Second method: HCl was added to 250 mg of sample, followed by toluene extraction, clean-up with cysteine solution and back-extraction into toluene; the recovery of (96 ± 2)% was verified by spiking a CRM extract; an internal standard was added (ethyl mercury). Separation was by capillary gas chromatography (S.P.B. 608 Supelchem; 30 m length; internal diameter of 0.53 mm; film thickness of 0.5 μm; injector temperature of 250 °C; detector temperature of 350 °C; column temperature of 100 °C). Final detection was by cold vapour atomic absorption spectrometry. Calibration was by calibration graph and standard additions, using a MeHgCl calibrant in water verified with different MeHgCl solution.

Microwave Digestion/GC-QFAAS

A subsample of 500 mg was digested by microwave digestion with HNO_3, followed by derivatization with $NaBEt_4$ and cryogenic trapping in an U-tube filled with chromatographic material; the recovery of 95–100% was verified by spiking the CRM before extraction and cross-checking with the RM used in the intercomparison. Separation was by packed column gas chromatography and the final detection was by quartz furnace atomic absorption spectrometry. Calibration was by calibration graph with MeHgCl calibrant in HNO_3.

Distillation/GC-CVAFS

Distillation was performed on a previously wetted sample of *ca.* 200 mg by addition of 0.7 mL of H_2SO_4/KCl mixture, heating the distillation vials at 145 °C; the distillation was carried out under a nitrogen flow rate of 60 mL min^{-1} for 3–6 h until 85–90% of the distillate was collected; distillation recovery of (98 ± 6)% was verified by spiking the CRM before distillation (standard additions). After pretreatment, derivatization was performed by aqueous phase ethylation ($NaBEt_4$) in an acetate buffer solution, after which MeHg was converted into methyl(ethyl)mercury and inorganic mercury was converted to ethylmercury. These volatile species were purged out of the solution by N_2 bubbling through the reaction vessel and collected on Carbotrap columns. Separation was by gas chromatography (column of 2 m length, internal diameter of 4 mm; Chromosorb WAW-DMCS with 15% OV as stationary phase; column temperature of 100 °C; He as carrier gas at 30 mL min^{-1}). After separation, the Hg forms were passed through a pyrolytic column consisting of a 20 cm quartz glass tube partly filled with quartz wool and heated at *ca.* 800 °C. All species were consequently converted to Hg^0 and detected by cold vapour atomic fluorescence spectrometry. Calibration was by calibration graph using a MeHgCl solution in isopropanol.

Distillation/GC-CVAAS

Distillation was performed on a sample of *ca.* 250 mg by addition of a mixture of H_2SO_4/NaCl/H_2O, followed by $NaBEt_4$ derivatization; the recovery (80%) was verified by spiking the CRM before distillation. Separation was by gas chromatography (Chromosorb WAW-DMCS with 15% OV-3 as stationary phase; column length of 0.45 m; 60–80 mesh; column temperature ranging from −196 °C to 180 °C). The final detection was by cold vapour atomic absorption spectrometry. Calibration was by calibration graph and standard additions with MeHgCl in water verified with pure MeHg in isobutanol.

Distillation/HPLC-CVAAS

Ca. 250 mg of sample were mixed with 1 mL of 10% NaCl and 1 mL of 50% H_2SO_4 and diluted to 10 mL with deionized water, then distilled with a distillation rate of 8–10 mL min^{-1} at 140 °C; the distillate was made up to 10 mL with deionized water and an aliquot of 3 mL was buffered with 5% NH_4OAc to pH 6. Complexation was performed by addition of 200 μL of 0.5% SPDC (sodium pyrrolidinedithiocarbamate); after 5 min, the solution was concentrated on a C_{18} column at a flow rate of 1.5 mL min^{-1}; the recovery of (103 ± 7)% was verified by spiking the CRM before distillation and equilibrating overnight (standard additions). Separation was by high performance liquid chromatography (Hypersil ODS, type Chromosphere ODS; 6 cm length; internal diameter of 4.6 mm; 3 μm particle size; MeCN/H_2O as mobile phase at a flow rate of 1 mL min^{-1}). MeHg was reduced to Hg^0 by addition of $NaBH_4$ and detected by cold vapour atomic absorption spectrometry at 253.7 nm. Calibration was by calibration graph and standard additions using a solution of MeHgCl in H_2O as calibrant (verified with a $HgCl_2$ solution).

Distillation/HPLC-ICP-MS

200 mg of the sample were distilled with 500 μL H_2SO_4 (8 mol L^{-1}) and 200 μL 20% NaCl; the solution was diluted to 10 mL with bidistilled water.

Distillation was carried out at a rate of 6–8 mL h^{-1} under an argon flow of 60 mL min^{-1} at 145 °C for 75–90 min; the distillate was collected in a glass vial which was kept in an ice-cooled water bath. The distillate was diluted up to 10 mL and 5 mL were buffered with 0.1 mmol L^{-1} ammonium acetate up to pH 6; the recovery of (104 ± 6)% was verified by standard additions at three spiking levels on wet sediment and equilibration for 5 h. After adding a fresh solution of 0.5 mmol L^{-1} sodium pyrrolidinedithiocarbamate, the 5 mL sample was preconcentrated with a flow of 1 mL min^{-1} on a RP-C_{18} ODS Hypersil column (acetonitrile/water as mobile phase at a flow rate of 1 mL min^{-1}). The enriched and complexed mercury compounds were separated by high performance liquid chromatography, oxidized on-line by UV irradiation, reduced by $NaBH_4$ (in 0.1 mol L^{-1} NaOH) and swept out by argon into the torch of an ICP-MS for final detection. Calibration was by calibration graph and standard additions with a solution of MeHgCl in water.

Toluene Extraction/HPLC-CVAFS

1000 mg of sample were extracted in 7 mL HCl (6 mol L^{-1}), twice with 5 mL toluene, and back-extracted into 1 mL $Na_2S_2O_3$, and an acetate buffer (NH_4OAc) was added; the recovery of (103 ± 2)% was verified by spiking the CRM before extraction. Complexation by mercaptoethanol was performed on-line with HPLC separation (Chromosphere C_{18}; 20 cm length; internal diameter of 3 mm; 5 μm particle size; MeOH/H_2O as mobile phase at a flow rate of 0.5 mL min^{-1}); the mercaptoethanol complexes were eluted from the column and destroyed by on-line oxidation (H_2SO_4 and $CuSO_4$ as catalyst); the resulting Hg^{2+} was reduced to Hg^0 by on-line reaction with $SnCl_2$ (1.5% in 1 mol L^{-1} NaOH) and purged from the eluent with argon at a flow rate of 50 mL min^{-1}. The argon stream carrying the Hg^0 was dried with a membrane before detection by atomic fluorescence spectrometry. Calibration was by calibration graph and standard additions with MeHgCl in water.

Supercritical Fluid Extraction/GC-MIP

About 500 mg of sample were extracted with supercritical CO_2 and eluted with toluene; the extraction recovery was (58 ± 6)% as verified by single MeHg addition to the CRM. Derivatization was carried out by butylation with Grignard reagent [butyl magnesium chloride (2 mol L^{-1}) in tetrahydrofuran]. Separation was by capillary gas chromatography (column of 15 m length, internal diameter of 0.53 mm; stationary phase DB-1 (dimethylpolysiloxane), 1.5 μm film thickness; direct injection of 2 μL; temperature of the injector of 180 °C, column temperature of 50 °C for 1 min, then programmed to 180 °C at 40 °C min^{-1}; He as carrier gas at 18 mL min^{-1}). Detection was by microwave-induced plasma (atmospheric pressure) atomic emission spectrometry at 253.7 nm. Calibration by calibration graph and standard additions using MeHgCl in Milli-Q water and $HgCl_2$ in HCl as calibrants.

4.3 Interlaboratory Studies

From the beginning it was clear that it was essential to examine each step critically in the methods for MeHg determination. To do so, several series of samples were prepared to (i) test the detection methods with solutions of known composition of Hg compounds, (ii) test the performance of the separation using cleaned extracts, (iii) verify the clean-up procedures by performing analysis of raw extracts and (iv) test the total analytical procedure by analyzing real samples. These evaluations were carried out in three intercomparisons, the first one of which started in 1987 (simple solutions), the second one in 1988 (cleaned and raw extracts) and the third round-robin exercise was carried out in 1989 (extracts and real samples) [81]. Youden plots [83] in some cases supported the technical conclusions.

First Interlaboratory Study

Samples

Three solutions were studied in the first interlaboratory study. Solution 1 contained about 10 mg kg^{-1} of MeHgCl in toluene and solution 2 was a mixture of *ca.* 10 mg kg^{-1} each of MeHgCl, Et-HgCl and Ph-HgCl. These solutions were prepared by the Danish Isotope Centre (Denmark) by dissolving the mentioned mercury compounds in 5 mL DMSO and in 10 L toluene. Samples were provided in 250 mL bottles protected against light. Solution 3 was an aqueous solution containing 2 mg kg^{-1} of MeHgCl and $HgCl_2$. The optimal NaCl and HCl concentrations to avoid adsorption were studied. This sample was prepared at the KFA in Jülich (Germany).

The stability of solutions 1 and 2 was verified on the content of 10 bottles stored at 4 °C in the dark over a period of 8 weeks at the National Food Agency. Analyses were performed by packed column gas chromatography followed by electron capture detection. Determinations were performed in triplicate on each of three bottles.

Stability tests of solution 3 (aqueous solution) were performed at the KFA by ion exchange/CVAAS. In a first stage, storage experiments at 0 °C and at ambient temperature were carried out but no measurable effects of temperature were observed. However, significant losses of Hg were observed after a 100-fold dilution of the solution containing approximately 2 mg L^{-1} of MeHgCl and ionic Hg; therefore it was recommended to dilute the solution only just prior to the determination. The stability was verified over three months on the content of one bottle and no significant changes were observed for both MeHg and ionic Hg at ambient temperature. All determinations were performed in seven replicates.

Results

The first intercomparison of the solutions did not reveal any major discrepancies in the results and hence in the final methods of determination used. The

Table 4.1 *Results of the three intercomparisons*[a]

Comparison	*CV%*	*Range*
Solution 1 (first round)	8.0	1.3
Solution 2 (first round)	8.9	1.2
Solution 3 (frst round)	12.3	1.4
Raw extract (second round)	16.6	1.6
Spiked extract (second round)	17.4	1.6
Cleaned extract (second round)	12.5	1.5
Aqueous solution (second round)	8.4	1.3
Raw extract (third round)	11.3	1.7
Spiked extract (third round)	8.8	1.4
Mussel tissue (third round)	17.4	1.7
Tuna fish (third round)	13.7	1.6

CV: coefficient of variation (%) between laboratories.
Range: ratio of higher value/lower value.

mean of laboratory means was in all cases close to the value expected upon preparation [(10.1 ± 0.8) mg kg^{-1} as MeHgCl for solution 1, (12.7 ± 1.2) mg kg^{-1} as MeHgCl for solution 2 and (2.13 ± 0.26) mg kg^{-1} as MeHgCl for solution 3, respectivel]). Table 4.1 lists the coefficient of variation (*CV*) obtained between laboratories: both for solutions 1 and 2, the *CV*'s obtained (8.0 and 8.9%, respectively) were considered to be acceptable. In the case of the aqueous solution (solution 3), a *CV* of 12.3% was found to be too high for the present state of the art. On the basis of these results, it was decided to organize a second interlaboratory exercise [81].

Second Interlaboratory Study

Samples

Approximately 4 kg of flounder was purchased at Sletten Havn located in Niva Bay, 30 km north of Copenhagen, where high levels of total Hg in fish tissues had been reported previously. The fish sample was mixed with redistilled water, homogenised and stored at −20 °C. Six sub-samples of homogenate each of 0.2 g were analysed for total Hg by RNAA. The total content was found to be (191 ± 20) μg kg^{-1} (as Hg) on wet mass basis. Extracts were then prepared by the Danish Isotope Centre and the stability of

MeHg was verified by the National Food Agency (Denmark). Another batch of aqueous solutions as described under the first intercomparison was prepared by the KFA and sent to the participants. The preparation of the samples was the following.

Raw extract. Subsamples of 30 g of fish homogenate were mixed with 80 mL HCl and 20 mL $CuSO_4$. This mixture was shaken, left to react for 15 min and extracted 3 times with toluene to obtain *ca.* 37 L of extract which was dried by addition of anhydrous Na_2SO_4 and stored at 5–10 °C. The samples were bottled in 500 mL light-protected borosilicate bottles with PTFE gaskets in the screw cap.

Raw extract spiked with MeHg. Approximately 2500 mL of the raw extract were spiked with MeHgCl dissolved in DMSO to obtain a concentration of about 11 $\mu g\ L^{-1}$. Samples were bottled in 250 mL light-protected borosilicate bottles.

Cleaned extract. Sub-samples of the raw extract were extracted twice with 150 mL of cysteine acetate. After acidification with HCl, the mixtures were back-extracted twice with toluene. This cleaned extract was dried by addition of anhydrous Na_2SO_4 and samples were distributed in 100 mL bottles.

Aqueous solution. 1 mg of MeHgCl and $HgCl_2$ were dissolved in water containing 30 g NaCl and 25 mL of HCl per liter. Samples were bottled in 250 mL borosilicate glass bottles with screw caps and stored in the dark at ambient temperature.

All the samples were provided with solid MeHgCl calibrants (purity >99.9%). The stability of the extracts stored at 4 °C in the dark was verified 5 months after the preparation at the National Food Agency. Analyses were performed by packed column gas chromatography followed by electron capture detection either by direct injection (cleaned extract) or after clean-up with cysteine/toluene (raw and spiked raw extracts) in triplicate in each of three bottles.

Stability tests of the aqueous solution were performed at the KFA. The stability was verified over 2 months on the content of one bottle (10 replicate analyses) and no significant changes were observed for MeHg.

Results

Analyses of extracts led to difficulties, mainly due to a lack of good long-term reproducibility for many laboratories. Capillary GC was found to offer good possibilities but its use was hampered by the absence of commercially available columns. Furthermore, sources of error were likely due to losses of MeHg. A Youden plot of raw and spiked extract demonstrated systematic errors (Figure 1) which was illustrated by the high *CV*'s found between laboratories (Table 4.1: 16.6 and 17.4% for raw and spiked raw extracts, respectively). Better results were obtained for the cleaned extract (12.5%) but the spread was still considered to be too high. However, a consequent improvement was obtained for the aqueous solution analysis (*CV* of 8.4%).

Owing to the high spread of results obtained for the MeHg determinations

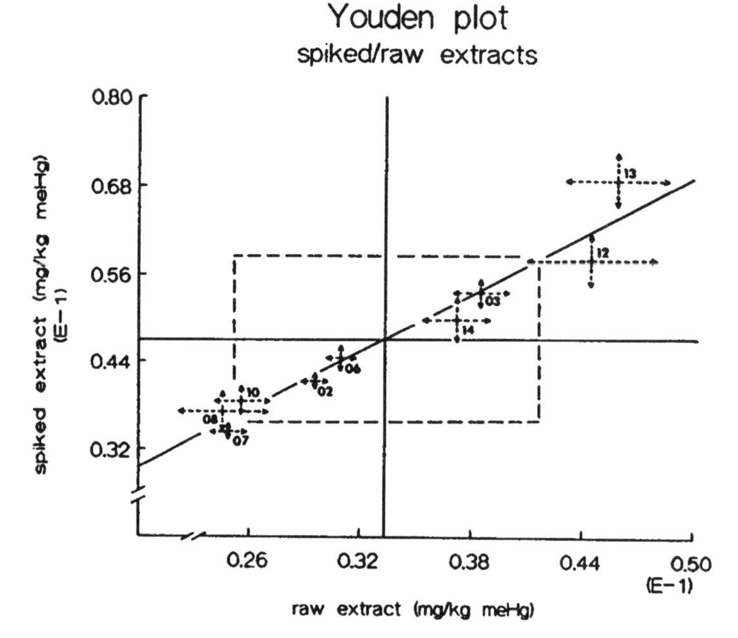

Figure 4.1 *Youden plot. MeHg in spiked extract versus MeHg in raw extract. Horizontal and vertical continuous lines are the means of the laboratory means, dotted lines being the standard deviations of these means. The length of the bars are equal to the standard deviation (5 replicates) of the laboratories.*

[Reproduced by permission from P. Quevauviller *et al.*, *Appl. Organomet. Chem.*, **7**, 413 (1993). © John Wiley & Sons Limited]

in the extracts, it was decided to organize a third round-robin exercise to improve the situation.

Third Interlaboratory Study

Samples

4.6 kg of cod homogenate was prepared as described before from samples fished in the Koege Bay, south from Copengagen. The total Hg concentration determined by CVAAS in triplicate in the homogenate was *ca.* 190 μg kg^{-1}.

Raw extract. Portions of 30 g of fish homogenate were treated with 80 mL HCl and 20 mL $CuSO_4$ solution, shaken and left to react for 15 min. This mixture was extracted three times with about 80 mL of toluene per extraction. Portions of 1500 mL of the toluene extract were then extracted with 400 mL of cysteine acetate solution. After separation, the cysteine acetate solution was acidified and extracted twice with 95 mL of toluene. A batch of 4000 mL of toluene was produced. A method and reagent blank solution was produced alternating with the actual extraction using the same glassware and reagent.

Calibrant solution. MeHgCl was dissolved in toluene in order to obtain a stock solution with a concentration of 44.43 mg kg^{-1}. 10.5 g of this solution was diluted with toluene to obtain a calibrant solution of 266.9 µg kg^{-1}.

Spiked extract. 4.4 g of the stock solution was added to the toluene extract (raw extract) to obtain a concentration in the spiked extract of (0.9972X + 123.77) µg kg^{-1} of MeHgCl where X is the concentration in the toluene extract. The four toluene extracts were bottled in 50 mL light-protected borosilicate bottles with PTFE gaskets in the screw cap.

In addition to these samples, mussel and tuna samples were prepared at the Joint Research Centre of Ispra (Italy). Mussel flesh (wet weight) was collected and minced whereas the tuna fish was filleted; the samples were frozen, freeze-dried, ground and homogenized. The two samples were bottled in brown borosilicate bottles each containing 15 to 20 g.

The stability of the calibrant solution and extracts was verified over a period of 5 months at the National Food Agency (Denmark). Analyses were performed by packed column gas chromatography followed by electron capture detection in five replicates on each of three bottles.

Stability of freeze-dried fish sample. The stability of MeHg in freeze-dried biological samples is an important figure to be known when preparing a certification campaign. Preliminary experiments have been performed at the Danish Isotope Centre on a laboratory reference material of freeze-dried tuna fish. No instability of the MeHg content in the RM stored at ambient temperature could be observed after three years. Both total organic mercury and methyl/phenylmercury were determined at regular intervals by RNAA [81].

Results

Sources of discrepancies were found for the analysis of the matching calibrant, the most important one being the inadequacy of the use of packed chromatographic columns. The *CV* obtained for raw data was 13.7% (Table 4.1); however, the results improved to 6.3% after outliers had been removed (on technical grounds). The results obtained with CP-SIL 8 capillary columns appeared in most of the cases better than the ones using packed columns. It was stated that a precision (as *CV*) of *ca.* 3–4% can be achieved with CP-SIL 8 capillary columns lasting for 1–2 years using proper optimization. Extensive work was carried out to evaluate the CP-SIL 8 columns [83]; the results showed that the use of an on-column insert is recommended to avoid losses of Hg due to contact with hot metal surfaces in the injector. The experiments suggested also that it is important to use capillary columns with a thick film, considering that such film reduces the contact between the volatilized mercury and a possibly not entirely deactivated silica column.

An additional source of error was calibration, which should be performed systematically using the compound to be determined (*e.g.* MeHgCl) and not with, for example,$HgCl_2$.

The *CV*'s between laboratories (Table 4.1) showed, however, that the state of the art improved in comparison with the second round-robin exercise

(11.3% instead of 16.6% for the raw extract, and 8.8% instead of 17.4% for the spiked raw extract).

With respect to mussel and tuna analyses, interferences were systematically higher with mussel tissue but the higher *CV* obtained (17.4% in comparison with 13.7% for tuna fish) could also be due to the fact that the MeHg content in mussel was much lower than that in tuna fish (0.14 ± 0.01 mg kg^{-1} as MeHgCl in mussel and 4.33 ± 0.11 mg kg^{-1} as MeHgCl in tuna fish). The use of cysteine paper was also recommended in order to remove impurities by washing repeatedly with toluene whilst MeHg is immobilized on the cysteine paper.

4.4 Certification of Fish Reference Materials

It was recognized that there was a strong need for controlling the quality of methylmercury determinations in fish by using certified reference materials (CRM's) representative of samples moderately to highly contaminated in mercury. The existing CRMs of tissues produced by the National Research Council of Canada (DOLT-1 and DORM-1) [84,85] were considered to be suitable for controlling the quality of analysis of low levels of contamination (certified values of 0.080 mg kg^{-1} and 0.731 mg kg^{-1} of MeHg, respectively), but CRMs containing more than 2 mg kg^{-1} of MeHg were deemed necessary for quality control of the analysis of contaminated fish tissues. Consequently, the BCR was requested to produce fish materials with MeHg levels representative of moderate to high contamination. These materials (CRMs 463 and 464) were prepared in 1989 and have been certified for their content of total mercury and methylmercury [86,87]. This section presents the certification work performed.

Preparation of the Candidate CRMs

Collection

The two candidate reference materials were collected in the Adriatic Sea; they were produced from tuna fish that were rejected from the normal trade because their total mercury content excedeed 0.8 mg kg^{-1}. Samples of 302 kg and 322 kg of tuna fish, respectively for CRM 463 and CRM 464, were sliced, frozen (−25 °C) and transported to Ecoconsult in Gavirate (Italy). The dorsal fish muscles of each material were minced using a Quick Mill 2300 mincer with tungsten carbide blades. After mincing, the material obtained was stored frozen in high density polyethylene containers. The material was then freeze-dried until reaching a moisture mass fraction below 2.5%. The resulting material (*ca.* 36 kg of each candidate CRM) was immediately frozen.

Homogenization and Bottling

The freeze-dried materials were sent to the Joint Research Centre of Ispra where they were ground using a mill equipped with zirconium dioxide balls. The ground materials were sieved using a vibrating stainless steel sieve. The fractions with particles larger than 125 μm were discarded and the remaining materials were stored in polyethylene boxes in an argon atmosphere. The two materials were then homogenized in a mixing drum for 16 days and bottled in brown borosilicate glass bottles. A total of 1000 bottles each containing *ca.* 15 g of material were produced for each candidate reference material. Both materials were stored at 4 °C.

Homogeneity Control

The between-bottle homogeneity was verified by the determination of total mercury and methylmercury on sample intakes of 0.2 g taken from 20 bottles which were set aside at regular intervals during the whole period of bottling. The within-bottle homogeneity was assessed by 10 replicate determinations on the content of one (re-homogenized) bottle.

For the determination of total mercury, the samples were mineralized by microwave digestion using nitric acid. The final determination was performed by CVAAS. Methylmercury was determined by CGC/ECD after extraction of 0.2 g fish powder in toluene, back extraction with a cysteine acetate solution and further extraction with toluene. Calibrations were performed by standard additions. The uncertainties of the methods of final determination were assessed by the analysis of 10 aliquots from one digest solution (nitric acid for total mercury, toluene for methylmercury). The *CV* of the method, therefore, does not comprise the *CV* introduced by the extraction procedure.

The *CV*'s for total mercury and methylmercury in CRMs 463 and 464 are presented in Table 4.2. An *F*-test at a significance level of 0.05 did not reveal significant differences between the within- and between-bottle variances for MeHg in both materials and total Hg in CRM 464. The within-bottle *CV* is very close to the *CV* of the method and, therefore, no inhomogeneities of the material were suspected. In the case of total Hg in CRM 463, the within- and between-bottle *CV*'s were slightly higher than the *CV* of the method; this difference is likely due to an increase in uncertainty related to the mineralization step with a microwave oven, which may differ from one material to another. In the case of MeHg, it is supposed that the extraction step is less vulnerable to variations. The fact that the within- and between-bottle *CV*'s overlap proves that there is no difference of inhomogeneity between bottles.

It was concluded that the two materials are suitable for use as CRMs and are homogeneous at an analytical portion of 0.2 g and above for total mercury and methylmercury

Stability Control

The stability of the total mercury and methylmercury content was tested at −20 °C, +20 °C and +40 °C. Total mercury and methylmercury were determined at the beginning of the storage period and after 1, 3, 6 and 12 months. Samples were analysed using the same procedures as for the homogeneity study. Total mercury and methylmercury were each determined five times (one

Table 4.2 *Homogeneity study: within- and between-bottle homogeneity for CRM 463[a] and CRM 464*

Component	*Between-bottle*	*Within-bottle*	*Method of final determination*
CRM 463			
Total Hg	6.8 ± 1.1	6.0 ± 1.3	2.8 ± 0.9
MeHg	3.6 ± 0.6	4.4 ± 1.0	2.6 ± 0.8
CRM 464			
Total Hg	3.9 ± 0.6	3.4 ± 0.8	2.8 ± 0.9
MeHg	5.3 ± 0.8	2.3 ± 0.5	2.6 ± 0.8

[a] $(CV \pm U_{CV})$%.
Uncertainty on the *CV*s: $U_{CV} \approx CV/2\sqrt{n}$.
[b] Single determination on the content of each of 20 bottles.
[c] 10 replicate determinations on the content of one bottle.
[d] 10 replicates of a digest/extract solution.

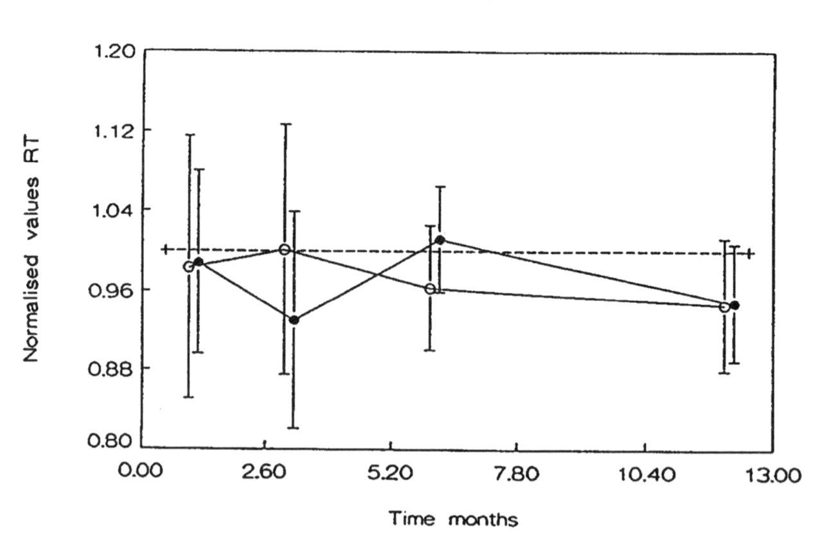

Figure 4.2 *Stability tests of MeHg in tuna fish for CRMs 463 and 464.* $R_T = X_T/X_{-20°C}$, *where* X_T *is the mean of 10 replicates at temperature* T *(+20 °C or +40 °C), and* $U_T = (CV_T^2 + CV_{-20°C}^2)^{1/2} R_T$, *where* CV_T *is the coefficient of variation of 10 replicates at temperature* T

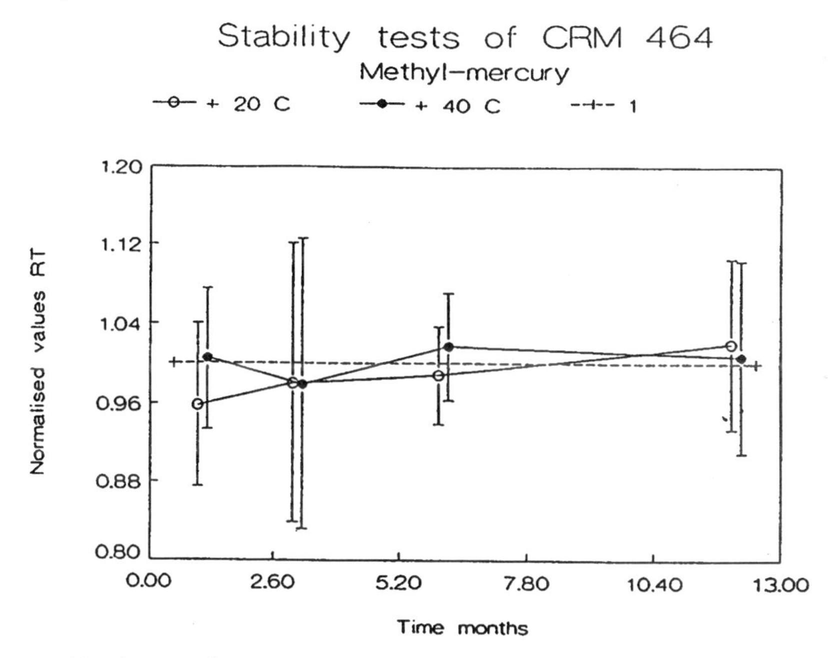

Figure 4.2 (*continued*)

replicate analysis in each of five bottles stored at different temperatures) at each occasion of analysis. The stability was monitored, following the procedure described in Chapter 3. On the basis of the results (see Figure 4.2), it was concluded that no instability could be demonstrated [86].

Technical Evaluation

The results submitted in the certification were discussed amongst all the participants in a technical meeting; they were presented in the form of bar graphs showing the laboratory codes and the abbreviation of the method used, the individual means and standard deviations and the mean of laboratory means with its standard deviation. Figures 4.3a,b show the sets of results for MeHg (after technical scrutiny) in the CRMs 463 and 464, respectively.

Low results were observed for a technique using distillation: the recovery of the distillation procedure had been verified by a spike (distillation stopped on the basis of the distilled spike solution). Usually, the recovery ranged from 90 to 95% but in the case of CRM 463 it varied between 68 and 95% which was due to a too low MeHg spike content compared to the methylmercury present in the sample. This procedure was not considered to be acceptable and the results were therefore rejected.

Too high results provided by one participant were due to the fact that the calibrants were added after the digestion step (with HCl), which resulted in different treatment between calibrants and samples, and possible different

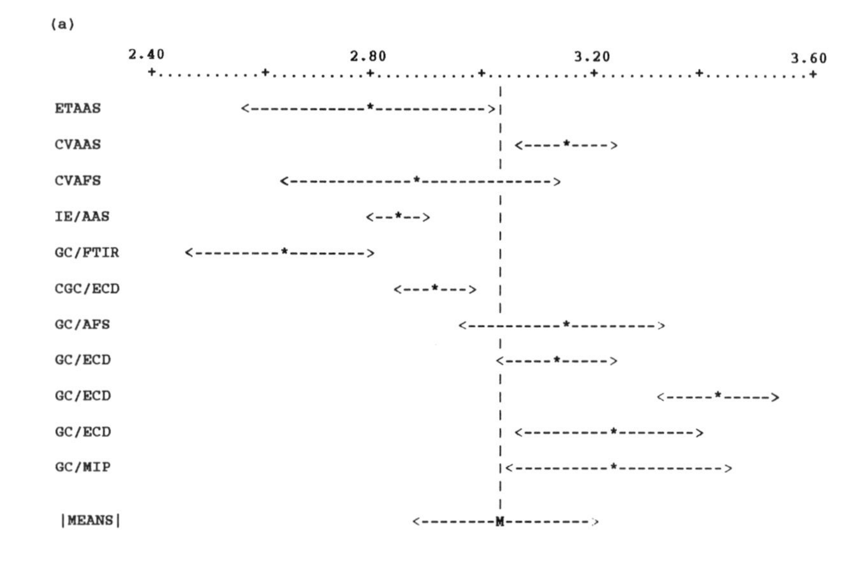

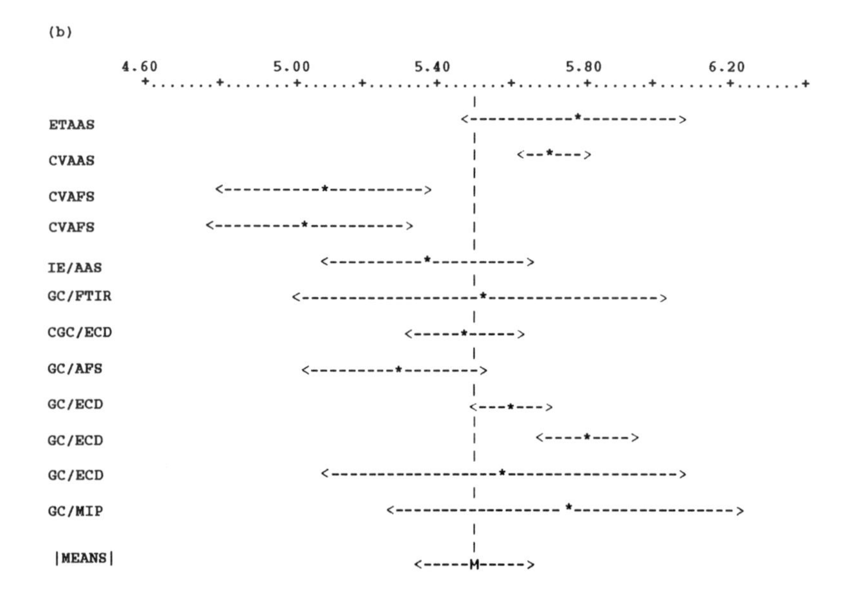

Figure 4.3 *Bar-graphs of MeHg certified values in CRMs 463 and 464. The results correspond to five replicate determinations, "MEANS" is the mean of laboratory means with 95% confidence interval.*

Table 4.3 *Certified mass fractions of total mercury and methylmercury in CRMs 463 and 464*[a]

Compound	*Certified value*	*Uncertainty*	*Unit*	*p*
CRM 463				
total Hg	2.85	0.16	$mg\,kg^{-1}$ (as Hg)	8
MeHg	3.04	0.16	$mg\,kg^{-1}$ (as MeHg)	11
CRM 464				
total Hg	5.24	0.10	$mg\,kg^{-1}$ (as Hg)	8
MeHg	5.50	0.17	$mg\,kg^{-1}$ (as MeHg)	12

[a] The certified values are the unweighted mean of *p* accepted sets of results.

extraction efficiencies. The laboratory withdrew its results, which were not used in the calculation of the certified value.

Certified Values

Following the technical evaluation, the sets of accepted results were submitted to statistical tests (Kolmogorov–Smirnov–Lilliefors, Nalimov, Bartlett and Cochran tests, and one-way analysis of variance) which are described in detail in the certification report [86]. The certified values (unweighted mean of *p* accepted sets of results) and their uncertainties (half-width of the 95% confidence intervals) are given in the Table 4.3 as mass fractions (based on dry mass). Total and methylmercury are certified as mass fractions ($mg\,kg^{-1}$) of Hg and $MeHg^+$ respectively.

4.5 Interlaboratory Study on Sediment

Preparation of the Material

Characterization of the Bacterial Flora

The main difficulty in preparing a sediment reference material for interlaboratory studies on chemical species is to achieve the stability of the relevant compounds [12]. With respect to methylmercury, the main source of instability is due to bacteria, either by demethylation [88,89] or formation of volatile dimethylmercury [90]. The conversion is indirectly provoked by the biological activity of various types of bacteria, such as (i) aerobic mesophilic heterotrophic microorganisms, (ii) anaerobic sulfate-reducing bacteria and (iii) anaerobic spore-forming bacteria. In order to control the remaining bacteria present after different irradiation treatments, a bacterial enumeration was performed on samples which were dehydrated after homogenization and irradiated at various γ-ray doses (0, 4, 8, 12, 25 and 50 kGy). The determination of the sulfate-reducing bacteria was performed using the Most Probable Number (MPN) technique [91], anaerobic spore forming bacteria were enumerated by plate count on nutrient agar; and the enumeration of aerobic mesophilic heterotrophic microorganisms was performed by the plate count method.

The cultivable bacteria of the three groups were high in sediment samples analysed immediately after drying and homogenization processes (non-irradiated samples). In the sample irradiated at the 4 kGy dose, the bacteria content was significant for the heterotrophic and the spore-forming bacterial group, whereas the sulfate reducers were inhibited.

As a conclusion of these tests, the γ-irradiation of sediment at 8 kGy was found to be able to sterilize the sediment completely.

Sample Collection and Treatment

Sediment samples were collected in the Ravenna Lagoon at two locations, namely close to a petrochemical plant water discharge (high Hg levels) and downstream. Two batches of *ca.* 250 kg were made available. The wet materials were dried at ambient temperature under an air stream. The two batches of dry sediment were then mixed and homogenized in order to obtain around 80 kg of dry homogenized sediment to be used as candidate certified reference material. A composite sediment was prepared for the purpose of the interlaboratory study by mixing two parts of contaminated sediment with one part of sediment containing low levels of methylmercury; around 100 bottles each containing *ca.* 50 g were prepared and distributed to the participating laboratories.

Homogeneity and Stability Tests

The homogeneity of the sediment material was verified by taking one bottle out of 20 bottles during the bottling procedure and determining the total mercury and methylmercury content at 50, 100 and 250 mg levels of intake (between-bottle homogeneity). The within-bottle homogeneity was verified by 10 replicate determinations performed in one bottle. The methodological uncertainty was determined by five replicate analyses of a digest solution. For methylmercury determination, 0.2 g of sediment was digested in 5 mL of H_2SO_4 by heating for 30 min in a water bath at 100 °C; this was followed by toluene extraction and back-extraction into thiosulfate. After $NaBH_4$ derivatization, methylmercury hydrides were cryogenically trapped and purged for GC-FTIR separation. The final detection was by CVAAS. Total mercury was determined by CVAAS after $SnCl_2$ reduction and *aqua regia* digestion. Both homogeneity and stability tests demonstrated that the sediment reference material was suitable for use in this interlaboratory study.

Effects of γ-Irradiation on MeHg Content

The effects of various γ-doses on the stabilization of MeHg were studied by determining the content in five bottles submitted to doses of 4, 8, 12.5, 25 and

50 kGy. No significant differences were observed in the results, which varied from (54.3 ± 3.3) to (56.4 ± 1.8) μg kg^{-1} as MeHg. It was hence concluded that methylmercury is not affected by γ-irradiation and that this procedure could be used safely for the stabilization of the material used in this study and for the candidate reference material [92].

Results

As an introductory remark to the discussion, at first glance the very high results submitted by one method (obtained with a technique involving extraction/CVAFS with no separation step) clearly showed its inadequacy for sediment analysis due to the extraction of inorganic mercury at the same time as MeHg. The overall set of data showed three outliers: (596 ± 48) μg kg^{-1}, (317 ± 43) μg kg^{-1} and (184 ± 21) μg kg^{-1}, which were first discussed.

One set obtained by HPLC demonstrated that this technique was not optimized for sediment analysis. Doubts were thrown on the extraction as well as on the separation; in addition, complexation with mercaptoethanol was not considered to be well adapted. This technique was not considered to be suitable for certification at this stage. Considerable improvements were desirable.

The coupled GC-QFAAS technique (involving cold trapping and derivatization with tetraethylborate) was faced with difficulties in the analysis of this organic-rich sediment. It was suspected that alkaline digestion was not suitable in this case and that the technique should be tested using other alternative digestion procedures.

The selected data showed a better agreement but the spread was, however, still considered to be too high (*CV* of 47% between laboratory means). This spread was mainly due to some apparent outliers, *i.e.* (13.9 ± 1.9) μg kg^{-1}, (19.1 μg kg^{-1}, only one replicate), (21.5 ± 1.1) μg kg^{-1}) and (30.9 ± 2.5) μg kg^{-1}.

Problems were experienced with thiosulfate back-extraction followed by DDTC complexation. It is likely that problems of complexation occurred between thiosulfate and SPDC which would explain a low recovery. The participant stressed that this problem was not observed for water analysis in which a one-step complexation procedure is used. Another laboratory similarly reported low results; an additional reason for explaining these was that the results were not corrected for recovery. Recommendations were given that experiments be performed to optimize the technique prior to certification.

The technique based on toluene extraction, derivatization and thiosulfate back-extraction followed by CVAAS detection was questioned since no separation is involved; it was suspected that inorganic Hg might have been detected beside MeHg. Proof should be given that only MeHg is determined.

Low results were suspected to be due to a low efficiency of the distillation procedure which should be verified by spiking samples. Further improvements were recommended. Problems at the distillation step were also experienced by another participant which had a rather large standard deviation, suspected to be due to interfering compounds when the distillation was too long. It was

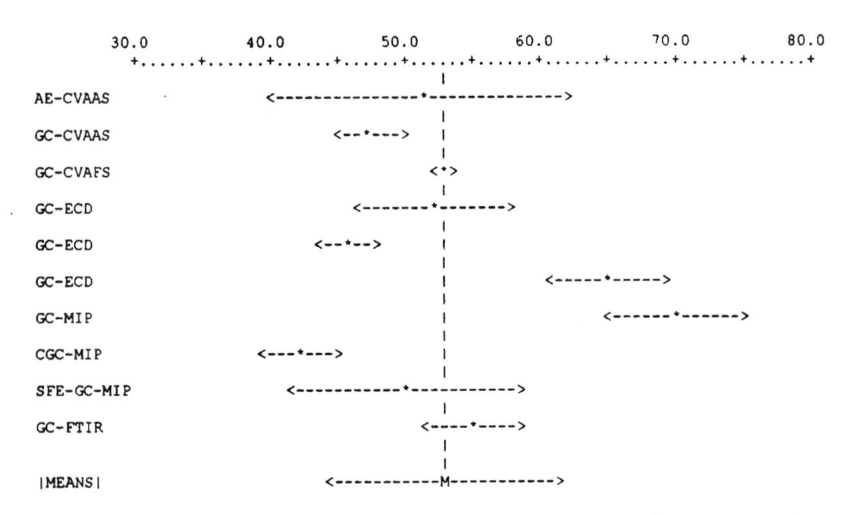

Figure 4.4 *Bar graph of MeHg in sediment. The laboratory codes are given along with the methods used. The results plotted correspond to five replicate determinations. M is the mean of the laboratory means =* (53.1 ± 8.5) *µg kg*$^{-1}$*. The results given in this graph correspond to the sets of data selected after technical scrutiny.*

recommended to stop the distillation after 85% to avoid this interference, which led to higher results (and hence high standard deviations). Furthermore, since the results are based on calculations by differences (measurement before and after UV destruction), it was requested that proof be given that only MeHg is actually determined.

Other technical remarks were made on the selected set of data (Figure 4.4), *e.g.* the precision of supercritical fluid extraction (SFE) could be improved by extending the extraction time in order to achieve a better recovery. Finally, interferences from sulfur extracted at the same time as MeHg were noted in MIP, stressing the importance of possible clean-up.

4.6 Certification of sediment reference material

Preparation of the candidate CRM

Collection

The candidate CRM was collected (30–40 cm top sediment layer) in the Ravenna Lagoon (Italy), sieved to pass apertures of 2 mm and air-dried at 25 °C in a drying chamber. The residual humidity was measured periodically; when a constant humidity level was reached (3.5%), the material was manually crushed and sieved again at 2 mm (the fraction higher than 2 mm was discarded). The resulting material was stored in polyethylene bags and transported at the Joint Research Centre of Ispra (Italy).

Homogenization and Bottling

The dry sediment was passed through a tungsten carbide-bladed hammermill and sieved to pass apertures of 90 μm. The <90 μm fraction was collected in a PVC mixing drum (filled with dry argon) and homogenized for 14 days at about 48 rpm.

The bottling procedure was performed manually: after an additional period of mixing of 2 days, a first series of 20 bottles was filled and immediately closed with screw caps and plastic inserts. Several series of 20 bottles were hence successively filled, alternating with re-mixing of the powder for 2 minutes. All bottles were stored at ambient temperature in the dark.

Homogeneity Control

The between-bottle homogeneity was verified by the determination of total mercury and methylmercury on sample intakes of 50, 100 and 250 mg taken from 20 bottles which were set aside at regular intervals during the whole period of bottling. The within-bottle homogeneity was assessed by 10 replicate determinations of total mercury and methylmercury on sample intakes of 50, 100 and 250 mg on the content of one bottle which was first re-homogenized.

For the determination of total mercury, 100 mg dry sediment was mineralized by addition of *aqua regia*. The final determination was performed by CVAAS after $SnCl_2$ reduction. Methylmercury was determined by CGC followed by HGAAS. A portion of 0.1–0.5 g dry sediment was extracted by addition of 5 mL H_2SO_4 (2.5 mol L^{-1}), followed by addition of 5 mL toluene; after centrifugation, the supernatant was placed into a 10 mL vial containing 4 mL thiosulfate solution. After shaking for 30 s, the toluene layer was discarded and the thiosulfate solution was transferred into a 25 mL beaker and heated on a hot plate (200 °C) to evaporate the solution to 2–3 mL. After cooling, the remaining solution was diluted to 10 mL with water and stored for analysis. An aliquot of this extract was transferred into a purge trap vial and 100 μL of 1% $NaBH_4$ was added to the purged solution. The methylmercury hydride formed was trapped at 120 °C in a capillary GC column, separated and detected by cold vapour AAS. The uncertainties of the methods of final determination were assessed by the analysis of 5 aliquots from one digest solution (nitric acid for total mercury, toluene for methylmercury). The *CV* of the method, therefore, does not comprise the *CV* introduced by the extraction procedure.

The *CV*'s for total mercury and methylmercury in CRM 580 are presented in Table 4.4. An *F*-test at a significance level of 0.05 did not reveal significant differences between the within- and between-bottle variances for MeHg and total Hg in the candidate CRM. The within-bottle *CV* for total mercury is very close to the *CV* of the method; with respect to methylmercury, the higher value of the within-bottle *CV* in comparison to the method *CV* relates to the additional uncertainty related to extraction which is not taken into account in the method *CV* calculation (analysis of extracts). On the basis of these results,

Table 4.4 *Within- and between-bottle homogeneity for CRM 580*[a]

Component	*Between-bottle*[b]	*Within-bottle*[c]	*Method of final determination*[d]
50 mg intake			
Total Hg	–	4.5 ± 1.2	–
MeHg	–	7.2 ± 2.2	–
100 mg intake			
Total Hg	–	5.1 ± 1.6	–
MeHg	–	6.2 ± 1.9	–
250 mg intake			
Total Hg	3.6 ± 0.6	5.0 ± 1.5	4.1 ± 1.3
MeHg	7.2 ± 1.2	5.6 ± 1.8	2.5 ± 0.8

Uncertainty on the *CV*'s: $U_{CV} \approx CV\sqrt{2n}$
[a] $(CV \pm U_{CV})$%.
[b] Single determination on the content of each of 20 bottles.
[c] 10 replicate determinations on the content of one bottle.
[d] 5 replicates of a digest/extract solution.

no inhomogeneities of the material were suspected. It was concluded that the material is suitable for use as a CRM and is homogeneous at an analytical portion of 50 mg and above for total mercury and methylmercury.

Stability Control

The stability of the total mercury and methylmercury content was tested at −80 °C, +20 °C and +40 °C over a period of 15 months. Total mercury and methylmercury were determined at the beginning of the storage period and after 1, 3, 6, 11 and 15 months. Samples were analysed using the same procedures as for the homogeneity study. Total mercury and methylmercury were each determined five times (one replicate analysis in each of five bottles stored at different temperatures) at each occasion of analysis.

The stability was evaluated, following the procedure described in section 4.3. On the basis of the results (Table 4.5), it was concluded that no instability of the material could be demonstrated.

Technical Evaluation

The determination of MeHg in sediment is prone to a range of possible sources of errors that have to be carefully controlled, *e.g.* insufficient extraction, insufficient derivatization or distillation, or interferences in detection. The first aspect considered in the technical discussion was related to the verification of the calibrants. Most of the participants verified their calibrant using alternative calibrant solutions; other participants preferred to check

Table 4.5 *Normalized results of the stability study*

Species	*Time (months)*	$R_t \pm U_t$	
		20 °C	*40 °C*
Total Hg	1	1.06 ± 0.07	1.01 ± 0.09
	3	1.10 ± 0.05	1.04 ± 0.03
	6	0.96 ± 0.04	0.98 ± 0.04
	11	1.02 ± 0.03	1.10 ± 0.06
	15	1.06 ± 0.07	1.00 ± 0.03
Methylmercury	1	1.05 ± 0.10	1.13 ± 0.08
	3	1.00 ± 0.06	1.14 ± 0.06
	6	0.99 ± 0.06	1.05 ± 0.08
	11	0.98 ± 0.09	0.99 ± 0.07
	15	1.07 ± 0.15	1.09 ± 0.14

their techniques using a certified reference material of sediment (the one produced by the IAEA [93]).

Discussions arose on the verification of the extraction recoveries. At present, there is no standardized procedure to check the extraction efficiency; the methods used differed widely from one laboratory to another. The participants recognized that it would be necessary to find out the most suitable recovery test to propose a standardized procedure in order to avoid possible discrepancies; the technique which was most supported was standard addition (*e.g.* three levels on wet sediment), equilibrating the spiked mixture overnight.

High results were observed with a technique involving distillation and UV destruction (separation of MeHg by water steam distillation, removal of inorganic mercury by anion exchange resin, decomposition of MeHg to ionic mercury by UV irradiation and detection by CVAAS). Although the results were confirmed by HPLC, it was suspected that this technique did not remove all inorganic mercury. This doubt had already been expressed at the first interlaboratory study and the set was not accepted for certification.

Another set of high results was also rejected. The results were obtained by hexane extraction, derivatization with $NaBEt_4$, CGC separation and MIP-AES detection. Although this technique was recognized to be suitable for MeHg determination, it appeared that its application in the laboratory was not sufficiently under control to provide accurate results.

The SFE technique gave a rather low recovery which was nevertheless accepted for certification. The laboratory had submitted a second set of data obtained by distillation and CGC-MIP which was on the low side; the efficiency of the distillation was shown to be much lower (70%) than the other laboratories using the same technique, which justified the data to be withdrawn. Although doubts were expressed on distillation-based techniques during the Conference "Mercury as a Global Pollutant" [94], these techniques (except one set of data) were generally in good agreement with alternative

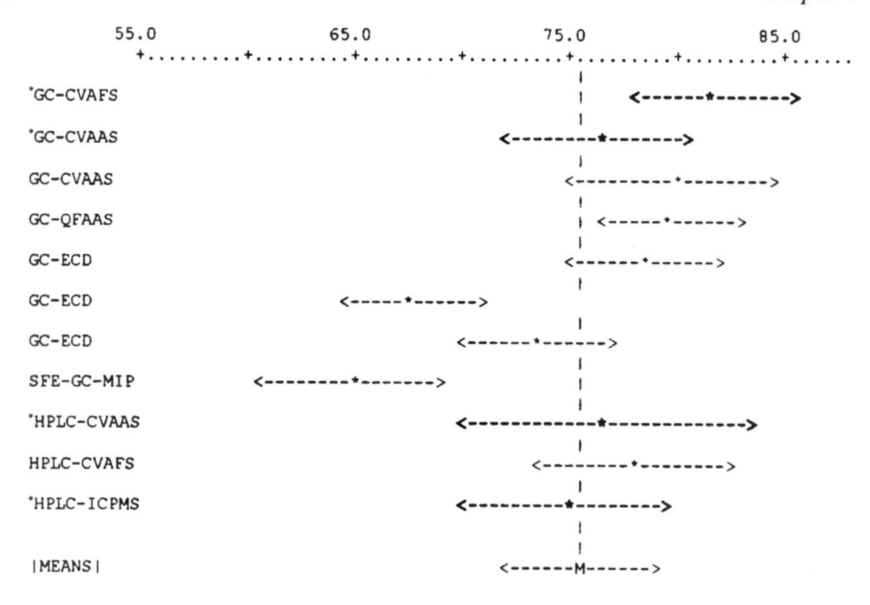

Figure 4.5 *Bar graph of MeHg in CRM 580*

techniques (distillation-based sets of data are indicated in bold and marked by an asterisk in Figure 4.5) and there was no reason to suspect a systematic bias [95].

Certified Values

The certified values (unweighted mean of *p* accepted sets of results) and their uncertainties (half-width of the 95% confidence intervals) are given in Table 4.6 as mass fractions (based on dry mass). Total mercury and methylmercury are certified as mass fractions of Hg and $MeHg^+$, respectively.

Table 4.6 *Certified mass fractions of total and methylmercury in CRM 580*

Component	Certified value	Uncertainty	Unit	Number of sets
Total Hg	132	3	mg kg^{-1}	13
MeHg	75.5	3.7	μg kg^{-1}	11

CHAPTER 5

Tin Speciation

5.1 Aim of the Project and Coordination

Justification

Butyl- and phenyl-tin compounds, particularly trisubstituted species, are known to be very toxic to marine organisms at very low concentrations [96,97]. Tributyltin (TBT) is released in the marine environment from the leaching of TBT-based antifouling paints used on boats and ships, whereas triphenyltin (TPhT) is used both as an antifouling agent and in herbicide formulations. The monitoring of tin compounds is required by the EC legislation, *e.g.* under Council Decisions 75/437/EEC (Marine Pollution from Land-based Sources), 77/585/EEC (Mediterranean Sea) and 77/586/EEC (Rhine River), and Council Directive 80/68/EEC (Groundwater). The compliance to this legislation and the need for a comparison of data produced worldwide require that the analyses are accurate. However, as said before, the methods currently used for the determination of organotin compounds involve various analytical steps such as extraction, derivatization, separation and final detection, which multiply the risks of analytical errors [11] since some of these techniques are far from being validated. In order to improve and ensure good quality control of tin speciation analysis, a series of interlaboratory studies (including certifications) have been organized in the past few years.

The Programme and Timetable

The programme to improve the quality control of organotin determinations in environmental matrices was started in 1988 [98] by a consultation of European experts. It was decided to follow a stepwise approach for the evaluation of the performance of methods used in butyltin analyses. The first exercise (simple solutions) was conducted in 1988–89 and a second interlaboratory study (spiked sediment) was organized in 1989–90. The first certification campaign (harbour sediment) was carried out in 1990–92, whereas the certification of a coastal sediment was conducted in 1992–93. The most recent certification focused on organotin determinations in mussel samples (1994–97).

Coordination

Interlaboratory Studies

The preparation of the solutions was carried out by TNO (Zeist, NL), which also provided the participants with butyltin calibrants of verified purity. The preparation of the TBT-spiked sediment was carried out by the EC Joint Research Centre (Environment Institute) of Ispra (Italy) in collaboration with TNO.

Certification of Sediment Reference Materials

The two sediment samples used as candidate CRMs were collected by the University of Pau (France) and prepared at the Institute for Reference Materials and Measurements (IRMM, Geel, Belgium). The verification of the homogeneity and stability was carried out by the University of Pau as described below. The certifications were coordinated by the BCR.

Certification of Mussel Reference Material

The candidate mussel CRM was collected by the ENEA, Casaccia Research Centre (Rome, Italy). The sample freeze-drying was carried out at the Biostarters Company (Parma, Italy) and the homogenization and bottling were performed jointly by Ecoconsult (Gavirate, Italy) and the EC Joint Research Centre of Ispra (Italy).

Participating Laboratories

The following laboratories participated in the interlaboratory studies: CIBA-GEIGY, Bensheim (Germany); CID-CSIC, Departamento de Química Ambiental, Barcelona (Spain); De Monfort University, School of Applied Sciences, Leicester (United Kingdom); ENEA, Divisione di Chimica Ambientale, Rome (Italy); Vrije Universiteit, Inst. Milieuvraagstukken, Amsterdam (The Netherlands); IFREMER, Nantes (France); Institut für Spektrochemie, Dortmund (Germany); Max-Planck Institut für Chemie, Mainz (Germany); Ministry for Agriculture, Fisheries and Food, Burnham-on-Crouch (United Kingdom); Rijkswaterstaat, Haren (The Netherlands); Schering AG, Bergkamen (Germany); Technical University of Athens, Dept. of Chemical Engineering, Athens (Greece); TNO, IJmuiden (The Netherlands); Universiteit Antwerp, Dept. Scheikunde, Antwerp (Belgium); Université de Bordeaux I, Lab. Photophysique et Phtotochimie Moleculaire, Talence (France); Université de Pau, Lab. Chimie Analytique, Pau (France); University of Plymouth, Department of Environmental Sciences, Plymouth (United Kingdom); Universidad de Sevilla, Departamento de Química Analítica, Sevilla (Spain).

The following laboratories participated in the certification campaigns on sediment: CIBA-GEIGY, Bensheim (Germany); CID-CSIC, Departamento de

Química Ambiental, Barcelona (Spain); ENEA, Divisione di Chimica Ambientale, Rome (Italy); Ministry for Agriculture, Fisheries and Food, Burnham-on-Crouch (United Kingdom); Universiteit Antwerp, Dept. Schiekunde, (Belgium); Université de Bordeaux I, Lab. Photophysique et Phtotochimie Moleculaire, Talence (France); Université de Pau, Lab. Chimie Analytique, Pau (France); University of Plymouth, Department of Environmental Sciences, Plymouth (United Kingdom); Universidad de Sevilla, Departamento de Química, Sevilla (Spain)

The following laboratories participated in the certification campaign on mussel: CID-CSIC, Departamento de Química Ambiental., Barcelona (Spain); ENEA, Divisione di Chimica Ambientale, Rome (Italy); Vrije Universiteit, Inst. Milieuvraagstukken, Amsterdam (The Netherlands); GALAB, Geesthacht (Germany); International Tin Research Institute, Uxbridge (United Kingdom); Ministry for Agriculture, Fisheries and Food, Burnham-on-Crouch (United Kingdom); Rijkswaterstaat, RIKZ, Haren (The Netherlands); Universiteit Antwerp, Dept. Scheikunde, Antwerp (Belgium); Universidad de Barcelona, Lab. Química Analítica, Barcelona (Spain); Universidad de Bordeaux, I. Lab. Photophysique et Photochimie Moleculaire, Talence (France); Università de Genova, Dipartimento di Chimica, Genova (Italy); Universidad de Huelva, Departimento de Química Analítica, Huelva (Spain); Universidad de Oviedo, Departmeto de Física y Analítica, Oviedo (Spain); Université de Pau, Lab. Chimie Analytique, Pau (France); University of Plymouth, Department of Environmental Sciences, Plymouth (United Kingdom); Universidad de Santiago de Compostella, Lab. de Química Analítica (Spain).

5.2 Techniques Used in Tin Speciation

Techniques used in the interlaboratory exercises and certification campaigns are described below. Discussions of the sources of discrepancies detected in the various exercises are reported in the section related to the technical evaluation of the results.

Butyltin Determination in Sediment

The techniques described below have been tested for sediment analysis at the occasion of three interlaboratory studies (TBT-spiked sediment, harbour sediment and coastal sediment). As detailed in Section 5.3, they were first tested with simple solutions to verify that no systematic error could be attributed to the detection step. Since calibration is of paramount importance, participants were provided with pure calibrants of tributyltin chloride synthesized by TNO for the verification of their own calibrants. In all the methods, the storage of the extracts (when necessary) was always at 4 °C in the dark. The moisture correction was carried out by drying a separate sediment portion of 1 g at (105 $\pm$ 2) °C for 2 h; the moisture content obtained by the different laboratories ranged from 1.2 to 2.3%.

Hydride Generation/GC-QFAAS

First method: about 1 g sediment was extracted with 20 mL acetic acid. The extract was centrifuged at 4000 rpm during 15 min. Derivatization was carried out by addition of 10% $NaBH_4$ in 1% NaOH in milli-Q water. The generated hydrides were cryogenically trapped in a U-tube filled with chromatographic material (column of 0.5 m, internal diameter of 4 mm; OV-101 as stationary phase loaded with 10% Chromosorb WHP 80–100 mesh; He as carrier gas at 100–300 mL min^{-1}; column temperature ranging from −196 to 220 °C). Detection was by QFAAS with a H_2/O_2 mixture (respective flow rates of 300 and 40 mL min^{-1}). The calibration was carried out with $MBTCl_3$, $DBTCl_2$ and TBTCl calibrants by standard additions. The extraction recovery was verified by spiking the sediment sample with butyltins before extraction; the recoveries obtained were (106 ± 13)% for MBT, (85 ± 4)% for DBT and (118 ± 15)% for TBT. The method was cross-checked with the PACS-1 material.

Second method: 2 g sediment were extracted with 20 mL pure acetic acid, stirring with a Teflon magnetic stirrer for 12 h. The mixture was centrifuged at 4000 rpm for 15 min and 2 mL supernatant was used for analysis. Triethyltin bromide was added as internal standard to the extraction mixture prior to stirring. Hydride generation was carried out by addition of 5% $NaBH_4$ in a 50 mL Milli-Q water flask. The generated hydrides were cryogenically trapped in a U-tube filled with chromatographic material (Chromosorb WHP 60–80 mesh coated with 10% SP-2100). The U-tube GC column was removed from liquid nitrogen and gently warmed up to desorb volatile butyltin hydrides (carried by helium at a flow rate of 300 mL min^{-1}). Detection was by QFAAS (wavelength of 224.6 nm) with addition of H_2 and O_2 at respective flow rates of 150 and 20 mL min^{-1}; the atomization temperature was 900 °C. Calibration was performed by standard additions using a mixed solution of $MBTCl_3$, $DBTCl_2$ and TBTCl.

Third method: *ca.* 2 g sediment were extracted (mechanical shaking during 16 h) with 10 mL acetic acid in a 25 mL screw-capped centrifuge tube; the tube was subsequently centrifuged at 4000 rpm for 5 min. A 2 mL aliquot of the extract was transferred to a 250 mL reaction flask filled with clean seawater. After a 2 min purge with N_2 flow, *ca.* 30 mL of 4% $NaBH_4$ were added *via* a peristaltic pump. Hydrides were carried by a N_2 flow to an U-tube filled with chromatographic material (Chromosorb GNAW 60/80 mesh, loaded with 3% SP-2100) cooled in liquid nitrogen. After being removed from liquid N_2, the column was heated to *ca.* 200 °C by gentle electric heating. Tin hydrides were separated and carried by a helium flow of 300 mL min^{-1} to the detector; detection was by QFAAS at 224.6 nm (electrothermally heated at 900 °C) with addition of H_2 and O_2 at respective flows of 250 and 7 mL min^{-1}. Calibration was by standard additions, using $MBTCl_3$, $DBTCl_2$ and TBTCl as calibrants.

Hydride Generation/CGC-FPD

Ca. 2 g sediment were wetted with 2 mL water, extracted with 8 mL of 0.1% NaOH in methanol and back-extracted in 2 mL hexane. Derivatization was by hydride generation, using $NaBH_4$, followed by back-extraction into hexane. Extraction recoveries were assessed using TBTCl, $DBTCl_2$ and tripropyltin as internal standard; they ranged from (59 $\pm$ 3)% for TPrT, (87 $\pm$ 3)% for DBT and (98 $\pm$ 5)% for TBT. Separation was by CGC (column of 25 m length, 0.32 mm internal diameter, 5% phenylmethylsilicone as stationary phase, 0.52 µm film thickness; H_2 as carrier gas at 2 mL min^{-1}; N_2 as make-up gas at 30 mL min^{-1}; injector temperature of 40 °C; column temperature ranging from 40 to 250 °C) and detection was by FPD (detector temperature of 250 °C).

Ethylation/CGC-FPD

Ca. 1 g sediment was extracted with 5 mL HCl and 10 mL toluene, mechanically shaking for 15 h. The recovery was verified by spiking a marine sediment and was (98.3 $\pm$ 1.3)% for TBT. Ethylation was performed with 150 µL of 2% $NaBEt_4$. Separation was by CGC (column of 25 m length, 0.32 mm internal diameter, CPSIL-5 as stationary phase, 0.14 µm film thickness; H_2 as carrier gas at 10 mL min^{-1} and make-up gas at 40 mL min^{-1}; injector temperature at 200 °C; column temperature ranging from 70 to 200 °C). Detection was by FPD (detector temperature at 250 °C). Calibration was by calibration graph, using TBTAc, $DBTCl_2$ and $MBTCl_3$ as calibrants.

SFE/Ethylation/CGC-FPD

This technique was developed in the frame of a RTD project funded by the SM&T programme to serve as reference for evaluating extraction procedures [99].

Ca. 3.5 g sediment were extracted by supercritical fluid extraction (CO_2 and 20% methanol) after HCl addition. The extraction recovery for TBT (verified with a TBT-spiked sediment) was (82 $\pm$ 6)%. Grignard derivatization was performed, using 2 mol L^{-1} EtMgCl in tetrahydrofuran. Separation was by CGC (column of 30 m length, internal diameter of 0.25 mm, DB-5 as stationary phase, 0.1 µm film thickness; He as carrier gas at 2 mL min^{-1}; injector temperature at 40 °C; column temperature ranged from 40 to 290 °C). Detection was FPD (temperature of 225 °C). Calibration was carried out with butyltin chloride calibrants; calibration graph and standard additions of tripropyltin as internal standard.

Pentylation/CGC-FPD

Ca. 5 g sediment were stirred for 1 h with 50 mL of a HBr–water mixture, followed by extraction for 2 h with tropolone/pentane and centrifugation at 1000 rpm for 10 min. The liquid phases were transferred to a 250 mL funnel in which the aqueous phase was separated and removed. The organic phase was dried with Na_2SO_4 and reduced in volume in a rotary evaporator to 0.5 mL. A 4 mL aliquot of pentylmagnesium bromide solution (1 mol L^{-1}) in ether was added to the extract and left standing for 1 h with occasional shaking. The excess of Grignard reagent was destroyed by dropwise addition of H_2SO_4. The organic layer was collected and extracted twice with 5 mL pentane; all the extracts were combined and dried over anhydrous Na_2SO_4. The resulting extract was evaporated to 0.5 mL and purified by chromatography on a 7 $\times$ 1 cm i.d. column of Florisil (elution with pentane). The pentane extract was concentrated to 0.5 mL under a N_2 stream after addition of a previously derivatized Me_2SnPe_2 internal standard. Separation was by CGC (column of 15 m length, 0.53 mm internal diameter, SPB-1 as stationary phase, 1.5 μm film thickness). Detection was by FPD.

Pentylation/CGC-QFAAS

Ca. 2 g sediment were extracted with a mixture of 4 mL deionized water, 2 mL acetic acid, 2 mL diethyldithiocarbamate (DDTC) in pentane and 25 mL hexane. The mixture was ultrasonically treated for 30 min. After phase separation, the hexane phase was collected and the sediment was back-extracted with 25 mL hexane. The combined hexane extracts were evaporated to dryness on a rotary vacuum evaporator. The extracted butyltin compounds were resolved in 250 mL *n*-octane with Pr_3SnPe as internal standard. Recovery experiments were performed with the CRM 462 and were (109 $\pm$ 7)% for TBT and (100 $\pm$ 5)% for DBT. A Grignard reaction was performed with 1 mL of *n*-PeMgBr (2 mol L^{-1}) in diethyl ether. The pentylated butyltins were separated by CGC (fused silica megabore column of 15 m, internal diameter of 0.53 mm, polydimethylsiloxane as stationary phase, 1.2 μm film thickness; Ar as carrier gas at 6 mL min^{-1}; air and H_2 as make-up gases at 145 mL min^{-1} and 350 mL min^{-1}; injector temperature at 230 °C; column temperature ranging from 100 to 270 °C). The detection was by QFAAS (quartz cell electrothermally heated to *ca.* 900 °C; wavelength of 286.4 nm). Pentylated calibrants (Bu_3SnPe, Bu_2SnPe_2 and $BuSnPe_3$) were prepared from commercial salts which were pentylated with 2 mol L^{-1} *n*-PeMgBr in diethyl ether in a solution of nonane, and checked by GC-QFAAS and GC-MS.

Pentylation/CGC-MS

First method: *ca.* 1 g sediment was extracted ultrasonically with 20 mL diethyl ether/HCl in tropolone (the recovery ranged from 97 to 108% for the three butyltin compounds as assessed by spiking the CRM with the respective compounds). Derivatization was performed by addition of 2 mol L^{-1} PeMgBr in diethyl ether. The final extract was cleaned up with silica gel. Separation was by CGC (column of 25 m length, 0.32 mm internal diameter, methylsilicone as stationary phase, 0.8 μm film thickness; He as carrier gas at 1 mL min^{-1}; injector temperature at 280 °C; column temperature ranging from 80 to 280 °C;

detector (transfer line) temperature at 280 °C). Detection was by mass spectrometry. Calibration was by standard additions, using TBTCl, $DBTCl_2$ and $MBTCl_3$ as calibrants.

Second method: *ca.* 0.5 g sediment was extracted with methanol/tropolone after addition of HCl. Tripropyltin was added as internal standard. Derivatization was performed by addition of pentylmagnesium bromide (2 mol L^{-1}) followed by clean-up with silica gel. Separation was by CGC (column of 25 m length, 0.2 mm internal diameter, methylphenylsilicon as stationary phase, 0.11 µm film thickness; He as carrier gas at 130 mL min^{-1}; injector temperature at 260 °C; column temperature ranging from 80 to 280 °C; detector (transfer line) temperature at 280 °C). Detection was by mass spectrometry. Calibration was by calibration graph, using butyltin chloride compounds as calibrants.

HPLC/ICP-MS

This technique was developed in the frame of a RTD project funded by the SM&T programme to serve as a reference method in the certification campaigns of butyltins in sediment and mussel reference materials [77,100].

2 g sediment were extracted with acetic acid and back-extracted into toluene. The extract was preconcentrated by solvent evaporation. Separation was by HPLC (Partisil SCX cation exchange, 10 µm particle size, methanol/water as mobile phase with ammonium citrate buffer). The final detection was by ICP-MS of mass 120. Calibration was by standard additions, using TBTAc and $DBTAc_2$.

Polarography

25 g sediment were leached with 150 mL HCl (0.5 mol L^{-1}) for 24 h at room temperature. After centrifugation of the slurry at 1500 rpm for 30 min, the solid phase was washed with HCl (0.5 mol L^{-1}) followed by a second centrifugation. The combined supernatant was filtered and extracted with dichloromethane in a separating funnel. The organic phase was transferred into a 5 mL polarographic cell and 5 mL electrolyte LiCl electrolyte (0.2 mol L^{-1}) in ethanol was added. The mixture was deoxygenized by nitrogen bubbling for 10 min and the alternating current polarograms were scanned directly in the organic phase from -0.4 to -1.3 V. The working electrode was a dropping mercury electrode; the reference electrode was an Ag/AgCl electrode filled with 0.2 mol L^{-1} LiCl in ethanol and the auxiliary electrode was a Pt electrode. The TBTCl peak was at -0.90 ± 0.05 V and the $DBTCl_2$ peak was at -0.68 ± 0.03 V. Calibration was by standard additions, spiking the leaching solution in the polarographic cell.

Butyl- and Phenyl-tin Determinations in Mussel

Following the above-mentioned certification, it was considered of paramount importance to improve the quality control of various analytical steps such as extraction and derivatization. Spiking procedures were discussed for the evaluation of extraction recoveries and it was generally accepted to spike the material (after wetting) at three levels and to calculate the mean of the recovery means obtained; results of organotin determinations were requested to be corrected for extraction recovery. With respect to derivatization, the only limitation for the evaluation of the derivatization yield was the lack of available calibrants; consequently, the project coordinator has carried out the preparation of a series of calibrants of verified purity and stoichiometry for distribution to the participants in the certification campaign. The purity of commercially available organotin compounds is often not adequate, while many of the alkylated derivatives need to be prepared in-house. It was felt that by providing ultrapure calibrants, synthesized and purified in large quantities by an expert laboratory, some errors related to calibration could be avoided. In order to ensure the best conditions for achieving accurate results in the certification of CRM 477, it was hence decided to prepare a set of organotin calibrants to be distributed to the certifying laboratories for quality control checks. The task was performed by the Department of Organic Chemistry of the Free University of Amsterdam; the purpose was to prepare highly purified butyltin and phenyltin compounds (in the form of salts) as well as their ethylated and pentylated derivatives for use as calibration and recovery tests.

Commercial organotins were used as starting materials; purification was carried out by a number of recrystallization and/or distillation steps. For the alkylated derivatives, the purified organotin salts were used as the starting materials of a Grignard reaction. The resulting products were purified again by repeated recrystallizations and/or distillations. The purity was checked by elemental analysis, by ^{1}H and ^{13}C NMR and, in case of the alkylated derivatives, by GC-MS. In total, 18 compounds were prepared, representing more than 5 g of alkylated derivatives and more than 10 g of purified salts. The list below gives the calibrants related to butyltins (*i.e.* used in the certification of MBT, DBT and TBT):

MBT:	$(C_4H_9)SnCl_3$	purity >99%
DBT:	$(C_4H_9)_2SnCl_2$	purity >99%
TBT:	$(C_4H_9)_3SnCl$	purity >98%
MBT-Pe:	$(C_4H_9)SnPe_3$	purity >98%
DBT-Pe:	$(C_4H_9)_2SnPe_2$	purity >98%
TBT-Pe	$(C_4H_9)_3SnPe$	purity >98%
MBT-Et:	$(C_4H_9)SnEt_3$	purity >99%
DBT-Et:	$(C_4H_9)_2SnEt_2$	purity >99%
TBT-Et:	$(C_4H_9)_3SnEt$	purity >98%

Corrections for moisture were performed by drying a separate mussel subsample of 1 g at $(105 \pm 2)\,^{\circ}C$ for 2 h; the moisture content obtained by the different laboratories ranged from 4.2 to 6.4%.

Beside the preparation of pure calibrants for the purpose of certification, independent studies were carried out to investigate possible sources of error existing at the extraction and derivatization steps. The study on derivatization demonstrated that the yields obtained by Grignard reactions were generally acceptable and that no systematic error could be suspected [101]; the comparison of extraction methods also showed that techniques currently used gave satisfactory results and could hence safely be applied to certify reference materials [102].

Hydride Generation/GC-QFAAS

An aliquot of 0.1 g mussel was digested with 10 mg protease and 10 mg lipase (enzymatic digestion) and 5 mL phosphate buffer during 4 h at 37 °C. Extraction recoveries were evaluated by spiking a similar matrix with a butyltin mixture; the recoveries ranged from 92% (TBT) to 99% (DBT) to 100% (MBT). Hydride generation was carried out in acetic acid using 10% $NaBH_4$ in 1% NaOH. The generated hydrides were cryogenically trapped in a U-tube filled with chromatographic material (Chromosorb WHP 80–100 mesh as stationary phase loaded with 10% OV-101; 0.5 mm length, 4 mm internal diameter; He as carrier gas at 100–250 mL min^{-1}; column temperature ranging from −196 to 220 °C). The organotin hydrides were detected by QFAAS electrothermally heated at 950 °C at a wavelength of 286.3 nm; a mixture of H_2 and O_2 was added to the furnace at respective flow rates of 200 and 45 mL min^{-1}.

Ethylation/CGC-AAS

Ca. 0.1 g sample was digested with 10 mL of 20% tetramethylammonium hydroxide (TMAH), buffering with 1 mL of HCl and 2 mL sodium acetate/acetic acid (pH 4). Extraction was carried out by addition of 10 mL methanol (ultrasonic shaking for 1 h and mechanical shaking for 1 h). Recoveries were evaluated by spiking the mussel matrix with the different organotin compounds: MBT (98%), DBT (85%), TBT (99%), MPhT (95%), DPhT (94%) and TPhT (98%). Derivatization was performed with 2% $NaBEt_4$ in 2 mol L^{-1} sodium acetate/acetic acid mixture (shaking for 1 h). Clean-up was with an alumina column. Separation was by CGC (column of 30 m length, 0.32 mm internal diameter, DB-5 as stationary phase, 0.25 µm film thickness; He as carrier gas at 5 mL min^{-1}; air and H_2 as make-up gases at 90 and 350 mL min^{-1}; column temperature ranging from 80 to 250 °C). Detection was by AAS (detector temperature of 750 °C). Calibration was by standard additions, using organotin chloride calibrants in methanol.

Hydride Generation/GC-ICP-MS

A 0.2 g sample was extracted with 20 mL glacial acetic acid (stirring overnight with a magnetic stirrer); the extract was centrifuged for 20 min. 50 µL of extract were injected into a reaction flask containing 50 mL Milli-Q water, 2 mL glacial acetic acid and 10 mL of 2% $NaBH_4$. The generated hydrides were stripped out with helium for 3 min and cryogenically trapped in a U-tube filled with chromatographic material (Chromosorb WHP 60–80 mesh as stationary phase loaded with 10% SP-2100; the column was electrically heated with a Ni–Cr wire). Final detection was by ICP-MS. Extraction recoveries were calculated by spiking the material (previously wetted with methanol and mixed with spiking solution) at three levels; the recoveries ranged from 97% (TBT) and 98% (MBT) to 112% (DBT). Calibration was by standard additions, using $MBTCl_3$, $DBTCl_2$ and TBTCl in methanol.

Hydride Generation/CGC-FPD

First method: *ca.* 0.5 g sample was digested with 0.1% NaOH in methanol/H_2O, mechanically shaking for 45 min, followed by extraction with 2 mL hexane. TPrT was added as internal standard. Hydride generation was carried out with 100 mg (96% pure) $NaBH_4$ (solid), mechanically shaking for 10 min and centrifuging at 2000 rpm for 5 min. Separation was by capillary GC (column of 25 m length, 0.32 mm internal diameter, 5% phenylmethyl-silicone as stationary phase, 0.52 μm film thickness; He as carrier gas at 170 mL min^{-1}; N_2 as make-up gas at 28 mL min^{-1}; injector temperature of 20 °C; column temperature ranging from 40 to 225 °C). Detection was by FPD (detector temperature at 225 °C). Recovery was evaluated by spiking (losses were observed due to methanol evaporation which was hence not recommended); recoveries were (105 $\pm$ 4)% for MBT, (106 $\pm$ 4)% for DBT, and (98 $\pm$ 4)% for TBT. Calibration was by standard additions, using $MBTCl_3$, $DBTO_2$, TBTO and TPhTCl as calibrants.

Second method: 0.5 g sample was digested with a 5 mL mixture of 10% HBr and 60% methanol and stirred for 1 h. Extraction was carried out by addition of 10 mL of 0.05% tropolone in methanol, stirring for 30 min and then centrifuging for 5 min. The organic layer was derivatized with 0.2 mL of 4% $NaBH_4$ solution. Clean-up was performed with Florisil followed by elution with hexane. Separation was by GC (column of 5 m length, 0.32 mm internal diameter, DB-1 as stationary phase; He as carrier gas at 5 mL min^{-1}; injector temperature of 250 °C; column temperature ranging from 80 to 250 °C). Detection was by FPD (detector temperature of 350 °C).

Ethylation/CGC-FPD

First method: 0.5 g sample was extracted with 2.5 mL methanol (mechanically shaking for 2 h) and 12.5 mL of HCl (0.12 mol L^{-1}) (ultrasonic mixing for 1 h). 0.5 mL of the methanolic extract was buffered with 100 mL acetate buffer (pH 4.8), derivatized with 0.2 mL of 2% $NaBEt_4$ solution in deionized water, and back-extracted with 0.3 mL isooctane (extraction into the organic solvent at 420 rpm for 45 min). Separation was by capillary GC (column of 30 m length, 0.25 mm internal diameter, DB-1 as stationary phase (polydimethylsoloxane), 0.25 μm film thickness; N_2 was used as carrier gas at 0.7 mL min^{-1} and as make-up gas at 30 mL min^{-1}; injector temperature of 290 °C; column temperature ranging from 80 to 270 °C). Detection was by FPD (detector temperature of 290 °C). Recoveries were assessed by wetting 0.25 g sample with 2 mL methanol followed by addition of 1 mL methanol solution containing the spiking compounds (four additions); results were 82% for MBT, 89% for DBT, 85% for TBT, 75% for MPhT and 77% for TPhT. Calibration was by standard additions, using organotin chloride compounds in methanol.

Second method: *ca.* 1 g sample was treated with 25% tetramethylammonium hydroxide followed by hexane extraction and clean-up with an alumina column. TPeT was added as internal standard. Ethylation was performed with 0.6% $NaBEt_4$ solution in acetate buffer (pH 5). Clean-up was carried out on alumina, with hexane elution. The derivatization yield was verified by comparison between the derivatized compounds provided by SM&T and underivatized compounds (chloride). Separation was by capillary GC (column of 30 m length, 0.25 mm internal diameter, DB-17 as stationary phase, 0.25 μm film thickness). Recoveries were assessed by spiking (standard additions); results were (66 $\pm$ 1)% for DBT, (74 $\pm$ 2)% for TBT and (55 $\pm$ 1)% for MPhT. Calibration was by calibration graph and standard additions, using the calibrants provided by SM&T.

Third method: 0.2 g sample was extracted with 8 mL acetic acid/H_2O. Ethylation was performed with 0.75% $NaBEt_4$ solution, simultaneously extracting with 1 mL nonane during a 3 min microwave exposure (40 W). TPrT was added as internal standard. The derivatization yield was verified with ethylated compounds. Clean-up was performed with an alumina column followed by elution with diethyl ether. Separation was by capillary GC (25 m column length, dimethylpolysiloxane as stationary phase, 0.17 μm film thickness; N_2 as carrier gas; injector temperature at 250 °C; column temperature ranging from 120 to 280 °C). Detection was by FPD (detector temperature of 350 °C). Recoveries were evaluated by spiking the reference material at three different levels; results were (71 $\pm$ 6)% for MBT, (104 $\pm$ 7)% for DBT, (94 $\pm$ 7)% for TBT and (81 $\pm$ 8)% for TPhT. Calibration was by standard additions.

SFE/Ethylation/CGC-FPD

As mentioned above, this technique has been developed for the purpose of certification (under SM&T funding) and was used as reference method [99,102].

Ca. 1 g sample was extracted by supercritical fluid extraction with CO_2 and a mixture of acetic acid and 0.2% tropolone with hexane as solvent (pressure of 50 atm, temperature of 50 °C). Ethylation was carried out with 2 mol L^{-1} ethylmagnesium chloride. TPeT was added as internal standard. Separation was by capillary GC (column of 30 m length, 0.25 mm internal diameter, DB-17 as stationary phase, 0.25 μm film thickness; H_2 as carrier gas at 5 mL min^{-1}; N_2 as make-up gas at 30 mL min^{-1}; injector temperature of 250 °C; column temperature ranging from 60 to 280 °C). Detection was by FPD (detector temperature of 300 °C). Recoveries were assessed by standard additions; results were (82 $\pm$ 4)% for DBT, (75 $\pm$ 2)% for TBT and (65 $\pm$ 3)% for TPhT. Calibration was by calibration graph and standard additions, using the calibrants provided by SM&T.

Pentylation/CGC-FPD

First method: *ca.* 0.2 g sample was pre-treated with 50 mL HBr/H_2O mixture for 1 h and extracted into 50 mL of 0.05% tropolone in dichloromethane for 2 h under manual and mechanical shaking. Pentylation was carried out with 1 mol L^{-1} pentylmagnesium chloride for 1 h; the derivatization yield ranged from 85 to 119% (verified with pentylated compounds). Clean-up was performed with Florisil, evaporating the resulting extract by rotary evaporation under gentle flow of N_2 until dryness; redissolution was carried out in a Pe_2SnMe_2 hexane solution. Separation was by CGC (column of 15 m length, 0.53 mm internal diameter, SPB-1 as stationary phase, 1.5 μm film thickness; He as carrier gas at 1.2 mL min^{-1}; injector temperature of 250 °C; column temperature ranging from 80 to 250 °C). Detection was by FPD (detector temperature of 300 °C). Recoveries were evaluated by three-level spiking of the mussel material; results were 131% for MBT, 114% for DBT, 99.4% for TBT, 166% for MPhT, 77% for DPhT and 99% for TPhT. Calibration was by standard additions.

Second method: *ca.* 0.1 g sample was extracted with 15 mL of 0.03% tropolone in methanol solution and 1 mL of 12 mol L^{-1} HCl by ultrasonic shaking for 15 min and centrifuging at 3000 rpm for 10 min; this was followed by an addition of 15 mL dichloromethane and 100 mL of 5% NaCl, manually shaking for 3 min. The resulting extract was reduced to 1 mL by rotary evaporation to near dryness under N_2 flow. Pentylation was performed with 1 mL of 2 mol L^{-1} pentylmagnesium bromide in ethyl ether; the derivatization yields ranged from 85 to 119% (as verified with pentylated compounds). Tripropyltin was added as internal standard. Separation was by CGC (column of 30 m length, 0.53 mm internal diameter, methylphenylsilicon as stationary phase, 1.5 μm film thickness; He as carrier gas at 9 mL min^{-1}; injector temperature at 240 °C; column temperature ranging from 80 to 280 °C). Detection was by FPD (detector temperature of 240 °C). Recoveries were evaluated by three-level spiking; results were 114% for MBT, 111% for DBT, 115% for TBT, 88% for MPhT, 104% for DPhT and 78% for TPhT. Calibration was by calibration graph, using the calibrants supplied by SM&T.

Ethylation/CGC-MIP-AES

First method: 0.1 g sample was digested with 5 mL of 25% tetramethylammonium hydroxide, stirring magnetically for 4 h at 50 °C. Derivatization was by addition of 20 mL of 0.1 mol L^{-1} acetate buffer (pH 5), 1.3 mL of acetic acid, 1 mL of 0.6% $NaBEt_4$ solution and 2 mL of hexane containing Pe_3SnEt as internal standard, shaking for 5 min and centrifugating at 3500 rpm for 3 min; the derivatization yield was verified by analysis of a spike (without mussel) and found to be 100% for all compounds except DPhT (44% only). Clean-up was performed with a column filled with alumina, eluting with 0.5 mL hexane and 1 mL diethyl ether. The combined eluate was reduced to 0.5 mL using a gentle stream of N_2. Separation was by CGC equipped with programmed temperature vaporization (HP-1 column of 25 m length, 0.32 mm internal diameter, 0.17 μm film thickness; He as carrier gas; injector temperature of 15–20 °C; column temperature ranging from 45 to 280 °C). Final detection was by MIP-AES at 303.42 nm. Recoveries were assessed by spiking 0.1 g tissue with methanol solution, leaving overnight and evaporating under N_2 flow (three-level spiking); results were (77 $\pm$ 7)% for MBT, (86 $\pm$ 6)% for DBT, (88 $\pm$ 5)% for TBT, (68 $\pm$ 8)% for MPhT and (75 $\pm$ 8)% for TPhT. Calibration was by calibration graph using ethylated butyl- and phenyltin compounds as organic salts in methanol.

Second method: *ca.* 0.2 g sample was digested with 15 mL of 25% tetramethylammonium hydroxide by microwave leaching in pressurized vessels at 120 °C for 3 min. The pH was adjusted to 5 by addition of 10 mL of 1 mol L^{-1} acetic/acetate buffer, 3.8 mL of glacial acetic acid, 2 mL isooctane and 2 mL of 1% $NaBEt_4$ (shaking for 5 min); this was followed by centrifugation at 2500 rpm for 5 min. Tripropyltin was used as internal standard. Clean-up was performed with alumina. Separation was by CGC, followed by MIP-AES detection.

Ethylation/CGC-MS

Ca. 0.3 g sample was extracted by addition of 15 mL methanol, 1 mL concentrated acetic acid and 10 mL hexane. Derivatization was by addition of 5% $NaBEt_4$, simultaneously extracting into hexane for 15 min; a small constant flow of reagent was added during the 15 minute extraction; water was added to facilitate the transfer of the derivatized organotins into the organic phase. Clean-up was carried out with aluminium oxide (10% water) and elution with hexane. The eluate was washed with 6 mol L^{-1} to remove by-products of the reaction, and then evaporated to near dryness. Tripropyltin was added as internal standard. Separation was by CGC (column of 50 m length, 0.20 mm internal diameter, HP-methylsilicon as stationary phase, 0.33 μm film thickness; He as carrier gas; injector temperature at 250 °C; column temperature ranging from 70 to 270 °C). Calibration was by calibration graph.

Pentylation/CGC-MS

Ca. 0.2 g sample was digested with 5 mL diluted HCl, 12 mL diethyl ether and 0.3 g NaCl. Extraction was by adding (twice) 12 mL of diethyl ether with 0.25% tropolone, followed by evaporation under a N_2 flow and drying with Na_2SO_4. Derivatization was performed with 2 mol L^{-1} pentylmagnesium bromide in diethyl ether. Clean-up was carried out with 5 g of 100% active alumina, eluting with 6 mL hexane/diethyl ether mixture. Ph_2SnEt_2 and Ph_3SnEt were added as internal standards. Separation was by CGC (column of 30 m length, 0.25 mm internal diameter, 5% phenylmethylpolysiloxane as stationary phase, 250 μm film thickness; injector temperature ranging from 60 to 200 °C; column temperature ranging from 60 to 280 °C). Detection was by MS (ion trap MS). Recovery experiments were performed by spiking 0.2 g tissue at three levels;

results were (68 ± 3)% for MBT, (67 ± 4)% for DBT, (49 ± 13)% for TBT, (78 ± 6)% for MPhT, (80 ± 26)% for DPhT and (58 ± 4)% for TPhT. Calibration was by calibration graph, using pentylated organotin compounds.

HPLC-ICP-AES

Ca. 0.5 g sample was extracted into 30 mL methanol and 0.05% tropolone mixture. 200 μL of extract were injected onto a cationic exchange HPLC column (Partisil, cation exchange, 25 cm length, 4.6 mm internal diameter, 10 μm particle size, mobile phase consisting of 80% methanol–water, 0.1 mol L^{-1} ammonium acetate and 0.1% tropolone). The eluent was mixed with 0.3 mol L^{-1} HCl and 0.25 mol L^{-1} $NaBH_4$ in 1 mol L^{-1} NaOH. The organotin hydrides were drained to a gas–liquid separator. Final detection was by ICP-AES. Recoveries were evaluated by spiking the material; results were (108 ± 3)% for MBT, (79 ± 8)% for DBT and (102 ± 11)% for TBT. Calibration was performed by standard additions.

HPLC-ICP-MS

As stated above, this technique was developed as a reference method to be used in the certification campaigns; a full description is given elsewhere [100].

Ca. 1 g sample was digested by enzymatic digestion with 0.05 g of lipase and 0.05 g of protease in a 0.1 mol L^{-1} citrate/phosphate buffered medium (*ca.* 40 mL) at pH 7.5; the mixture was kept overnight at 37 °C under mechanical shaking. Extraction was performed by three additions of 10 mL dichloromethane. The extract was preconcentrated by rotary evaporation until dryness. Dilution of the dried extract was performed with 66% MeOH and 33% mobile phase (70% MeOH and 30% of 0.03 mol L^{-1} aqueous citrate buffer). Separation was by HPLC (gradient elution, cation exchange with Partisil-10 SCX, 2 × 25 cm length, 4.6 mm internal diameter, 10 μm particle size). Final detection was by ICP-MS. Recoveries were assessed by spiking the CRM; results were (12 ± 4)% for DBT, (65 ± 15)% for TBT and (27 ± 3)% for TPhT. Calibration was by calibration graph, using $DBTCl_2$, TBTCl and TPhTCl as calibrants.

HPLC/Fluorimetry

Ca. 0.25 g sample was digested with 20 mL of 0.6 mol L^{-1} HCl and 2.5 mol L^{-1} NaCl aqueous solution, with addition of 10 mL ethyl acetate (twice), followed by mechanical shaking for 30 min and centrifugation at 10 000 rpm for 20 min. The combined extracts were washed with 10 mL of 0.5 mol L^{-1} $NaHCO_3$ and 1 mol L^{-1} NaCl, manually shaking for 2 min and centrifugating at 2000 rpm for 5 min. 5 mL of ethyl acetate was added to the washing aqueous phase and the organic phase was evaporated until dryness by rotary evaporation at 35 °C, followed by addition of 2.5 mL methanol. The filtered methanolic phase was injected into the HPLC column (cation exchange,

Partisil SCX, 25 cm length, 4.6 mm internal diameter, 10 μm particle size). Post-column derivatization was carried out with 0.02 mol L^{-1} Triton X-100 and 3,3′,4′,7-tetrahydroxyflavone at a flow rate of 3 mL min^{-1}. Final detection was by fluorimetry. Recoveries were assessed by spiking the CRM at one level (3 replicates); the results were (82 ± 3)% for TBT and (89 ± 3)% for TPhT.

Parallel Research Developments

In parallel to the organization of the interlaboratory studies and certification campaigns, considerable work has been carried out either to optimize existing techniques or to develop novel analytical methods. Firstly, a systematic study (coordinated by the University of Bordeaux I, Laboratory of Molecular Photophysics and Photochemistry, Talence, France) has been performed to investigate the factors inhibiting the hydride generation reaction in sediment containing high organic carbon content (see Section 5.4). Other systematic studies (coordinated by ENEA, Rome, Italy and the CID-CSIC in Barcelona, Spain) to test and optimize either extraction or derivatization have also been carried out in parallel to the certification of organotins in the mussel CRM 477 (see Section 5.6).

A research project (carried out under the former BCR and coordinated by the CID-CSIC in Barcelona, Spain) also enabled development of a supercritical fluid extraction procedure which was successfully applied to the certification of TBT in the CRM 462 [99]. The effect of extraction variables, such as extraction time, temperature and extraction agent composition, in supercritical fluid extraction have been optimized by using a factorial-fractional experimental design. Under the optimum conditions (T = 60 °C, P = 35 MPa, 5.1 mol L^{-1} methanol in CO_2, t = 30 minutes), the TBT extraction efficiency was 82% with a coefficient of variation of 9.2% (after determination by GC-FPD).

Another successful project (coordinated by the University of Plymouth, Department of Environmental Sciences, United Kingdom) was aimed at developing a novel methodology for the determination of organotins employing isotope dilution high performance liquid chromatography inductively coupled plasma mass spectrometry (ID-HPLC-ICP-MS) [100]. The parameters for isotope dilution analysis have been studied and optimized and isotopically enriched ^{116}Sn has been prepared for this purpose. The method was successfully used for the certification of TBT in CRMs 462 and 477.

5.3 Interlaboratory Studies

Interlaboratory Study on Organotin in Solutions

Four solutions (HCl-acidified) containing different possible interferents were used in the first round: solution A with TBTAc only and solutions B, C and D containing TBTAc along with Sn(IV) and $DBTCl_2$ (sol. B), Sn(IV) and $MBTCl_3$ (sol. C) and TPhTAc (sol. D). The results of this intercomparison

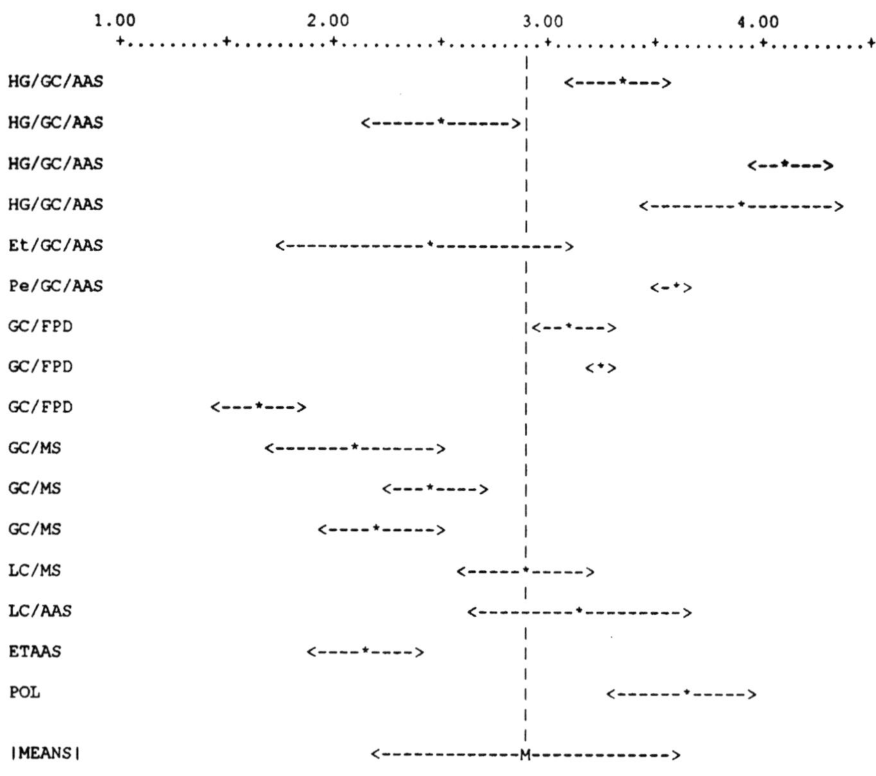

Figure 5.1 *Intercomparison of TBT in a spiked sediment (content as TBTAc mass fraction in mg kg^{-1}). The laboratory codes are indicated along with the methods used. The results plotted correspond to five replicate determinations. MEANS is the mean of laboratory means with its standard deviation. M = (2.9 ± 0.7) mg kg^{-1} (as Sn)*

were found to be in good agreement, which indicated that possible systematic errors did not arise from the techniques of final determination [43].

Interlaboratory Study on TBT in a TBT-spiked Sediment

The second exercise was undertaken in 1989 on the determination of TBT in a spiked sediment (collected in the Laggo Magiore, Italy, and prepared at the EC Joint Research Centre of Ispra) [98]. The results of this interlaboratory trial (Figure 5.1) did not reveal any systematic errors in the different analytical methods that were compared. The coefficient of variation (*CV*) obtained between laboratories (25%) was considered to reflect the state of the art at that stage and the group of experts recommended proceeding with the organization of a certification campaign. The participants recognized, however, that a better agreement should be achieved for certification. No particular source of error due to a method could be detected and the analytical techniques used in this intercomparison were therefore found suitable for certification. The need to allow spikes to equilibrate at least overnight to obtain a realistic assessment of extraction recoveries was highlighted [43].

5.4 Certification of Harbour Sediment, RM 424

Preparation

Collection and Preparation

The harbour sediment (RM 424) was collected in the Sado Estuary (Portugal) in the vicinity of the harbour of Setúbal [103]. The first 10 cm sediment layer was collected with a shovel and stored at ambient temperature in plastic containers. After decantation, the sediment sample was air dried for 7 days at ambient temperature on a cotton sheet in a well ventilated room, and stored at −20 °C in polythene bags. The material (about 180 kg dry mass) was then stored in ice boxes for transportation to the Institute for Reference Materials and Measurements (Geel, Belgium). The material was dried at 80 °C in air for 100 h; it was then ground and put through a sieve of 75 μm mesh size. Before homogenization, the powder obtained was sterilized by heating at 120 °C for one hour and homogenized in a mixer for one hour. The material was finally stored in 60 mL well-cleaned brown glass bottles with polyethylene inserts and plastic screw caps, each containing *ca.* 25 g of powder.

The moisture content of the material was determined by Karl Fischer titration on 10 samples selected during the bottling procedure. The mean moisture mass fraction measured was (2.24 ± 0.36)%.

Homogeneity Control

The between-bottle homogeneity was verified by the determination of TBT on intakes of 2 g taken from 20 bottles which were set aside at regular intervals during the whole period of bottling. The within-bottle homogeneity was assessed by 10 replicate determinations on the well mixed content of one bottle. The technique used for this study was hydride generation/cold trapping–gas chromatography/atomic absorption spectrometry. No significant differences were observed between the *CV* of the method (8.7 ± 2.8%) and the *CV* within-bottles (8.2 ± 1.8%). The *CV* between-bottles appeared to be significantly higher than the *CV* within-bottles (18.3 ± 2.9%); however, the value was found to correspond to determinations performed over a period of six months and, therefore, included the long term reproducibility of the analytical method used, which was realistic at this mass fraction level. Any inhomogeneity of the material would have actually been detected in the stability study.

Stability Control

The stability of the TBT content was tested at −20 °C, +20 °C and +40 °C over a period of 12 months and TBT was determined at the beginning of the

Table 5.1 *Stability tests of coastal sediment RM 424*[a]

Compound	*Time* (month)	*Temperature*	R_T	U_T
TBT	1	+20 °C	0.984	0.141
	3	+20 °C	1.036	0.227
	6	+20 °C	1.059	0.293
	12	+20 °C	0.967	0.139
	1	+40 °C	0.938	0.160
	3	+40 °C	1.000	0.227
	6	+40 °C	0.922	0.138
	12	+40 °C	0.883	0.123

[a] $R_T = X_T/X_{-20\,°C}$ X_T = mean of 10 replicates at temperature T (+20 °C or +40 °C); $U_T = (CV_T^2 + CV_{-20\,°C}^2)^{1/2}R_T$; CV_T = coefficient of variation of 10 replicates at temperature T.

storage period and after 1, 3, 6 and 12 months. The stability was assessed, following the method described in Chapter 3; samples stored at −20 °C were used as reference for the samples stored at +40 °C and at +20 °C. No instability was detected for TBT at +20 °C (Table 5.1). In the case of storage at +40 °C, slight TBT losses were, however, suspected to occur after 6 months storage. Although the CRM was shown to be stable at +20 °C, a recommendation was given to store this material at +4 °C for long term storage.

Technical Discussion

The original bar graph presentation showed a very large scatter of data which ranged from less than 10 μg kg^{-1} of TBT to more than 150 μg kg^{-1}. Detailed discussions were necessary to explain the sources of discrepancies. Some laboratories reported "not detected" values ranging from less than 15 μg kg^{-1} to less than 60 μg kg^{-1}.

The laboratories reported their results of extraction recovery, which were generally acceptable (from 80 to 100%). It was assumed that the main problems were not due to extraction but to possible interferences in the derivatization and/or detection steps. Aromatic compounds were suspected to have inhibited the hydride generation but not the ethylation reaction; methanolic HCl extraction did not extract the oil present in the sample but a 35% suppression of hydride generation was still observed. It was felt, however, that the interferences were occurring at the atomization stage rather than at the hydride generation step; high contents of Fe and Cr could have led to a suppression of the TBT signal, as demonstrated with spiking experiments. An extensive study was carried out by the University of Bordeaux on interference effects from inorganic (metals) and organic substances (organic solvents, PCB, pesticides, *n*-alkanes, humic substances) on the yield of hydride generation. The major effects observed were mostly linked to high levels of trace metals; organic compounds had generally negligible effects on the signal suppression [104].

The methods using hydride generation and AAS as final determination were, however, in considerable difficulty with this complicated matrix, due to unknown interferences either at the derivatization step or in AAS detection, or determination limits which were too low. As observed by the participants, the laboratories using gas chromatographic separation and detection either by FPD or MS tended to agree, which would confirm that these methods would be more suited to the determination of TBT in this particular material.

Although the analytical methods involving hydride generation were successfully applied in an interlaboratory study on TBT in spiked sediment [43] as well as in the certification of TBT and DBT in the CRM 462 (coastal sediment) [70], it was concluded that the low TBT mass fractions and the complicated matrix did not allow the use of these methods for an accurate TBT determination in this material. It was stressed that, although certification was contemplated at the start of the interlaboratory study, this material would probably not be suited as a CRM for the following reasons:

- The low content of TBT is very close to the limits of determination of most of the techniques used in this exercise
- The high level of interferences makes this material difficult to analyse, which would render quite doubtful its use as a CRM for routine analysis
- Both the low TBT content and the complicated matrix do not resemble sediments usually analysed for TBT monitoring. This unrepresentativeness is another justification for not proposing this material as a CRM

Consequently, the RM 424 was not certified and will be considered as a research material for laboratories willing to evaluate techniques such as GC-FPD, GC-MS or HPLC-ICP-MS. Figure 5.2 presents the results obtained by GC-FPD, GC-MS, GC-MIP-AES and HPLC-ICP-MS. The good agreement found between these methods gave a good confidence that no systematic error was left undetected. Owing to the potential risks of interferences, calibration by standard additions is a prerequisite.

The reference value (unweighted mean of 8 accepted sets of results) and its standard deviation is (20 ± 5) $\mu g\ kg^{-1}$ as mass fractions (based on dry mass) of TBT cation. Indicative values were given for DBT and MBT, respectively (53 ± 19) $\mu g\ kg^{-1}$ as DBT cation (6 sets of results: GC-FPD, GC-MS and GC-AAS) and (257 ± 54) $\mu g\ kg^{-1}$ as MBT cation (3 sets of results: GC-FPD, GC-MS).

5.5 Certification of Coastal Sediment, CRM 462

Feasibility Study on Material Preparation

One of the major requirements for the preparation of a CRM is its stabilization. This should be achieved in such a way that the pattern of compounds is not changed (*i.e.* no degradation) and the long term stability of the material is achieved. γ-Irradiation is a procedure which has been used for this purpose;

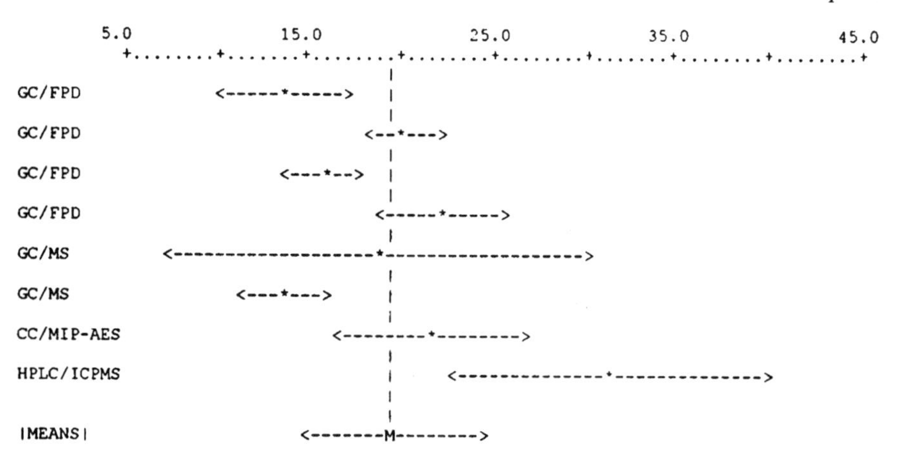

Figure 5.2 *Bar graph of TBT. The laboratory codes are indicated along with the methods used. The results plotted correspond to five replicate determinations. MEANS is the mean of laboratory means with its standard deviation*

however, it has been mainly carried out to stabilize materials to be certified for their total element content. In the case of organotin compounds, it was suspected that γ-irradiation would lead to a degradation of TBT, which was confirmed by a literature search performed within the BCR [105]. Consequently, heating procedures were tested on a sediment portion to investigate the feasibility of stabilizing the material and the possible effects on the stability of butyltin compounds. The aim of this study was to stop possible microbial activity (risk of microbial degradation of TBT) by using a heating procedure which would preserve the organotin content of the sediment. A small amount of sediment sample was collected in the Arcachon Bay (France) and air dried at ambient temperature (*ca.* 20 °C) on a polyethylene film in a clean room for 7 days. The material was ground in a porcelain crucible and sieved at 80 µm mesh. The powder obtained was homogenized manually and bottled in 20 glass bottles each containing *ca.* 50 g. After bottling, 5 flasks were wrapped in an aluminium film to protect them against light and were placed in a cupboard at ambient temperature and kept as test samples. The other bottles were treated as follows:

- 5 bottles heated at 80 °C for 6 h
- 5 bottles heated at 120 °C for 2 h
- 5 bottles heated at 120 °C for 6 h.

The effect of heat on butyltin stability was assessed by determining TBT in duplicate in each bottle. The determination was carried out by acetic acid extraction, hydride generation (using $NaBH_4$), cryogenic condensation in a U-tube filled with chromatographic material (Chromosorb GAWHP 80–100 mesh) followed by thermal separation (based on the different organotin

Table 5.2 *Effects of heat on tributyltin stability*[a]

+20 °C	*6 h at 80 °C*	*2 h at 120 °C*	*6 h at 120 °C*
(102 ± 16) ng g^{-1}	(102 ± 17) ng g^{-1}	(105 ± 24) ng g^{-1}	(85 ± 17) ng g^{-1}

[a] The TBT contents are given as mass fractions (μg kg^{-1}) of Sn.

hydride sublimation points) and AAS detection in a quartz furnace electrically heated to 950 °C.

The results obtained are presented in the Table 5.2 as mass fractions of Sn. A statistical treatment (F-test) did not reveal any significant difference between the sets of results. It was therefore concluded that the heating procedures applied did not entail significant losses of TBT by degradation. Previous results obtained at the University of Pau had, however, shown that drying at 80 °C for an overlong period (*e.g.* overnight) could induce degradation of TBT. Hence, the procedure recommended for the preparation of the candidate CRM was drying at 60 °C for 48 h, grinding and sieving, followed by heat sterilization at 120 °C for 2 h.

Preliminary experiments were also performed in order to evaluate the stability of butyltins in the test material over a 3-month period in the dark. The test material was stored at different temperatures (−20 °C, + 20 °C and +40 °C) and MBT, DBT and TBT were determined in six replicates at each temperature after 1, 2 and 3 months. A statistical treatment of the results (T-test) did not reveal any significant changes in the butyltin contents with respect to the storage conditions tested. It was therefore concluded that the preparation of a coastal sediment for the certification of butyltin compounds could be undertaken safely.

Preparation of the Candidate CRM

Collection

The coastal sediment to be used as a candidate certified reference material (CRM 462) was collected in the southern part of the Arcachon Bay (France) in a small harbour (Larros) [70]. The first 10 cm sediment layer was collected with a shovel and stored in plastic containers. After decantation, the sediment sample was air dried for 7 days at ambient temperature on a cotton sheet in a well ventilated room, and stored at −20 °C in polyethylene bags. The material (about 180 kg dry mass) was then stored in ice boxes for transportation to the IRMM.

Homogenization and Bottling

The material was dried at 55 °C in air for 100 h. The moisture content determined at this stage was less than 3%. The material was then sieved

Table 5.3 *Within- and between-bottle homogeneity*[a]

Compound	*Between-bottle*[b]	*Within-bottle*[c]	*Method of final determination*[d]
DBT	9.6 ± 1.5	5.6 ± 1.2	6.0 ± 1.3
TBT	15.8 ± 2.5	10.0 ± 2.2	9.0 ± 2.0

[a] $(CV \pm U_{CV})$%.
[b] Single determination on the content of each of 20 bottles.
[c] 10 replicate determinations on the content of one bottle.
[d] 10 replicates of an extract solution.
Uncertainty on the CV's: $U_{CV} \approx CV/\sqrt{2n}$.

through a sieve of 1 mm mesh size and finely ground using a jet mill grinding device with a classifier. This procedure allowed a powder with a closely defined particle size distribution (less than 75 μm) and a sharp maximum size limitation (no oversized particles) to be obtained. The material was then sterilized by heating at 120 °C for two hours, homogenized in a mixer for two hours and finally stored in 60 mL well-cleaned brown glass bottles with polyethylene inserts and plastic screw caps, each containing *ca.* 25 g of powder.

The moisture content of the material was determined by Karl Fischer titration on 10 samples selected during the bottling procedure. The mean moisture mass fraction measured was (2.70 ± 0.40)%.

Homogeneity Control

The between-bottle homogeneity was verified by the determination of DBT and TBT on intakes of 1 g taken from 20 bottles which were set aside at regular intervals during the whole period of bottling. The within-bottle homogeneity was assessed by 10 replicate determinations on the well mixed content of one bottle. Each bottle was shaken manually for 5 minutes to eliminate segregation which might have occurred during transport and storage.

The determinations of DBT and TBT were carried out using the method previously described in the feasibility study. Calibrations were performed by standard additions. The uncertainty of the method of separation and final determination was assessed by five replicate determinations of each butyltin compound on one extract solution; the CV of the method does, therefore, not comprise the CV introduced by the extraction procedure.

The CV's for DBT and TBT are presented in Table 5.3. An F-test at a significance level of 0.05 did not reveal significant differences between the within- and between-bottle variances. The within-bottle CV is very close to the CV of the method and, therefore, no inhomogeneity of the material was suspected. It was hence concluded that the material is suitable for use as a CRM and is homogeneous at least at an analytical portion of 1 g and above for DBT and TBT.

Stability Control

The stability of the butyltin content was tested as carried out for RM 424 (Section 5.4). As shown in Figure 5.3a for DBT at +20 °C, the value 1 is contained in almost all the cases between $R_T - U_T$ and $R_T + U_T$. The uncertainty on the *CV* can account for the deviations observed. For TBT at +20 °C (Figure 5.3b), the content decreased after three months of storage and stabilized afterwards. This decrease was assessed to be due to a degradation of TBT directly to MBT as previously observed [103], because of a corresponding increase of MBT content after 3 months (from 66 to 93 μg kg^{-1} as Sn). In the case of storage at +40 °C, both DBT and TBT displayed strong losses after 3 months which was likely due to a degradation of these compounds to MBT and inorganic tin (Figures 5.3a and 5.3b). On the basis of the results, it was concluded that:

- DBT is stable at +20 °C in the dark
- TBT displayed losses at the start of the storage period but the content stabilized after 3 months. It is hence concluded that TBT remained stable at +20 °C in the CRM in the dark over 12 months; however, in order to avoid any risk of organotin degradation during long term storage, it was decided to store the material at +4 °C in the dark
- Both DBT and TBT are unstable at +40 °C. However, the two compounds are stable at this temperature for at least one month (Figures 5.3a and 5.3b), which indicates that the material could be transported safely under extreme conditions

Technical Evaluation

Tributyltin

Poorer extraction recoveries for TBT in CRM 462 (Figure 5.4a) were often observed in comparison to recoveries obtained from other sediment materials. Some laboratories found lower recoveries if the spike was allowed to equilibrate longer. The previous interlaboratory exercise on TBT-spiked sediment had identified the need to allow spikes to equilibrate at least overnight to obtain a realistic assessment of extraction recoveries.

The evaluation of polarographic methods showed that there were major problems with this sample. The surfactants present made it difficult to detect TBT and polarography was, therefore, a method not recommended for this certification.

In some cases, the uncertainty on extraction recoveries were not taken into account in the overall uncertainty of the laboratory means, which explained the occurrence of small standard deviations.

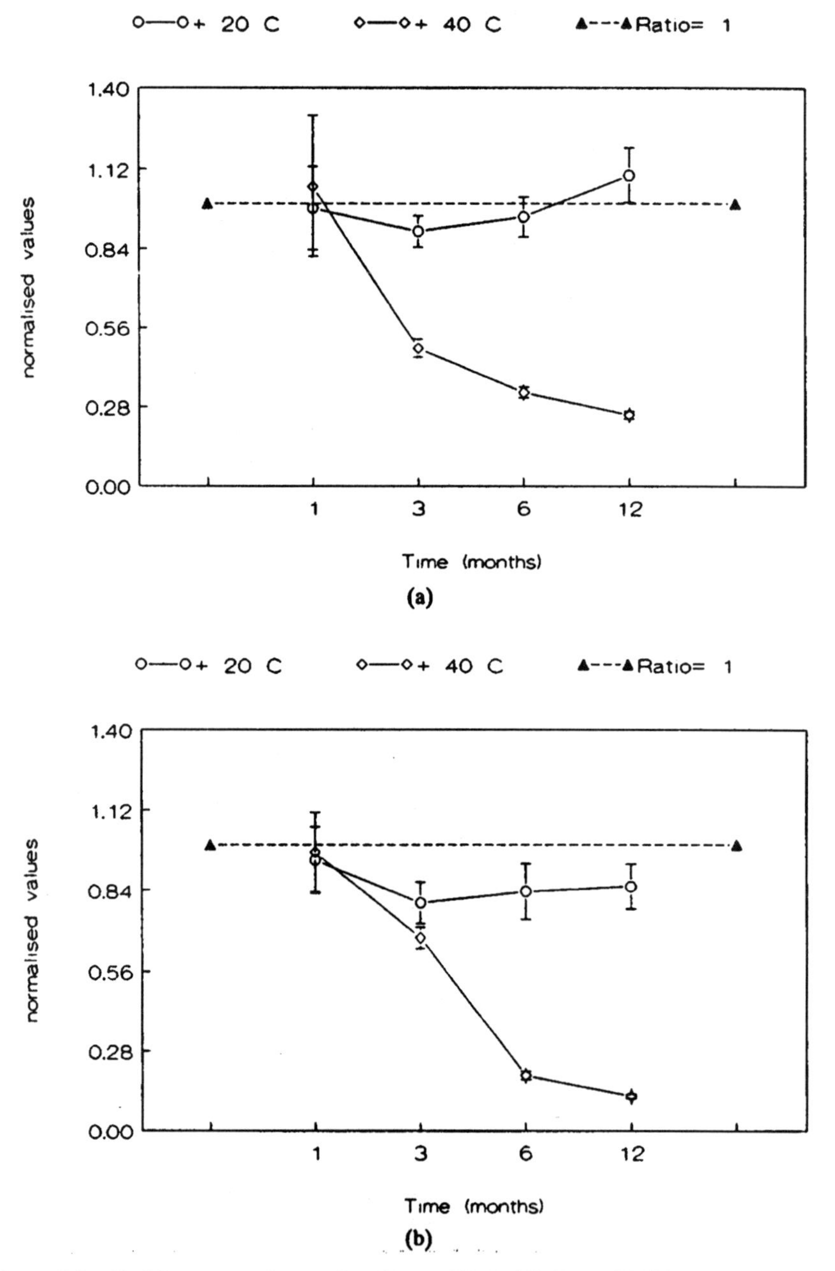

Figure 5.3 *Stability tests of coastal sediment CRM 462.* $R_T = X_T/X_{-20\,°C}$; X_T = *mean of 10 replicates at temperature* T *(+20 °C or +40 °C);* $U_T = (CV_T^2 + CV_{-20\,°C}^2)^{1/2} R_T$; CV_T = *coefficient of variation of 10 replicates at temperature* T. *(a) Stability tests of DBT at +20 and +40 °C. (b) Stability tests of TBT at +20 and +40 °C*

[Reproduced by permission from P. Quevauviller *et al.*, *Appl. Organomet. Chem.*, **8**, 629 (1994). © John Wiley & Sons Limited]

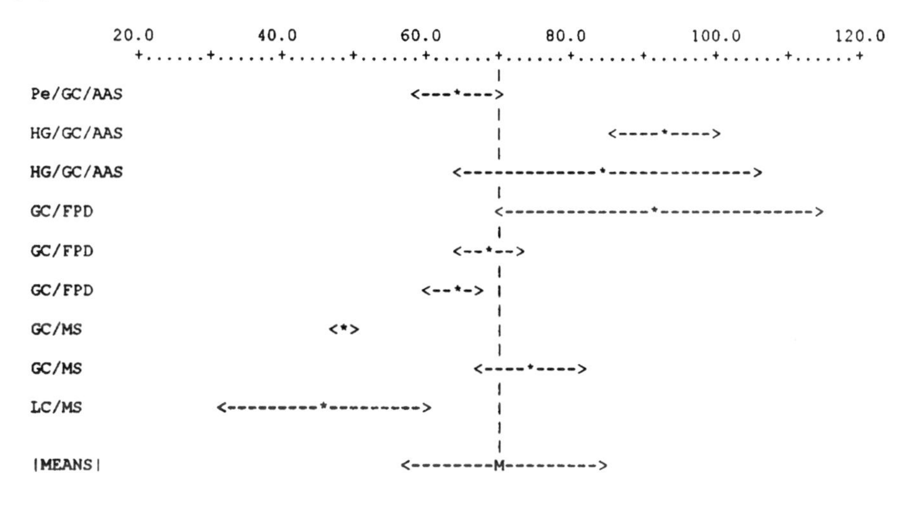

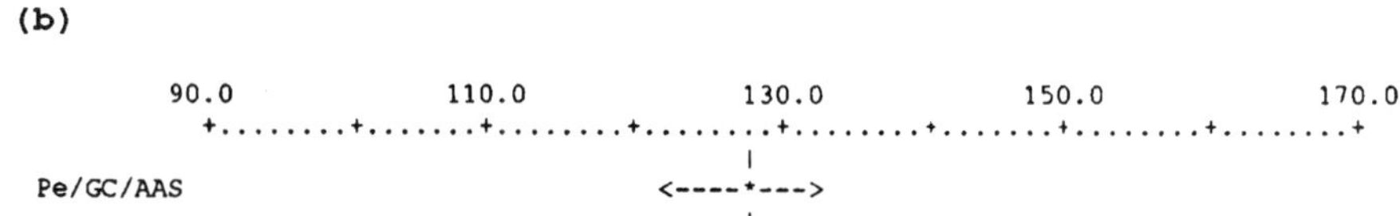

Figure 5.4 *Bar graphs of TBT and DBT (certified values in CRM 462). The results correspond to five replicate determinations. MEANS is the mean of laboratory means with 95% confidence interval. (a) Tributyltin in μg kg^{-1} (as TBT). (b) Dibutyltin in μg kg^{-1} (as DBT)*

Dibutyltin

As shown in Figure 5.4b, the overlap obtained between the different sets of results and the range of techniques used was found satisfactory and, as no doubts were thrown on the results presented, it was agreed to certify the DBT content.

Table 5.4 *Certified mass fractions of TBT and DBT in CRM 462*[a]

Compound	*Certified value*	*Uncertainty*	*Unit*	*p*
TBT	70	14	μg kg^{-1}	9
DBT	128	16	μg kg^{-1}	7

[a] The certified values are the unweighted mean of *p* accepted sets of results.

Monobutyltin

A high scatter of results was observed for MBT, which prevented certification. Many laboratories reported problems in the extraction step. The addition of complexing agents such as diethyldithiocarbamate or sodium dithiocarbamate may enhance the extraction recovery. However, the addition of complexing agents was found to prevent the hydride generation of a volatile species for MBT. Considering the high spread of results (13 to 244 μg kg^{-1} as MBT) and the doubts still remaining on the different techniques used, it was decided not to give any indicative values for MBT which could be misused by laboratories. Further efforts should be made to improve the state of the art of MBT determinations before contemplating certification of this compound at the μg kg^{-1} level. A similar lack of agreement on MBT results has already been shown in an interlaboratory study of analytical techniques [107]; this study confirmed, however, that the techniques used for DBT and TBT determinations were satisfactory.

Certified Values

Following the technical evaluation, the sets of accepted results were submitted to statistical tests (Kolmogorov–Smirnov–Lilliefors, Nalimov, Bartlett and Cochran tests, and one-way analysis of variance) which are described in detail in the certification report [108]. The certified values (unweighted mean of *p* accepted sets of results) and their uncertainties (half-width of the 95% confidence intervals) are given in the Table 5.4 as mass fractions (based on dry mass). The two compounds are certified as mass fractions (μg kg^{-1}) of, respectively, $Sn(C_4H_9)_3^+$ (TBT) and $Sn(C_4H_9)_2^{2+}$ (DBT).

The success of this certification campaign clearly illustrates the improvement achieved by the participating laboratories in the course of the interlaboratory programme. As mentioned in the other study on TBT in harbour sediment [103], the certification of butyltins in sediment could not be achieved in 1990 due to a lack of control of some techniques, *e.g.* involving hydride generation. The results obtained in this campaign show that all the techniques used (including hydride generation-based methods) were in the best obtainable agreement in relation to the state of the art. Recommendations were given to use standard addition procedures systematically and to check carefully the extraction recovery and the derivatization yield. Particular precautions were taken for the quality control of hydride generation in this exercise; this technique was accepted for certification, providing that all the proofs for a good QC were given.

Further efforts are needed to certify MBT in sediment. The certification of butyl- and phenyl-tins in both freshwater sediment and mussel material is now contemplated within the EC Measurements and Testing Programme.

Re-certification

The CRM 462 has been successfully used during two years (1994–1996) for the quality control of DBT and TBT in coastal sediment. Some doubts were, however, expressed on the long-term stability of the material stored at +4 °C, as notified by a laboratory currently using the material for quality control checks. These doubts justified that the material be withdrawn from the market and that new stability experiments be carried out. The new studies of the material confirmed that degradation of TBT (roughly 15%) had occurred. Consequently, a new stability study (at −20 °C) has been carried out and the material has been re-certified for its content of DBT and TBT by a group of selected laboratories [109]. This additional work clearly demonstrated that reference materials containing organotins have to be imperatively stored at −20 °C or below to avoid any risk of degradation.

5.6 Certification of Mussel Tissue, CRM 477

Preparation of the Material

Collection

In recent years, several monitoring campaigns were carried out in the La Spezia Gulf (Liguria, Italy) in order to study the environmental distribution and fate of organic micropollutants. The harbour of La Spezia is characterized by intense maritime traffic and dockyard activity (both civil and military) and by the presence of one of the most important Italian mussel farms. Analyses performed on mussels collected during these campaigns showed high contents of organotin compounds in mussel tissues as a consequence of maritime activities [110], which made samples from this area suitable as a candidate reference material for the certification of butyl- and phenyl-tin compounds.

1200 kg of mussels (*Mytilus edulis*) were purchased directly at the La Spezia mussel farm in July 1991 [111]. After collection, the mussel samples were washed with fresh water to eliminate matrix salts which could interfere in the preparation process or the analysis. The samples were immediately frozen by immersion in liquid nitrogen. Shelling could not be performed by cooking or using a vapour stream since this treatment could have caused degradation of organotin compounds. It was preferred to shell the frozen materials directly by using special mussel knives. The edible part was collected in thermally sealed polyethylene bags (*ca.* 4 kg per bag) and immediately stored at −25 °C.

Freeze-drying Procedure

The frozen material (*ca.* 325 kg) was transported to the Biostarters Company (Parma) where it was ground, using a PTFE-coated mill, and spread on sterilized flat trays for the freeze-drying treatment. The process consisted of dividing the homogenate into batches, leaving the material at *ca.* −55 °C for 6 h, then applying the vacuum for 48 h. Analyses were performed at the end of each freeze-drying process on samples collected from the top, the intermediate and the bottom levels of the flat trays, to evaluate the suitability of the process. The results showed that a moisture content of less than 4% was achieved. The freeze-dried material (final amount *ca.* 35 kg) was put into thermally sealed polyethylene bags, stored at −25 °C and transported to the Joint Research Centre at Ispra.

The freeze-dried material was ground for 15 days in a zirconia ball mill, taking all precautions to avoid contamination. It was then sieved at 125 μm mesh with a titanium sieve in order to separate the fibrous part of the bulk material and mixed for 15 days under argon atmosphere in a special polyethylene-lined mixing drum. The argon atmosphere was renewed after 5 and 10 days. The material was bottled under argon atmosphere in brown Pyrex-glass bottles, remixing the sample for 30 min after 40 bottles had been filled. Bottles were set aside during the bottling procedure for the homogeneity and stability studies. One thousand bottles, each containing *ca.* 15 g were obtained.

Homogeneity Control

The between-bottle homogeneity was verified by the determination of mono-, di- and tri-butyltin (MBT, DBT and TBT) and mono-, di- and tri-phenyltin (MPhT, DPhT and TPhT) on sub-samples of 500 mg taken from 20 bottles which were set aside at regular intervals during the whole period of bottling. The within-bottle homogeneity was assessed by 10 replicate determinations on the well-mixed content of one bottle.

The organotin determinations were carried out as follows [18]: the subsample (500 mg) was placed in a Pyrex vial (20 mL) and about 500 ng (as Sn) of internal standard (tripropyltin chloride in methanolic solution) was added. A mixture of methanolic tropolone and HCl was then added, and the vial, capped with a Teflon-lined screw cap, was placed in an ultrasonic bath for 15 min. The procedure was repeated twice, collecting the supernatent after centrifugation and transferring it to a 250 mL cylindrical separatory funnel. 25 mL of dichloromethane were added and the solution was shaken with 200 mL of a NaCl solution. This operation was repeated with another 25 mL of dichloromethane aliquot. The organic phases were collected through anhydrous sodium sulfate and concentrated down to few mL by a gentle stream of nitrogen and transferred to a 15 mL reaction vial. 2 mL of isooctane were added and the solution was brought near to dryness under moderate flow of

Table 5.5 *Recoveries of organotin compounds from 500 mg of non-spiked and spiked mussels*[a]

Compound	*Non-spiked /ng Sn*	*Spiked amount /ng Sn*	*Found /ng Sn*	*Recovery /%*
TBT	185 ± 24	160	314 ± 27	91 ± 9
DBT	61 ± 10	165	201 ± 23	89 ± 11
MBT	80 ± 13	150	195 ± 29	85 ± 15
TPhT	nd	150	138 ± 13	92 ± 9
DPhT	nd	157	133 ± 18	85 ± 14
MPhT	nd	147	120 ± 20	82 ± 17

[a] Results are the average of five different experiments [18]. nd: not detected.

nitrogen. 1 mL of an ethereal solution of pentylmagnesium bromide (2 mol L^{-1}) was then added, the vial was capped and the Grignard reaction was allowed to proceed by shaking in a thermostatic bath for 1 h, after which 2 mL of isooctane were added. The excess reagent was destroyed by carefully adding about 1–2 mL of water (dropwise) and then 5–10 mL of sulfuric acid (1 mol L^{-1}). The organic layer was removed with a Pasteur pipette and put on the top of a 3 g silica-gel column for clean-up. The column was eluted with hexane:benzene (1:1) until 5 mL were collected. The solution was finally reduced to an exact volume (usually 1 mL) and 1 μL was injected into the GC-MS apparatus. The gas chromatograph was a HP 5890 (capillary column HP Ultra 25 m × 0.2 mm × 0.11 μm, helium as carrier gas). The detector was a HP 5970 MSD (EI quadrupole).

As mentioned above, quality control of organotins in biological materials is limited, owing to the lack of CRMs available (with the exception of NIES 11, certified for TBT and DBT). To ensure the best traceability of measurements, pentylated organotin calibrant solutions were prepared at the beginning of the study and stored at −20 °C in the dark [112] to serve as independent control solutions, as the stability of fully derivatized organotins far exceeds that of the starting compounds, particularly for phenyltins. Freshly prepared stock and working calibrant solutions were checked for degradation products by GC-MS after pentylation and analysed with respect to the stored pentylated calibrant solutions. In order to control the method performance, organotin working calibrant solutions were regularly run through the whole analytical procedure. Single analytical steps were carefully checked for performance problems or relevant changes in the materials used (*e.g.* clean-up for new batches of silica gel). Recovery tests were performed using freeze-dried mussel samples analyzed before and after spiking with *ca.* 160 ng (as Sn) of each butyl- and phenyl-tin compound [18]. Organotins were added as solutions in methanol to the samples previously wetted with distilled water; after the addition, the mussel samples were shaken for at least 30 min and allowed to equilibrate overnight. The recoveries were calculated with respect to the sum of the contents of the incurred compounds and the spikes. Recoveries ranged from (80 ± 15)% to (90 ± 10)% (Table 5.5).

Table 5.6 *Within- and between-bottle homogeneity*[a]

Compound	*Between-bottle*[b]	*Within-bottle*[c]	*Method of final determination*[d]
MBT	4.0 ± 0.7	2.5 ± 0.6	1.9 ± 0.8
DBT	3.1 ± 0.5	1.9 ± 0.4	2.9 ± 0.7
TBT	2.9 ± 0.5	2.1 ± 0.5	2.0 ± 0.5
MPhT	4.4 ± 0.7	3.4 ± 0.8	3.0 ± 0.7
DPhT	11.0 ± 1.8	8.3 ± 1.9	5.2 ± 1.2
TPhT	6.5 ± 1.0	3.5 ± 0.8	3.0 ± 0.7

[a] $(CV \pm U_{CV})$%.
[b] Single determination on the content of each of 20 bottles.
[c] 10 replicate determinations on the content of one bottle.
[d] 10 replicates of an extract solution.
Uncertainty on the *CV*'s: $U_{CV} \approx CV/\sqrt{2n}$.

The uncertainty of the method of separation and final determination was assessed by seven replicate determinations of each butyltin compound on one extract solution; the *CV* of the method, therefore, does not comprise the *CV* introduced by the extraction procedure. The *CV*'s for the six organotin compounds are presented in Table 5.6. An *F*-test at a significance level of 0.05 did not reveal significant differences between the within- and between-bottle variances. The within-bottle *CV* is very close to the *CV* of the method and, therefore, no inhomogeneity of the material is suspected. It is hence concluded that the material is suitable for use as a CRM and is homogeneous at least at an analytical subsample of 500 mg for the six organotin compounds studied.

The ranges of organotin contents found in the mussel candidate CRM are representative of medium to high contamination levels, *i.e.* 700 to 1000 $\mu g\ kg^{-1}$ for each butyltin species, 1000 to 1500 $\mu g\ kg^{-1}$ for MPhT and TPhT and below 50 $\mu g\ kg^{-1}$ for DPhT.

Stability Control

This study followed the usual procedure used at the BCR, *i.e.* the stability was verified over a period of one year (after 1, 3, 6 and 12 months) at three different temperatures (−20, +4 and +40 °C), using the results obtained at −20 °C as reference to assess possible changes which might occurred at +20 and +40 °C.

The tests were repeated after 24, 36 and 44 months of storage at −20 °C. Samples were analysed using the procedures detailed above for the homogeneity study. MBT, DBT and TBT were each determined three times at each occasion of analysis (one replicate determination in each of three bottles stored at different temperatures).

A high rate of degradation (TBT and DBT degraded to MBT and probably inorganic tin) was observed at +40 °C; instability was also noted at +20 °C and to a lesser extent at +4 °C. With respect to stability tests carried out at −20 °C, a slightly higher uncertainty for MBT could be due to a change of extractability but this was not considered to affect the stability of this compound. On the basis of the results, it was concluded that butyltin compounds in mussel tissue are stable at −20 °C over a long term period (as verified over 44 months) and that the material is hence suitable to be used as a CRM (Figures 5.5a,b) [113].

Mono-, di- and tri-phenyltin were not found to be stable at the different temperatures tested. Consequently, it was decided not to certify and not to give any indicative values for these three compounds [114].

Technical Discussion

Two sets of results were withdrawn for all compounds owing to suspicion of systematic errors (systematic low values) and lack of quality control (large standard deviations).

Tributyltin

A difference in the standard deviation (SD) in the sets of HPLC-ICP-MS data was due to the fact that the dataset with the lower SD was obtained by an isotope dilution methodology; both sets of data were retained for certification. The bar graph for TBT is given in Figure 5.6.

Dibutyltin

Relatively high results with a large *CV* were obtained by HPLC-ICP-AES and doubts were raised about the specificity of the liquid/liquid extraction procedure used, which could have led to an over-estimate of the results due to uncertainties in recovery; the set of data was therefore withdrawn. The bar graph for DBT is shown in Figure 5.7.

Monobutyltin

One laboratory reported a (71 ± 6)% recovery which was lower than for other species and a possible over-estimation was suspected; the set of data was consequently withdrawn. The use of an ultrasonic stage in the extraction in the GC-MIP method (giving a 70% recovery) could be responsible for the apparent over-estimation of the results; another laboratory used a microwave method and did not observe such over-estimation.

One laboratory submitted two sets of data, one obtained with GC-FPD with a relatively low recovery (85 *cf.* 95%) and one obtained with GC-QFAAS involving enzymatic digestion; the former result was discarded. The bar graph for MBT is shown in Figure 5.8.

Triphenyltin

The effect of the use of methanol and drying (blowing down) the sample on the derivatization step (Grignard reaction) was discussed. It seemed to be clear that, under certain conditions, degradation products could be formed which

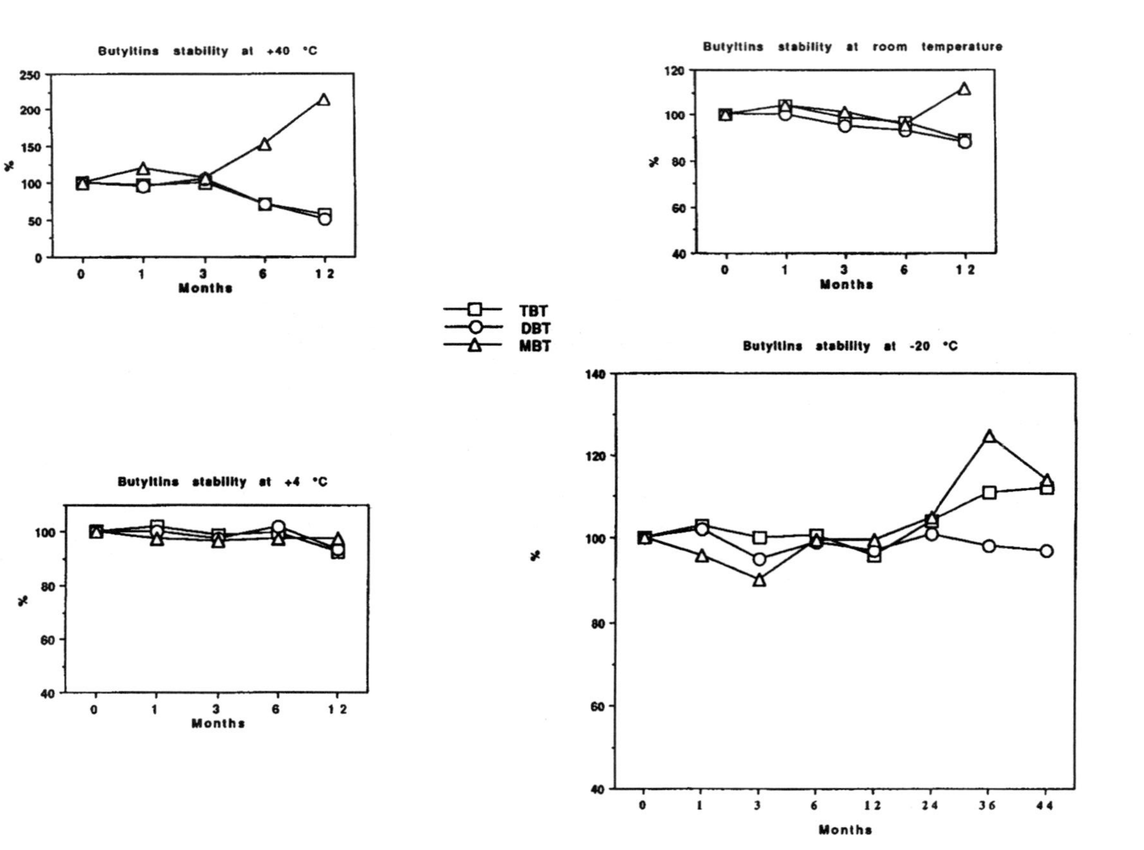

Figure 5.5 *Stability of (a) MBT, (b) DBT and (c) TBT in CRM 477 in mg kg*$^{-1}$

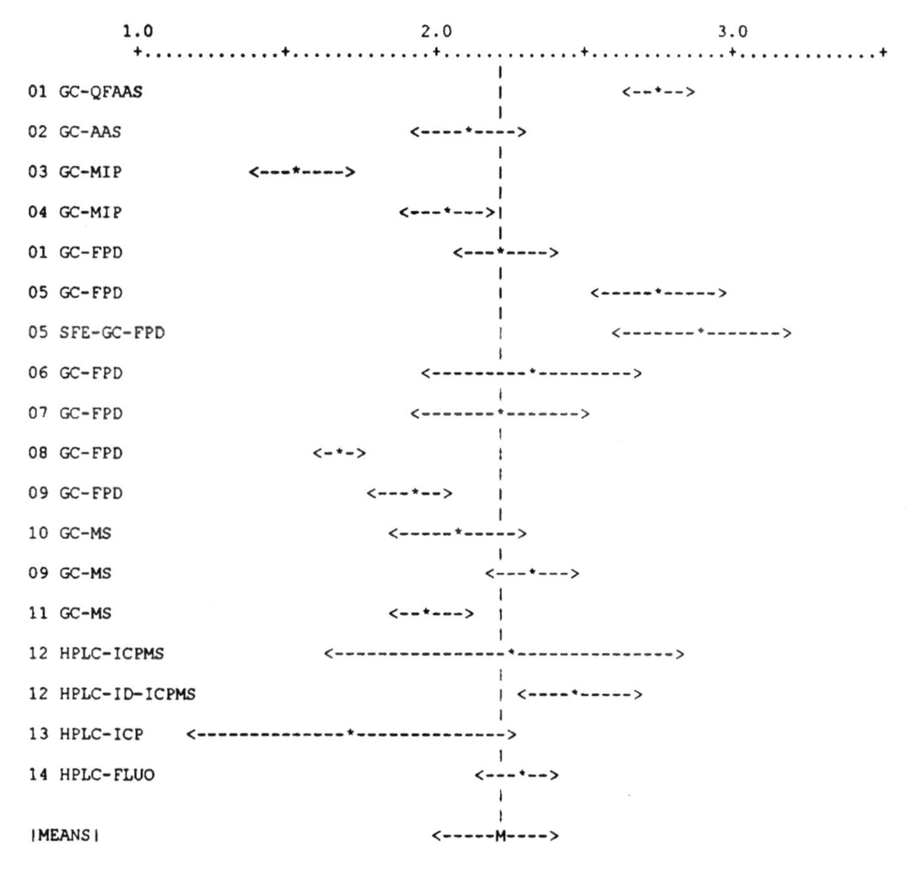

Figure 5.6 *Tributyltin in CRM 477 in mg kg^{-1} (as TBT)*

would lead to significant analytical uncertainties, particularly if the extraction/derivatization is too long. Degradation could be photoinduced. It was agreed that the extraction and derivatization were best achieved in a single step process.

The set of data obtained by SFE-GC-FPD was withdrawn since the technique was optimized for TBT but not for TPhT.

The relatively large SD obtained in HPLC-ICP-MS was discussed; difficulties in optimization of the HPLC conditions would have contributed to this poor precision. In order to safeguard against contributions from nearby peaks, peak heights had been used as the basis for quantification.

The relevance of the requested spike recovery test, practical aspects of the spiking [*e.g.* slurry (possible over-estimation of recovery) or dry down (better admixture but possible degradation risk)] and the problems of degradation of calibrants were discussed and will be the subjects of further studies. Apparent

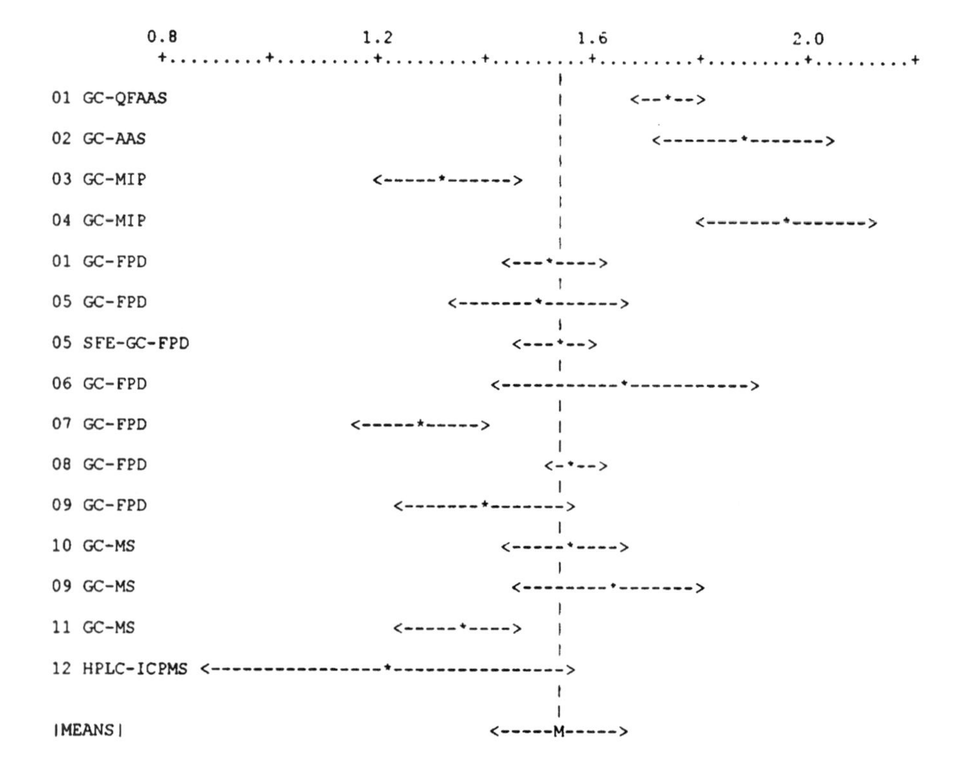

Figure 5.7 *Dibutyltin in CRM 477 in mg kg^{-1} (as DBT)*

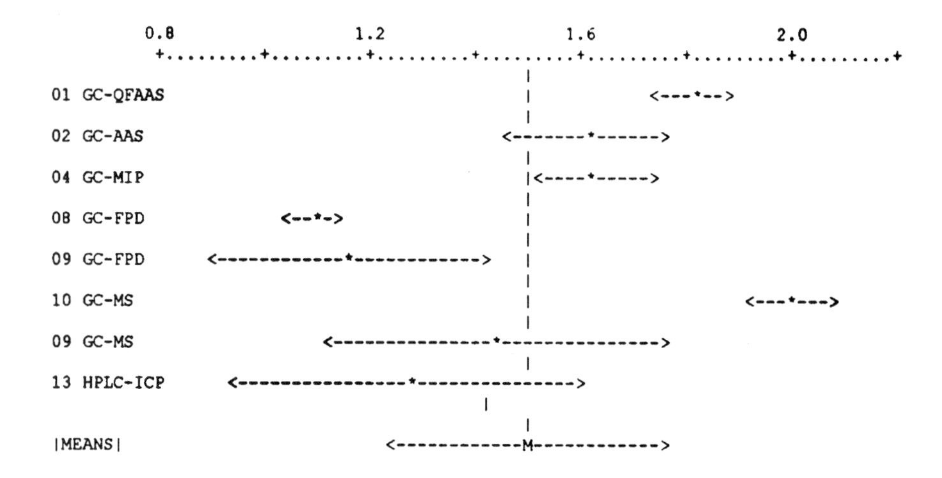

Figure 5.8 *Monobutyltin in CRM 477 in mg kg^{-1} (as MBT)*

Table 5.7 *Certified contents of MBT, DBT and TBT in CRM 477*

	Certified value ($mg\ kg^{-1}$ as cations) dry matter basis	*Uncertainty ($mg\ kg^{-1}$ as cations) dry matter basis*	*p*
MBT	1.50	0.27	8
DBT	1.54	0.12	15
TBT	2.20	0.19	18

p: number of accepted sets of results.

discrepancies in the propagation of errors used by some laboratories did not seem to reflect recovery efficiency variations in their final result.

Owing to the large spread of results and the demonstration of instability risks of this compound, it was decided not to give any value for TPhT.

Monophenyltin and Diphenyltin

A large spread of results, a poor between-bottle homogeneity (*CV* of *ca.* 48%) and a lack of stability precluded the certification of DPhT, for which no indicative value is proposed. The same situation was observed for MPhT in comparison to DPhT and a similar decision was taken.

Certified values

The certified values (unweighed mean of *p* accepted sets of results) and their uncertainties (half-width of the 95% confidence intervals) are given in Table 5.7 as mass fractions (based on dry mass). The three compounds are certified as mass fractions ($mg\ kg^{-1}$) of $Sn(C_4H_9)^{3+}$ (MBT), $Sn(C_4H_9)_2^{2+}$ (DBT) and $Sn(C_4H_9)_3^{+}$ (TBT), respectively.

CHAPTER 6

Lead Speciation

6.1 Aim of the Project and Coordination

Justification

Environmental contamination by lead is widespread, the major anthropogenic source of this element being the combustion of leaded gasoline. Although their use has been discontinued in some countries, the use of tetraalkyllead compounds as an antiknock agents remains the largest application of organolead compounds [115]. It is recognized that, owing to the ubiquity of lead and concern over the toxicity of organolead compounds in the environment, the monitoring of lead species will need to be continued over the next decade. Vehicular emissions of tetraalkyllead are subject to atmospheric breakdown to trialkyl- and dialkyl-lead and all three forms are scavenged from the atmosphere by rainfall [116]. Therefore, trimethyl- and triethyl-lead are found in road drainage and surface water [117]. As a consequence, a number of laboratories are performing analyses of rainwater and urban dust to monitor the levels of trialkyllead compounds in the environment. The techniques are generally based on a combination of different analytical steps including extraction, derivatization (*e.g.* ethylation, Grignard reaction), separation (*e.g.* gas chromatography or high performance liquid chromatography) and detection (*e.g.* atomic absorption or atomic emission spectrometry) [21].

In view of the urgent need to establish the state of the art of lead speciation analysis, a project has been discussed and designed with a group of European laboratories in the framework of the Measurements and Testing Programme (formerly the BCR) of the Commission of the European Communities. A feasibility study has been carried out to investigate the stability of alkyllead compounds in solution [118], showing that dialkyllead and triethyllead compounds were not sufficiently stable to be kept in water; the only compound which could be stored without significant degradation was trimethyllead and this compound was, therefore, selected for the preparation of lead-containing solutions. Interlaboratory exercises were designed as a first step of a larger project whose aim is to produce reference materials of simulated rainwater and urban dust certified for their content of trimethyllead.

The Programme and Timetable

The project to improve the quality control of lead speciation analysis was started in 1990 by a feasibility study on the stability of alkyllead species in solution [118], and was concluded in 1991. The first interlaboratory study was conducted in 1992 [119] and was followed by a second exercise carried out in 1993 [120]. The certification campaign of trimethyllead in artificial rainwater and urban dust was conducted in 1995–96.

Coordination

Feasibility Study

The feasibility study on the stability of lead species in solution was carried out by the University of Antwerp, Departement of Chemistry (Wilrijk, Belgium), which also verified the purity and stoichiometry of calibrants used in the interlaboratory studies.

Interlaboratory Studies

The artificial rainwater and urban dust samples used in the two interlaboratory studies were prepared by the School of Chemistry of the University of Birmingham (United Kingdom), which also verified their homogeneity and stability. The two exercises were coordinated by the BCR.

Certification

The candidate reference materials of artificial rainwater and urban dust were prepared by the School of Chemistry of the University of Birmingham (United Kingdom) which also verified their homogeneity and stability. The two exercises were coordinated by the Department of Environmental Sciences of the University of Plymouth (United Kingdom).

Participating Laboratories

The following laboratories participated in the interlaboratory studies: Associated Octel Company, South Wirral (United Kindgom); De Montfort University, Department of Chemistry, Leicester (United Kingdom); Health and Welfare Canada, Ottawa (Canada); University of Umeå (Sweden); Universität Münster, Institut für Chemo- und Biosensorik, Münster (Germany); University of Plymouth (United Kingdom); CNR-ICAS, Pisa (Italy); Universitaire Instelling Antwerpen, Departement Scheikunde, Wilrijk (Belgium); Philipps-Universität Marburg (Germany); Universidad de Zaragoza (Spain); National Institute of Occupational Health, Umeå (Sweden).

The following laboratories took part in the certification campaign: De Montfort University, Department of Chemistry, Leicester (United Kingdom);

GALAB, Geesthacht (Germany); Technische Universität Wien (Austria); University of Umeå (Sweden); Universität Mainz, Anorganische Chemie, Mainz (Germany); Universität Münster, Institut für Chemo- und Biosensorik, Münster (Germany); University of Plymouth (United Kingdom); CNR-ICAS, Pisa (Italy); Universitaire Instelling Antwerpen, Departement Scheikunde, Wilrijk (Belgium); Universidad de Zaragoza, Centro Politécnico Superior de Ingenerios, Zaragoza (Spain); Vrije Universiteit Amsterdam, Instituut voor Milieuvraagstukken, Amsterdam (The Netherlands).

6.2 Techniques Used in Lead Speciation

Techniques used in the interlaboratory studies and in the certification campaigns on trimethyllead in artificial rainwater and urban dust are described below.

Trimethyllead Determination in Artificial Rainwater

Each ampoule had to be made up to one litre of working solution with the purest water available in the laboratory.

Hydride Generation/ETAAS

1 mL intake was used for the analysis. HCl was added to pH 3 and complexation was performed with EDTA. Derivatization was carried out by addition of 0.5% $NaBH_4$; the hydrides were carried by an Argon flow at 75 mL min^{-1}. Detection was by ETAAS (with L'vov platform) with Zeeman background correction at 283.3 nm. Calibration was by calibration graph and standard additions, using Me_3PbCl as calibrant. This technique was considered to be acceptable when only trimethyllead is present in the sample since the method only differentiates between "inorganic" and "organic" lead species (*i.e.* there is no chromatographic separation of the organic species).

Ethylation/GC-QFAAS

A 20 mL solution was used for the analysis. Ethylation was carried out with 4% $NaBEt_4$ after addition of acetate buffer. The ethylated compound was cryogenically trapped in a U-tube filled with chromatographic material (0.5 m length, 3 mm internal diameter, OV-101 as stationary phase loaded with 10% silica 80/100 mesh; He as carrier gas at 120 mL min^{-1}; H_2 as make-up gas at 30 mL min^{-1}; column temperature at 75 °C). Detection was by QFAAS at 283.3 nm (detector temperature at 800 °C). Calibration was by standard additions, using the Me_3PbCl provided by SM&T.

Ethylation/CGC-QFAAS

A 100 mL intake was used for analysis. Extraction was performed with 50 mL NaOH and 2 mL sodium acetate/acetic acid (2 mol L^{-1}) in 10 mL hexane.

Clean-up was carried out using silica gel, followed by preconcentration over a N_2 stream to a volume of 1 mL. Derivatization was performed with 10% $NaBEt_4$ in acetic acid at pH 4. Separation was by CGC (column of 30 m length, 0.32 mm internal diameter, DB-5 as stationary phase, 0.25 μm film thickness; He as carrier gas; air/H_2 as make-up gases; injector temperature at 80 °C). Detection was by QFAAS at 283.3 nm (detector temperature at 750 °C). Calibration was by standard additions, using Me_3PbCl provided by SM&T.

Propylation/GC-QFAAS

First method: 150 mL were used for the analysis. NaOH was added up to pH 9, buffering with borax/HCl, followed by an extraction with 0.7 mL *n*-hexane, complexation with 10 mL 0.5 mol L^{-1} DDTC, and addition of 0.6 mol L^{-1} HCl. The extraction recovery was evaluated by spiking the same matrix: (97.3 ± 5)%. Derivatization was performed with 2 mol L^{-1} propylmagnesium chloride in diethyl ether. Separation was by CGC (column of 15 m length, 0.53 mm internal diameter, DB-1 (100% dimethylpolysiloxane) as stationary phase, 1.5 μm film thickness; He as carrier gas at 22 mL min^{-1}; H_2 as make up gas at 230 mL min^{-1}; injection temperature at 180 °C; column temperature ranging from 40 to 200 °C). Detection was by electrothermally heated QFAAS at 283.3 nm. Calibration was by calibration graph, using two different Me_3PbCl calibrants.

Second method: 500 mL of sample were extracted with 20 mL hexane after addition of NaCl and 5 mL of 0.5 mol L^{-1} sodium diethyldithiocarbamate. The mixture was mechanically shaken, evaporated under N_2 flow to 0.5 mL and 2 mol L^{-1} propylmagnesium chloride in diethyl ether was added. The extraction recovery was (98 ± 5)% (evaluated by spiking). Separation was by GC with a glass column packed with 10% OV-101 loaded with 10% Chromosorb W 80–100 mesh (column of 1.2 m length, 3 mm internal diameter; He as carrier gas at 130 mL min^{-1}; injector temperature at 150 °C; column temperature ranging from 85 to 185 °C). Detection was by electrothermally heated QFAAS (T-tube) at 283.3 nm with deuterium background correction (detector temperature at 950 °C). Calibration was by calibration graph, using Me_3PbCl as calibrant.

Ethylation/CGC-MIP-AES

10 mL intake were taken for the analysis. 500 μL of ammonium citrate/EDTA buffer solution was added at pH 8. Derivatization was performed with 1 mL of 0.8% aqueous $NaBEt_4$ and addition of 500 μL hexane containing Bu_4Pb as internal standard; the mixture was mechanically shaken for 5 min, decanted for 5 min and the hexane phase was collected. Separation was by CGC (column of 25 m length, 0.32 mm internal diameter, HP-1 as stationary phase, 0.17 μm film thickness; He was used as carrier gas at 300 mL min^{-1}; H_2 and O_2 were added as scavenger gases at 621 and 138 mL min^{-1}; column

temperature ranged from 45 to 280 °C). Detection was by MIP-AES at 405.8 nm. Calibration was by standard additions, using Me_3PbCl as calibrant. The recovery was estimated by spking the rainwater CRM and found to be 95%.

Propylation/CGC-ICP-MS

Ca. 20 mL was used for the analysis to which were added 2 mL EDTA/ammonia/citrate buffer solution to pH 8. 500 μL hexane were added, containing Bu_4Pb as internal standard. The mixture was mechanically shaken for 10 min, decanted for 5 min (phase separation) and 400 μL of the hexane phase was derivatized by addition of 40 mL of 2 mol L^{-1} propylmagnesium bromide in diethyl ether. The excess Grignard reagent was destroyed with 2 mL of 0.1 mol L^{-1} H_2SO_4. Separation was by CGC (column of 30 m length, 0.25 mm internal diameter, RSL-150 as stationary phase, 0.50 μm film thickness; He was used as carrier gas; the column temperature ranged from 60 to 230 °C). Detection was by ICP-MS. Calibration was by standard additions using ^{208}Pb isotope. The recovery (assessed by spiking) was 93%.

Pentylation/CGC-MS

5 mL intake was used for the analysis. The solution was buffered at pH 0.5 with citric acid/ammonia, followed by two extractions with 10 mL hexane. Complexation was carried out with sodium diethyldithiocarbamate. The extract was dried with Na_2SO_4 and preconcentrated under a N_2 stream. Derivatization was carried out by addition of 2 mol L^{-1} pentylmagnesium bromide in diethyl ether, followed by clean-up with 100% active alumina and elution with hexane/diethyl ether. Separation was by CGC (fused silica column of 30 m length, 0.25 mm internal diameter; 5% phenyl/methylpolysiloxane as stationary phase, 250 μm film thickness; He as carrier gas at 60 mL min^{-1}; injector temperature ranging from 60 to 280 °C; column temperature ranging from 60 to 200 °C). Detection by MS, monitoring ions 307 and 309 (detector temperature at 250 °C). Calibration was by standard additions, using Me_3PbCl as calibrant. Ph_2SnEt_2 was added as internal standard.

Butylation/CGC-MS

Ca. 75 mL solution was used for the analysis. Buffering was performed with ammonium citrate at pH 9, followed by addition of 0.5 mL of 0.25 mol L^{-1} diethyldithiocarbamate. Extraction was carried out with 5 mL pentane. The extract was dried with Na_2SO_4, evaporated to 0.5 mL and redissolved into hexane. Derivatization was by addition of 0.5 mL of 2 mol L^{-1} BuMgCl in THF, followed by addition of H_2SO_4. Separation was by CGC (fused silica column of 60 m length, 0.25 mm internal diameter, DB-1 as stationary phase, 0.25 μm film thickness; He was used as carrier gas at 110 mL min^{-1}; injector temperature at 250 °C; column temperature ranging from 50 to 260 °C). Detection was by MS (detector temperature at 280 °C), monitoring the ions

208, 223 and 253 (Me_3BuPb), 208, 237 and 295 (Et_4Pb), and 208 and 379 (Bu_4Pb). Calibration was by calibration graph, using Me_3PbCl as calibrant and addition of Et_4Pb as internal standard.

Trimethyllead Determination in Urban Dust

The participating laboratories had to apply moisture corrections according to the procedure recommended by the SM&T programme, as well as corrections for extraction recoveries using a trimethyllead-spiked urban dust material provided along with the reference material.

Ethylation/GC-QFAAS

First method: *ca.* 1 g intake (moisture of *ca.* 4%) was taken for analysis. NaCl/H_2O was added, the mixture was mechanically shaken and filtered at 0.2 μm. Recovery was evaluated with the spiked urban dust provided by SM&T and was found to be (75 ± 4)%. Derivatization was carried out with 4% $NaBEt_4$ in acetic acid/acetate buffer, using He as carrier gas at 120 mL min^{-1}. The derivatized compounds were cryogenically trapped and separation was performed by GC (U-tube glass column of 0.5 m length and 3 mm internal diameter, filled with OV-101 as stationary phase loaded with 10% silica 80–100 mesh; He was used as carrier gas at 120 mL min^{-1}; H_2 was added as make-up gas at 30 mL min^{-1}; the column temperature ranged from *ca.* −196 to 75 °C). Detection was by electrothermally heated QFAAS (T-tube) at 283.3 nm (detector temperature at 800 °C). Calibration was by standard additions, using Me_3PbCl as calibrant.

Second method: *ca.* 1 g sample was taken for analysis (moisture of *ca.* 7%) to which a mixture of NaAc/HOAc was added at pH 4, followed by addition of 10 mL H_2O and 10 mL methanol; the extraction was performed ultrasonically followed by a back-extraction into hexane. The recovery (assessed with the spiked dust provided by SM&T) was (97 ± 6)%. Derivatization was carried out by four additions of 1 mL of 10% $NaBEt_4$, shaking for 10 min at each occasion. Decantation of the organic phase was followed by a reduction of the resulting extract to 1 mL over a N_2 flow and clean-up with silica gel. Separation was performed by CGC (column of 30 m length, 0.32 mm internal diameter, DB-5 as stationary phase, 0.25 μm film thickness; He was used as carrier gas; air/H_2 were used as make-up gases; injector temperature at 80 °C). Detection was by electrothermally heated QFAAS at 283.3 nm (detector temperature at 750 °C). Calibration was by standard additions, using Me_3PbCl as calibrant and Me_3PbPr as internal standard.

Propylation/GC-QFAAS

Ca. 5 g of dust intake was used for the analysis (moisture of *ca.* 4%). The sample was wetted with H_2O and NaCl and the slurry was filtered through a glass microfibre filter. Complexation of Pb^{2+} was performed with EDTA, adjusting the pH to 8.5 with ammonia, following by complexation with 0.5 mol L^{-1} DDTC and hexane extraction. The extraction recovery (assessed with the spiked dust provided by SM&T) was (95 ± 7)%. Grignard derivatization was performed with 2 mol L^{-1} propylmagnesium chloride in diethyl ether, of which the excess was destroyed with H_2SO_4. Clean-up was performed on a silica solid-phase extraction column, eluting with hexane. Separation was by GC, using a glass column (U-tube) packed with 10% OV-101 loaded with 10% Chromosorb W (column of 1.2 m length, 3 mm internal diameter, 80–100 mesh; He was used as carrier gas at 130 mL min^{-1}; injector temperature of 150 °C; column temperature ranging from 85 to 185 °C). Detection was by electrothermally heated QFAAS (T-tube) at 283.3 nm with deuterium background correction (detector temperature at 150 °C). Calibration was by standard additions, using Me_3PbCl as calibrant.

Ethylation/CGC-MIP-AES

A *ca.* 1 g sample was taken for analysis (moisture of *ca.* 2.5%) to which 3 mL of ammonium citrate/EDTA was added, followed by 1 mL of 0.8% aqueous $NaBEt_4$ solution and 500 μL of hexane containing Bu_4Pb as internal standard, mechanically shaking for 5 min, leaving for decantation for 5 min and collecting the hexane phase. The extraction recovery (assessed with the spiked dust provided by SM&T) was 87%. Separation was by CGC (column of 25 m length, 0.32 mm internal diameter, HP-1 as stationary phase, 0.17 μm film thickness; He was used as carrier gas at 300 mL min^{-1}; H_2 and O_2 were added as scavenger gases at respective flows of 621 and 138 mL min^{-1}; the column temperature ranged from 45 to 280 °C). Detection was MIP-AES at 405.8 nm. Calibration was by standard additions, using Me_3PbCl as calibrant.

SFE/Propylation/CGC-MS

A *ca.* 1.5 g intake was used for the analysis (moisture of *ca.* 3%) to which NaOH was added to pH 9, buffering with borax/hydrochloric acid. Extraction was by supercritical fluid extraction (SFE) using CO_2 with 10% methanol as modifier (extraction temperature of 80 °C). Liquid–liquid extraction of the SFE eluate was performed with *n*-hexane after complexation with 0.5 mol L^{-1} DDTC. The extraction recovery (assessed with the spiked dust provided by SM&T) was (93 ± 3)%. Grignard derivatization was carried out with 2 mol L^{-1} propylmagnesium chloride in diethyl ether. Separation was by CGC (fused silica column of 25 m length, 0.20 mm internal diameter, HP-1 as stationary phase, 0.33 μm film thickness; He was used as carrier gas; injector temperature at 180 °C; column temperature ranging from 60 to 260 °C). Detection was MS (electron impact) of ions 223 and 253 (detector temperature of 265 °C). Calibration was by calibration graph, using Me_3PbCl as calibrant.

Pentylation/CGC-MS

A *ca.* 1 g sample was taken for analysis (moisture of *ca.* 2%) to which an aqueous buffer was added together with 0.5 mol L^{-1} DDTC. Extraction was performed twice with 10 mL pentane, mechanically shaking for 5 min. Clean-up was with 100% active alumina, followed by elution with hexane/diethyl ether. Grignard derivatization was carried out with 2 mol L^{-1} pentylmagnesium bromide. Separation was by CGC (fused silica column of 30 m length, 0.25 mm internal diameter, 5% phenyl/methylpolysiloxane as stationary phase, 250 μm film thickness; injector temperature at 60–280 °C; column temperature ranging from 60 to 200 °C). Detection was by MS (electron impact) of ions 307 and 309. Calibration was by standard additions, using Me_3PbCl as calibrant and Ph_2SnEt_2 as internal standard.

Butylation/CGC-MS

Ca. 1 g of sample was taken for analysis (moisture of *ca.* 2%). Extraction was performed with 5 mL pentane after buffering with ammonium acetate and DDTC complexation. Clean-up was carried out with deactivated alumina, followed by an evaporation of the extract to dryness and redissolution in hexane. The extraction recovery was assessed with the spiked dust provided by SM&T and was found to be (85 ± 4)%. Derivatization was performed with 0.5 mL of 2 mol L^{-1} butylmagnesium chloride in THF, followed by addition of 0.5 mol L^{-1} H_2SO_4. Separation was by CGC (fused silica column of 60 m length, 0.25 mm internal diameter, DB-1 as stationary phase, 0.25 μm film thickness; He was used as carrier gas at 110 mL min^{-1}; injector temperature at 250 °C; column temperature ranging from 50 to 260 °C). Detection was by MS in selective ion monitoring mode (detector temperature at 280 °C), monitoring ions 208, 223 and 253 (Me_3BuPb), 208, 237 and 297 (Et_4Pb), and 208 and 379 (Bu_4Pb). Calibration was by calibration graph, using Me_3PbCl as calibrant.

Development of HPLC-ID-ICP-MS

As part of the project on Pb speciation, a separate RTD project has been carried out at the University of Plymouth in the frame of a grant funded by the SM&T programme, of which the aim was to develop a novel isotope dilution HPLC-ICP-MS method for the determination of trimethyllead in environmental matrices [121]. One of the many advantages of ICP-MS, apart from its inherent sensitivity, is the possibility of measuring isotope ratios due to the MS component in the technique; a logical extension to isotope ratio measurement is to incorporate the procedure into an isotope dilution calibration strategy which has the advantage of increased accuracy and precision. Typically for lead the single measurement of *m/z* 206 or 208 may produce precision (RSD%) of 0.5–1.0%; however, the isotope ratio 206/208 would produce precision of 0.1–0.2% [121]. The project has extended the principle of isotope dilution to HPLC separation coupled to ICP-MS detection; this necessitated the synthesis of calibrants with verified purity and stoichiometry. An interface between HPLC and ICP-MS has been developed, as well as customized software for ID-HPLC-ICP-MS. The method development is fully described elsewhere [121]; the technique was successfully used in the certification campaign.

6.3 Feasibility Study

Prior to preparing reference materials for the interlaboratory study, a comprehensive study had been performed by the University of Antwerp to test the optimal preparation and storage conditions for solutions containing alkyllead species [118]. Laboratory experiments enabled the monitoring of the degradation of di- and tri-alkyllead compounds in water in the dark, in daylight and with UV irradiation. The findings showed that trialkyllead degradation did not give rise to appreciable formation of dialkyllead but directly yielded inorganic lead. Generally, ethyllead compounds were observed to be more sensitive towards degradation than the corresponding methyllead compounds. On the basis of this study, it was decided to consider only trimethyllead for the interlaboratory studies and to monitor the long term stability of this compound in both artificial rainwater and urban dust. This feasibility study is described extensively in the literature [118].

6.4 First Interlaboratory Study

Preparation and Verification of Calibrants

One of the most critical points in organometallic chemistry is the availability of calibrants of suitable purity and verified stoichiometry. This aspect was recognized at an early stage of the project and the purity of alkyllead compounds used in the feasibility study was carefully verified [118]. Additional experiments were performed on calibrants in the frame of the first interlaboratory exercise as described below [119].

Trimethyl- (TriML) and triethyl-lead (TriEL) compounds were obtained from Alfa products (Johnson Matthey) and their purity was verified as follows: carbon, hydrogen and chloride relative masses in the TriML and TriEL calibrants were determined by elemental microanalysis; the chloride concentration was determined by ion chromatography. Total lead was determined in the calibrants by electrothermal atomic absorption (ETAAS) using two different acid digestion procedures (concentrated nitric acid and a mixture of nitric acid/hydrogen peroxide). Calibrant solutions of TriML and TriEL at the 25 mg L^{-1} level were prepared in deionized distilled water (DDW) and analysed; a 200 μL aliquot of each of these solutions was added to a DDW solution (30 mL) containing NaCl (2 g), 0.5 mol L^{-1} NaDDTC (2 mL) and 0.1 mol L^{-1} EDTA (1 mL), and the mixtures were shaken manually in a separating funnel. Hexane (5 mL) was added to the funnel and the aqueous phase was removed after a 4 minutes shaking time. The extracted alkyllead compounds were then

Table 6.1

Components	*Expected (%)*	*Mean*	*Found (%)* σ	*n*	*Significance in* t-*test*
C[a]	12.5	13.0	0.36	3	—
H[a]	3.1	3.1	0.05	3	—
Cl[a]	12.3	10.5	0.02	2	S
Cl[b]	12.3	11.8	0.2	3	S
Total Pb[c]	72	73.5	5.0	4	—
Total Pb[d]	72	72.4	1.7	4	—
TriML[e]	100	98.8	3.9	4	—
TML[f]	0	1.8	0.08	2	S
$MeEt_3Pb$[f]	0	1.8	0.12	2	S
Other alkylPb[g]	0	0			
DML	0	0			

[a] Microanalysis.
[b] Ion chromatography (Dionex).
[c] HNO_3 digestion followed by ETAAS.
[d] HNO_3/H_2O_2 digestion followed by ETAAS.
[e] Hexane extraction, HNO_3/H_2O_2 back-extraction followed by ETAAS.
[f] Hexane extraction followed by GC-AAS.
[g] Hexane extraction, propylation followed by GC-AAS.

Table 6.2

Components	*Expected (%)*	*Mean*	*Found (%)* σ	*n*	*Significance in* t-*test*
C[a]	21.8	21.3	0.28	2	–
H[a]	4.6	4.4	0.09	2	–
Cl[a]	10.8	11.3	0.11	2	S
Cl[b]	10.8	10.2	0.20	3	S
Total Pb[c]	62.8	63.1	1.6	3	–
Total Pb[d]	62.8	62.6	1.4	3	–
TriEL[e]	100	97.7	4.7	3	–
TEL[f]	0	1.6	0.08	2	S
$MeEt_3Pb$[f]	0	0.2	0.06	2	–
Other alkylPb[g]	0	0			

[a] Microanalysis.
[b] Ion chromatography (Dionex).
[c] HNO_3 digestion followed by ETAAS.
[d] HNO_3/H_2O_2 digestion followed by ETAAS.
[e] Hexane extraction, HNO_3/H_2O_2 back-extraction followed by ETAAS.
[f] Hexane extraction followed by GC-AAS.
[g] Hexane extraction, propylation followed by GC-AAS.

re-extracted into dilute nitric acid and hydrogen peroxide, and determined by ETAAS.

An aqueous solution containing 500 ng L^{-1} of TriML and TriEL was prepared and extracted as above (except for the addition of nitric acid and hydrogen peroxide). The hexane extract was transferred to a 25 mL conical flask and 0.5 mL propylmagnesium chloride (Grignard) reagent was added, followed by gentle shaking for 8 min. The extract was then washed with 0.5 mol L^{-1} H_2SO_4 (5 mL) to destroy any excess Grignard reagent present. The organic phase was separated and dried with a minimum of anhydrous Na_2SO_4 and transferred to a 4 mL vial. TriML and TriEL were determined by gas chromatography (GC)-AAS. Student's *t*-tests were applied for comparing the experimental results obtained with those expected.

The results are listed in the Tables 6.1 and 6.2. Inorganic lead contamination of the calibrants can be estimated by subtracting the total organic lead from total lead concentration determined by acid digestion. However, it was found that the content of alkyllead in the calibrants was slightly less than 100%, and that extraction/analytical losses were likely causes for this, rather than inorganic lead (as this was not significantly different from the expected abundance). In addition, if there was contamination by inorganic $PbCl_2$ in the standards, the concentration of chloride as obtained by microanalysis and ion chromatography would be higher than the one expected. In the two chloride analyses it was found that the chloride concentration was slightly lower than that expected, probably due in part to the tetraalkyllead detected in the calibrants (around 2% of the total lead content). In the light of the other results (Tables 6.1 and 6.2), it was thought unlikely that significant amounts of either inorganic lead or other ionic alkyllead compounds were present. It was concluded, therefore, that the two trialkyllead calibrants were not less than 98% pure.

Preparation of Solutions

As mentioned in Section 6.3, trimethyllead was found to be stable in solutions kept at ambient temperature in the dark [118]. Consequently, a batch of solutions containing 40 mg L^{-1} of trimethyllead chloride (as Pb) and 100 mg L^{-1} of lead nitrate (as Pb) added as interferent was prepared and the stability was verified over a period of six months.

Results

Results are presented in the form of bar graphs (Figure 6.1). For the intercomparison exercise, participants were asked to dilute the solution 1000 times, *i.e.* to determine levels of TriML of *ca.* 40 μg L^{-1}. Some laboratories analysed the solutions after a 10 000 times dilution. The participants considered it impossible to correct the results for impurities in the calibrant matrix.

High results and small standard deviations obtained by DPASV were questioned. The small standard deviation is inherent to the technique used but interferences and adsorption problems on the Hg drop electrode were suspected, which could explain the higher value found (50.4 μg L^{-1}).

A double peak in the chromatogram was observed for the HRGC-ETAAS technique, eluting at the retention time of inorganic Pb which was not

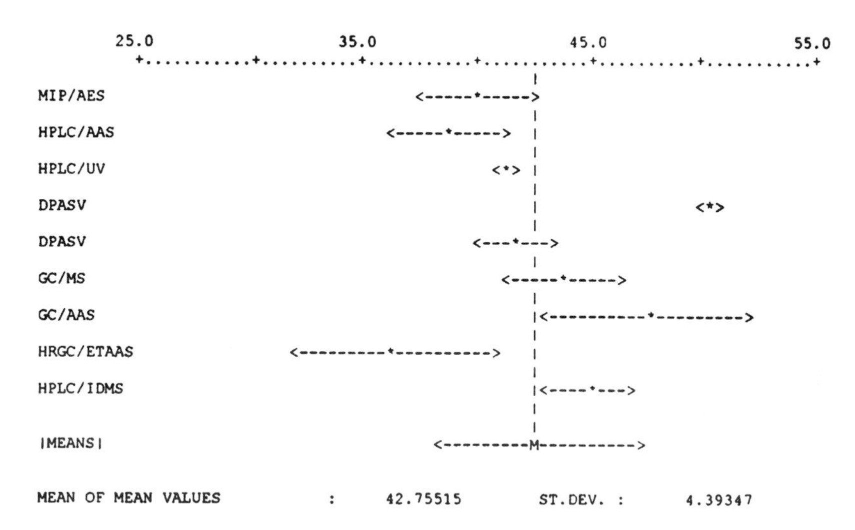

Figure 6.1 *Bar graph for laboratory means and standard deviations*

reproducible. It was suspected that the derivatization step was not fully under quality control, which could explain the low value found (36 μg L^{-1}).

A systematic error was suspected in the application of the HPLC-AAS method which could explain higher values found at the 4 μg L^{-1} level (10 000 dilution). This error could not be seen at the higher level (1000 dilution). Indeed, in the case of a small systematic error the calibration slope would be the same and standard additions could not correct for non-additive errors. The laboratory mentioned that a hydride generation procedure was successfully used in their laboratory for lead speciation analysis. No interferences were observed between inorganic Pb, Me_3PbCl and triethyllead using this method. However, this system is still under development and its ruggedness should be further tested. This method was based on earlier work by Blais and Marshall [122].

No problems due to interferences from inorganic Pb were mentioned by any of the participants.

6.5 Second Interlaboratory Study

Feasibility Study on Reference Material Preparation

Artificial rainwater solutions enriched with trimethyllead were prepared and diluted to obtain solutions containing three different levels: 50 ng L^{-1}, 500 ng L^{-1} and 5 μg L^{-1} [120]. These solutions were stored at room temperature and −20 °C. The composition of the artificial rainwater samples simulated the composition of real samples (addition of NH_4^+, K^+, Ca^{2+}, Mg^{2+}, Na^+, Cl^-, SO_4^{2-}, NO_3^- and H_3O^+). The pH of the solutions upon storage was 3.84. The stability was studied during six months with determinations performed every month. The results showed that the decomposition of trimethyllead was most rapid in the low concentration samples (50 ng L^{-1}). There was only a small degree of decomposition at the 500 ng L^{-1} level both in the dark at ambient temperature and at $-20\,°C$. The 5 μg L^{-1} samples were stable for at least 6 months under both storage conditions. On the basis of these results, the preparation of a candidate reference material of artificial rainwater was discussed. The participants agreed that the levels should be representative of natural concentrations but that the preparation of solutions containing 50 ng L^{-1} of trimethyllead would not be feasible. To overcome the instability problems it was decided to store the ten-fold concentrated artificial rainwater solution (500 ng L^{-1}) in 100 mL Nalgene bottles and dilute to 1 L prior to analysis. The pH corresponding to the storage conditions of the artificial rainwater solution should be 3.5 to arrive at a pH of 4.5 after dilution. Any further dilution would require use of artificial rainwater of similar composition.

Samples for the Interlaboratory Study

The participating laboratories received two sets of solutions containing respectively *ca.* 50 and 5 μg L^{-1} of TriML. They were requested to perform five replicate analyses of:

- 10 times dilution of the 50 μg L^{-1} concentrated solution (solution A)
- 10 times dilution of the 5 μg L^{-1} concentrated solution (solution B)
- 100 times dilution of the 5 μg L^{-1} concentrated solution (solution C)
- 1000 times dilution of the 5 μg L^{-1} concentrated solution (solution D)

In addition, the participants had to perform five replicate analyses of an urban dust sample.

Two sets of simulated rainwater solutions were prepared with the composition listed in Table 6.3. Aliquots of solution (100 mL) were transferred into 18 100 mL Nalgene bottles for each of the samples. The bottles were wrapped with aluminium foil, and then sealed in plastic bags.

Urban dust was collected from the Queensway road tunnel in Birmingham city centre. After being passed through a 500 μm sieve to remove large particles of debris, the dust was first treated by air-drying for several days (4–5 days) and then ground with a ballmill for three minutes. The ground dust was further sieved through a 80 μm sieve. Around 600 g of treated road dust was homogenized thoroughly in a 1 kg glass jar and then stabilized by freeze-drying for 20 hours. The bottling procedure followed immediately into 30 mL amber glass bottles containing approximately 25 g of the dust, which were sealed in plastic bags.

Results

The bar graphs obtained for solutions A, B and C, and for the urban dust sample are shown in Figures 6.2–6.5.

Table 6.3 *Composition of the solutions used in the interlaboratory study*

Composition	*Sample A*	*Sample B*
TriML (μg L^{-1})	54.8	6.39
Pb^{2+} (μg L^{-1})	100	100
10-fold concentrated rainwater (μmol L^{-1})		
NH_4^+	600	600
K^+	50	50
Ca^{2+}	120	120
Mg^{2+}	100	100
Na^+	600	600
Cl^-	900	900
SO_4^{2-}	300	300
NO_3^-	400	400

Solution A: 10 Times Dilution of 50 μg L^{-1} Solution

The bar graph presentation is shown in Figure 6.2.

The ETAAS technique did not include a separation step but the participant mentioned that EDTA extraction would only extract organic lead compounds; this technique was considered to be suitable for the analysis of a simple solution containing only one lead compound but would not be adapted to mixtures of lead species, *e.g.* the techniques would not allow for the separation of TriML and TriEL in a natural rainwater sample. In case organolead compounds should be determined in natural samples or solutions containing different lead compounds, ETAAS should be coupled to a separation technique, *e.g.* GC or HPLC.

Some doubts were expressed on the DPASV results obtained, considering the low standard deviation, and the analysis was repeated with a well conditioned electrode. The new set of data was (6.52 ± 0.22) μg kg^{-1}.

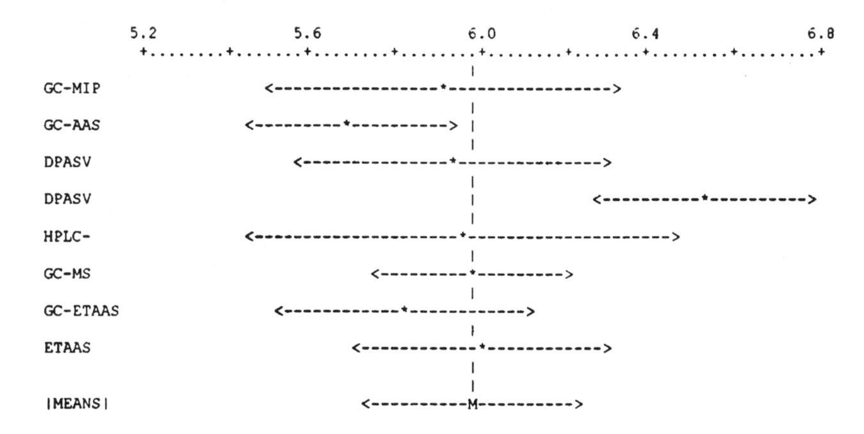

Figure 6.2 *Bar graph of the results of TriML in solution A. The mean of laboratory means was 5.98 ± 0.24 μg kg^{-1}*

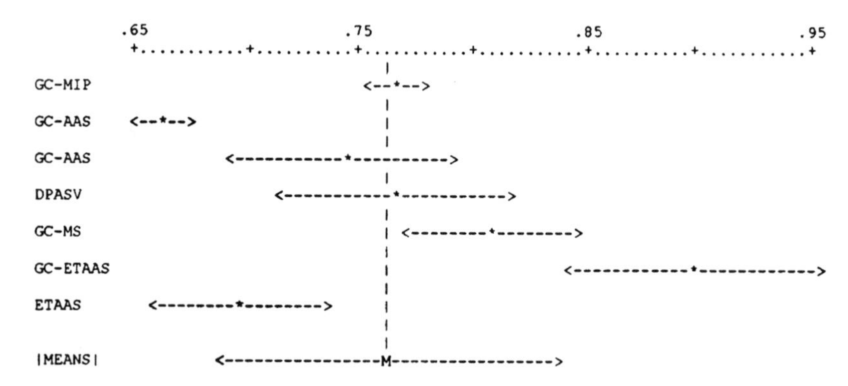

Figure 6.3 *Bar graph of the results of TriML in solution B. The mean of laboratory means was 0.76 ± 0.08* μg *kg*$^{-1}$

One laboratory (DPASV) repeated the analysis and observed a systematic difference due to two different sets of calibrants. The second set of data (5.94 ± 0.33 μg kg^{-1}) was obtained with a calibrant solution made with a newer calibrant from the same producer. This highlighted the need to verify the calibrant thoroughly, *i.e.* not to rely on calibrants from one producer of which the quality could vary from one set to another. Most of the laboratories actually used their own calibrants which were not verified for purity and stoichiometry. Only one laboratory used the calibrant previously verified and distributed in the first interlaboratory study. It was stressed that calibration was an important issue and that more effort should be put into the verification of calibrants in future exercises. It was agreed that the coordinator of the project would purchase calibrant from a chemical company and verify its purity; sets of verified primary calibrants would then be made available to participants in a further exercise to verify their own calibrants.

The verification of extraction recoveries was also questioned. Most of the laboratories performed standard addition procedures and hence did not need to correct for recovery. The coefficient of variation (*CV*) between laboratories was originally 20.9%. After technical scrutiny, the *CV* decreased to *ca.* 4%, which was found to be an excellent agreement.

Solution B: 10 Times Dilution of 5 μg L^{-1} solution

The bar graph for solution B is presented in Figure 6.3.

Low standard deviations were due, in two cases, to the fact that analyses were carried out on the same day and that, therefore, the day-to-day variability was not taken into account.

Reliable results were obtained by DPASV by doubling the deposition time (240 s instead of 120 s); the participant mentioned, however, that this concentration corresponded to the limit of determination of his technique.

The *CV* between laboratories was 14.8% before the evaluation and dropped to 10% after some sets of results were remove on technical grounds. This

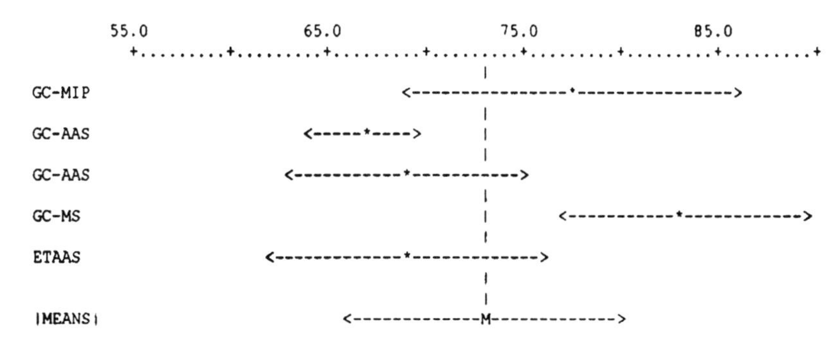

Figure 6.4 *Bar graph of the results of TriML in solution C. The mean of laboratory means was 73.1 ± 7.0 μg kg^{-1}*

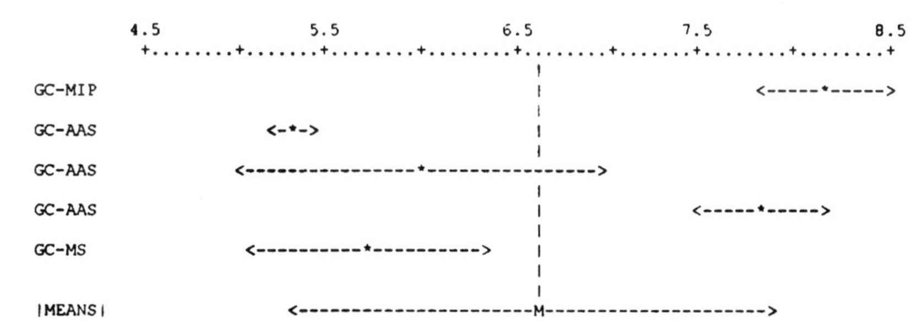

Figure 6.5 *Bar graph of the results of TriML in urban dust. The mean of laboratory means was 6.6 ± 1.3 mg kg^{-1}*

degree of agreement was found acceptable at this level of TriML concentration.

Solution C: 100 Times Dilution of 5 μg L^{-1} Solution

As shown in Figure 6.4, good agreement was obtained between laboratories at this level of concentration (*CV* of *ca.* 10% between laboratories).

Urban dust

A bar graph of the results is presented in Figure 6.5

The extraction recovery was considered to be the most critical point in urban dust analysis. Ultrasonic extraction used in GC-MIP gave poor recoveries and in extreme cases led to none or even highly negative recoveries. This was probably because such a treatment leads to the release of inorganic lead from the sample which consumes the reagent ($NaBEt_4$) despite the EDTA masking. In addition, the high amount of inorganic lead extracted creates important interferences at the detection step, which requires the addition of EDTA. Therefore, it was preferred to use a milder extraction procedure (shaking) to allow complete recovery of the TriML spike to be obtained, whereas inorganic lead was only partly extracted. One laboratory using GC-AAS verified its recovery by spiking a different road dust material and found values of 86 and 89% respectively; the results submitted were, however, not corrected and would then be slightly higher after correction. Here again, the analyses were performed on the same day which explained the small standard deviation.

Some doubts were expressed on the second GC-AAS procedure used, which filtered the suspension and took back TriML in water. Indeed, TriML is not stable in water and it was suspected that losses could have occurred by adsorption on the filter. The extraction recovery was not necessary as standard additions were performed prior to extraction; recovery values of 66 and 77% were obtained.

The extraction recovery for the GC-MS method was not verified, which again allowed some doubts on the possibly low results.

At this stage, it was hardly possible to confirm the doubts expressed on the extraction recovery of TriML in this material. The participants recommended that emphasis be put on the verification of extraction recovery in a further exercise, *i.e.* that a small batch of candidate reference material of urban dust be spiked with a known amount of TriML, left to equilibrate, homogenized and made available to the participants so that the extraction recovery may be verified. Recommendation was made to spike the dust in a slurry which should be freeze-dried, not dried, in order to avoid losses of TriML.

6.6 Certification of Trimethyllead in Rainwater

Preparation of the Candidate Reference Material

The composition of the artificial rainwater matrix was chosen to reflect that of natural rainwater which falls over continental land masses [123]. A stock solution was prepared at 100 times the required concentration of the artificial rainwater by the addition of inorganic compounds to deionized water. One litre of stock solution was prepared by the addition of the following compounds (Analytical reagent grade) to 1000 mL of Milli-Q deionized water:

$CaCl_2$	11.3 mg
$MgCl_2$	9.5 mg
$(NH_4)_2SO_4$	39.6 mg
$NaNO_3$	34.0 mg
NaCl	11.7 mg
KCl	3.7 mg
HCl	17.4 μL

The components were dissolved using an ultrasonic bath and the final solution was filtered through a 0.2 μm (47 mm diameter) cellulose acetate membrane in order to eliminate algal and bacterial particles. No problems were encountered with the solubility of any of the components of the stock solution.

Table 6.4 *Concentration of artificial rainwater components*

Component	*Concentration in stock solution/μmol L^{-1}*	*Concentration in final solution/μmol L^{-1}*
NH_4^+	600	60
K^+	50	5
Ca^{2+}	120	12
Mg^{2+}	100	10
Na^+	600	60
Cl^-	900	90
SO_4^{2-}	300	30
NO_3^-	400	40
$(CH_3)_3Pb^+$	500 ng kg^{-1}	50 ng kg^{-1}

Trimethyllead chloride (purity 98%) was used for spiking the candidate CRM. A large batch of artificial rainwater (100 L) was prepared in a rigid high-density polyethylene bin (0.14 m^3) with a lid. The solution was protected from light by wrapping the bin in aluminium foil. Trimethyllead chloride was added to give a concentration of 500 ng L^{-1} (as lead). The solution was thoroughly stirred for an extended period of time to ensure proper mixing by means of a glass rod which was passed through a hole in the lid of the container. The solution was dispensed into Nalgene bottles (125 mL) by syphoning it through a Teflon tube (6.4 mm diameter) which was passed through the hole in the container lid. The Nalgene bottles (800 units) were rinsed out with two small aliquots of the solution and 100 mL of the artificial rainwater matrix was dispensed to each bottle. The bottles were capped, wrapped in aluminium foil, sealed in polythylene bags and then stored in a cold room at +4 °C.

The concentrations of the components present in the stock solution and in the artificial rainwater are presented in Table 6.4.

Homogeneity Control

The between-bottle homogeneity of the reference material was verified by a single determination of trimethyllead in each of eight bottles randomly selected from the 800 bottles produced. The methodological uncertainty was determined by performing five replicate determinations of the trimethyllead content in one bottle. The artificial rainwater samples (100 mL) were diluted with 900 mL of Milli-Q deionized water prior to analysis. NaCl (20 g), 5 mL of 0.25 mol L^{-1} NaDDTC and 5 mL hexane were added to the diluted sample and the solution was shaken mechanically for 30 min. The organic fraction was transferred to a 25 mL conical flask, 0.5 mL propylmagnesium chloride was added and the flask was gently shaken for 8 min. The extract was washed with 5 mL of 0.5 mol L^{-1} H_2SO_4 to destroy any excess Grignard reagent present. The organic phase was dried over anhydrous Na_2SO_4 (*ca.* 100 mg) and then transferred to a 4 mL vial. A 50 μL extract was injected into a GC and

Table 6.5 *Between-bottle homogeneity and method CV for RM 604*[a]

Component	*Between-bottle*[b]	*Method of final determination*[c]
Trimethyllead	6.8 ± 1.7	4.3 ± 1.4

[a] $(CV \pm U_{CV})$%.
Uncertainty on the *CV*'s: $U_{CV} \approx CV \sqrt{2n}$.
[b] Single determination on the content of each of 8 bottles.
[c] 5 replicate determinations on the content of one bottle.

trimethyllead was determined by AAS. The extraction efficiency for trimethyllead was determined by spiking the sample with 100 ng L^{-1} of the compound and performing the analysis as described above. The mean recovery for four replicates was 90.2% with a relative standard deviation of 2.9%.

The *CV*'s for trimethyllead in the reference material are presented in Table 6.5. An *F*-test at a significance level of 0.05 did not reveal significant differences between the between-bottle and method variances. On the basis of these results, no inhomogeneities of the material were suspected.

Stability Control

The stability of the trimethyllead content was tested at 4 °C in the dark over a period of 12 months and trimethyllead was determined at the beginning of the storage period and after 1, 3, 6 and 12 months. Analyses were repeated after 37 months to verify the long term stability. In addition, a short-term stability study was carried out at +37 °C to simulate worst-case transport conditions. Samples were analysed using the same procedures as for the homogeneity study. Trimethyllead was determined in triplicate (one replicate analysis in each of three bottles stored at 4 °C) at each occasion of analysis. For the short-term stability at +37 °C, 15 bottles of RM 604 were stored at +37 °C and five of these samples were analysed after 5, 10 and 15 days. The evaluation of the stability followed the procedure described in Chapter 3, using the results obtained on the samples analysed at the beginning of the storage period as reference for the results obtained at each occasion of analysis. The results showed that no instability of the material could be demonstrated over a period of 12 months (Table 6.6). However, the *t*-test result ($p = 0.01$) indicated that the difference between the initial concentration of trimethyllead measured on the day of the rainwater preparation and the final concentration after 37 months storage was significant at the 95% confidence level. On average, more than 10% of trimethyllead in rainwater had decomposed at 4 °C after three years storage.

With respect to the data of the short-term stability study carried out at +37 °C (Table 6.7), an *F*-test (single-factor ANOVA) was used to determine whether or not a significant difference existed among the trimethyllead concentrations determined after 5, 10 and 15 days. According to this test, the

Table 6.6 *Normalized results of the stability study*

Compound	*Time /months*	$R_t \pm U_t$
Trimethyllead	1	1.00 ± 0.02
	3	0.99 ± 0.02
	6	0.97 ± 0.02
	12	0.96 ± 0.03
	37	0.86 ± 0.06

Table 6.7 *Stability study of trimethyllead in CRM 604 at +37 °C after 15 days*

Compound	*Time (days)*	*Mean ± SD/ ng*$^{-1}$ *L as Pb*
Trimethyllead	0	54.3 ± 4.0
	5	52.7 ± 3.0
	10	49.5 ± 3.2
	15	48.5 ± 3.0

stability of trimethyllead in the RM 604 may not be affected by the elevated temperature since the calculated *F*-value is less than the critical *F*-value (p-value > 0.05); however, this statistical conclusion must be viewed with caution since the calculated *F*-value is so close to the critical *F*-value. The difference between the initial trimethyllead in the material (day 0) and the final concentration (day 15) was also determined using a *t*-test; the statistical result (p = 0.0486) at the 95% confidence level showed that trimethyllead had decomposed significantly after 15 days storage at +37 °C [124,125].

Preparation of Calibrant

The techniques used by the participating laboratories are described in Section 6.2. A trimethyllead calibrant was prepared by the University of Plymouth for the purpose of the certification campaign in order to enable participating laboratories to verify their own calibrants. A portion of 39 g of tetramethyllead and toluene (80% w/w) was placed in a round bottom flask and hexane (250 mL) was added. Dried hydrogen chloride gas was bubbled through the mixture for 10 min at a flow rate of 150 mL min^{-1}. A heavy white precipitate was formed and was removed by filtration; it was first washed with hexane (300 mL) and finally rinsed with pentane before being dried under reduced pressure. The original reaction mixture was discarded, the apparatus was cleaned with hexane and a fresh preparation was undertaken in order to produce sufficient trimethyllead chloride for the purpose of the project. The purity of the product was assessed using NMR spectroscopy and was found to be greated than 99%. The product was also sent to an accredited external laboratory for carbon,

Table 6.8 *Carbon, hydrogen and chlorine analysis of the trimethyllead calibrant*

Element	*% Composition sample*	*% Composition theoretical*
Carbon	12.62	12.51
Hydrogen	3.24	3.14
Chlorine	12.04	12.30

hydrogen and chlorine analysis and the results are shown in Table 6.8. Good agreement was obtained between the observed percentage composition for these components and their theoretical values, providing confirmatory evidence of the purity of the product.

Technical Evaluation

The sets of results found acceptable after both the technical and statistical evaluation are presented in the Figure 6.6. Each set of results is identified by the code number of the laboratory.

The technical discussion focused firstly on the calibration methods used by the participants. All laboratories used the trimethyllead calibrant provided by the University of Plymouth, either for calibration or verification of their own calibrants. It was noted that some deterioration had been observed in a commercial calibrant over a two year period. No significant difference was observed by the laboratories between external calibration and standard addition. Some laboratories used tetraethyllead or tributyllead as internal standard.

With the exception of three laboratories which used a hydride generation method, DPASV or HPLC, all the participants had used gas chromatographic separation following a Grignard derivatization of the analyte. Some differences were observed in terms of precision by two laboratories using the same separation and detection but different Grignard reagents, namely pentylated Grignard reagent (some losses were suspected to occur at this stage) and butylation (resulting in a cleaner reaction, thereby accounting for the discrepancy in precision). In the subsequent discussion, it was agreed that the conditions of the Grignard reaction in terms of temperature, concentration and length of the alkyl chain were key factors which require careful control. The longer the alkyl chain of the Grignard reagent, the greater the risk of degradation product formation and peak broadening in the chromatographic stage.

Conclusions

While the results obtained by the participating laboratories were in good agreement (Figure 6.1) and illustrated the high quality of the measurements performed, the doubts expressed on the stability of the reference material did not encourage the SM&T programme to recommend this material for certifica-

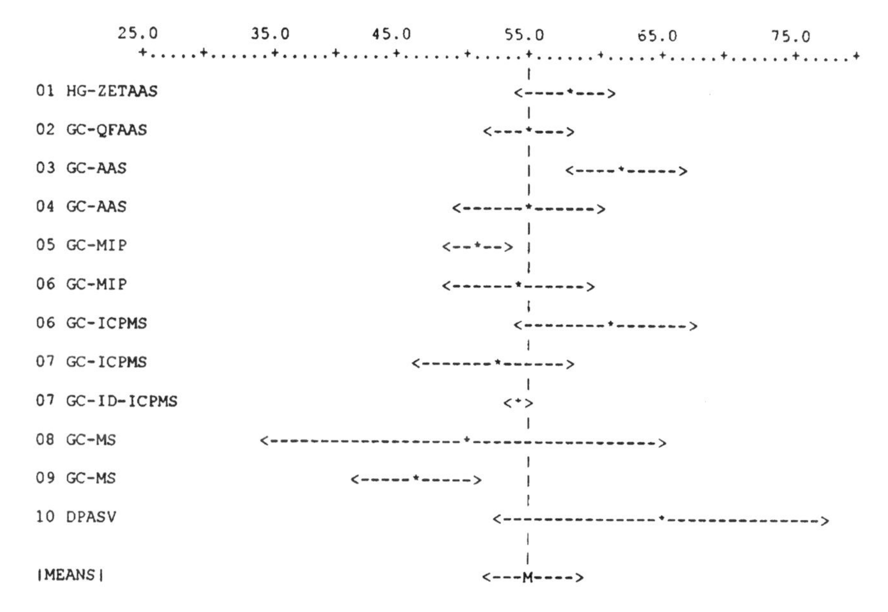

Figure 6.6 *Bar graph for laboratory mean and 95% CI. The results correspond to the trimethyllead content as mass fractions of Me_3Pb^+ (ng kg^{-1} as TrimL); the mean of laboratory means obtained was (55.2 ± 3.5) ng kg^{-1} as TrimL*

tion [124,125]. It is thought that such material could be certified, providing that storage at low temperature (+4 °C or below) in the dark would be constantly maintained; in the case of the RM 604, the good results obtained over 12 months at +20 °C were obviously not sufficient since the material was shown not to be stable after 37 months.

6.7 Trimethyllead in Urban Dust

Preparation of the Material

Around 15 kg of surface road dust was collected from sweeping a lay-by in the central section of the Queensway tunnel in Birmingham (UK). The tunnel is approximately 850 m in length and is a major traffic route through Birmingham from the motorway system. The dust was passed through a 500 µm sieve to remove large particles of debris. Then, the sample was treated by air drying at room temperature for five days, on a flat tray lined with clean paper, in a well ventilated and dark place. Further, the dust was ground in a ballmill for a period of three minutes and sieved through a 125 µm sieve.

About 10 kg of the pre-treated dust sample was homogenized by the following procedure: the whole amount of dust was divided into four subsamples of about 2.5 kg; each subsample was stored in a 5 L dark glass bottle and shaken on a mechanical shaker for a few hours (using different shaking directions). When all four sub-samples had been homogenized, they were redistributed again into a second batch of four sub-samples in such a way that each of the first four sub-samples contributed equally to the each of the second four sub-samples. This procedure was repeated three times after which the final four sub-samples were again combined in one sample. The homogenized dust sample was freeze-dried at $-50\,°C$ and 8 millibar pressure for 24 h. The bottling procedure followed immediately. Each of the 600 30 mL amber glass bottles, provided with screwcaps, was filled with *ca.* 15 g of the dust sample, sealed in a plastic bag and stored in the cold room at 4 °C.

A separate batch of material was set aside for the preparation of spiked samples to be used for the verification of extraction recoveries by the certifying laboratories. A solution (250 mL) of trimethyllead chloride calibrant containing 20 μg as Pb was first prepared, and then added to 400 g of candidate urban road dust. The slurry was placed in a 1 L wide-neck amber-glass bottle and was stirred manually for several hours with two glass rods, and then stored over 48 h at 4 °C. After freeze-drying over two nights, the dust was gently ground in a porcelain mortar, and then transferred into another 1 L wide-neck amber-glass bottle. The bottle was then shaken mechanically to homogenize the whole sample of the spiked road dust before it was distributed into 25 bottles, each containing *ca.* 15 g of the dust. The spiking level of trimethyllead in the urban road dust was 50 $\mu g\ kg^{-1}$ (as Pb).

Homogeneity Control

The between-bottle homogeneity of the candidate CRM 605 was verified by a single determination of trimethyllead in each of 15 bottles set aside during the bottling procedure (5 bottles out of each set of 200 bottles). The within-bottle homogeneity was verified by 10 replicate analyses from one bottle.

Samples (1 g) were transferred to 250 mL screwcap glass bottles containing Milli-Q water (100 mL) and NaCl (10 g). The bottles were shaken on a mechanical shaker for 30 min. The slurry was filtered through two pieces of glass microfibre filter (Whatman GF/C) and rinsed with 50 mL of Milli-Q water. The combined filtrates were then transferred to a second clean 250 mL glass bottle. After the pH was adjusted to 9.0 with an ammonia solution, EDTA (3 g), 5 mL of 0.5 mol L^{-1} NaDDTC and 15 mL hexane were added. After 30 min agitation, the organic phase was removed and the extraction was repeated with a second aliquot of hexane. The combined hexane extracts were then passed through anhydrous Na_2SO_4 with hexane rinses. The hexane extracts were transferred to a 25 mL conical flask and evaporated by purging with a N_2 stream in a water bath, set at 35 °C, until approximate 5 mL of the hexane extract remained. This was transferred into a 10 mL concentrator receiver tube with a hexane rinse and evaporated by purging with a N_2 stream until 0.5 mL of the extract remained. Grignard reagent (0.3 mL propylmagnesium chloride) was added and the concentrator tube was shaken in a ultrasonic bath for 5 min. 5 mL of 0.5 mol L^{-1} H_2SO_4 was added to destroy the excess

Table 6.9 *Between-bottle homogeneity and method CV for CRM 605*[a]

Component	*Within-bottle*[b]	*Between-bottle*[c]	*Method*[d]
Trimethyllead	8.2 ± 1.5	8.4 ± 1.9	8.2 ± 1.8

[a] $(CV \pm U_{CV})$.
Uncertainty on the *CV*'s: $U_{CV} \approx CV/\sqrt{2n}$
[b] 10 replicate determinations on the content of one bottle.
[c] Single determination on the content of each of 15 bottles.
[d] 10 replicate determinations on an extract solution.

Grignard reagent and the mixture was shaken in the ultrasonic bath for a further 5 min. The hexane layer was removed from above by means of a 200 μL pipette. A small amount of anhydrous Na_2SO_4 was placed within the pipette tip to dry the extract. This 25 μL of the final extract was injected into a GC-AAS system. The extraction recovery of trimethyllead in road dust was verified by spiking road dust material at a level of 4 μg kg^{-1}. The range of recovery was from 66.5% to 91.3% for four replicates [mean of (77.6 ± 9.5)%].

The *CV*'s for trimethyllead analysis in CRM 605 are presented in Table 6.9. An *F*-test at a significance level of 0.05 did not reveal significant differences between the within-bottle and the between-bottle variances and the method *CV*. On the basis of these results, no inhomogeneity was suspected and the material was considered to be homogeneous at a level of 1 g and above [126,127].

Stability Control

The stability of the trimethyllead content was tested in the dark at -20 °C, +20 °C and +37 °C and trimethyllead was determined at the beginning of the storage period and after 1, 3, 6, 12 and 37 months. Samples were analysed using the same procedures as for the homogeneity study. Trimethyllead was determined in triplicate (one replicate analysis in each of three bottles stored at -20, +20 and +37 °C) at each occasion of analysis. The classical evaluation of stability (see chapter 3) was carried out, using samples stored at -20 °C as reference for the samples stored at +20 and +37 °C respectively.

The results showed that no instability of the material could be demonstrated over a period of 37 months for the material stored at +20 °C (Table 6.9). However, a significant decrease in trimethyllead content was observed to occur at +37 °C. On the basis of these results, it was concluded that the material is stable at +20 °C whereas storage temperatures above this level should be strictly avoided.

Technical Evaluation

Results produced by hydride generation/ZETAAS were withdrawn; this method is capable of speciating between inorganic and organic lead but does

Table 6.10 *Normalized results of the stability study*

Component	*Time /months*	$R_T \pm U_T$ (+20 °C)	$R_T \pm U_T$ (+37 °C)
Trimethyllead	1	1.06 ± 0.10	0.83 ± 0.14
	3	0.98 ± 0.18	0.84 ± 0.14
	6	1.01 ± 0.14	0.83 ± 0.13
	12	1.00 ± 0.13	0.80 ± 0.10
	37	0.98 ± 0.12	0.73 ± 0.08

not have the specificity to differentiate between organolead species. The candidate CRM was likely to contain a variety of organolead compounds and, therefore, this result was withdrawn.

The recoveries obtained by the various laboratories were discussed and most were in the range 70–95% with standard deviations of 3–8%. It was agreed that the best practice was to conduct recovery studies alongside the analytical measurement. Trimethyllead in the spiked sample was determined as the difference between the spiked and the unspiked results.

The results of the statistical discussion are fully described in the certification report [126].

Certified Value

The certified value (unweighted mean of *p* accepted sets of results) and its uncertainty (half width of the 95% confidence interval) is given in the Table 6.11 as mass fractions. The bar graph of results obtained by the different laboratories is shown in Figure 6.7. Trimethyllead is certified as mass fractions of Me_3Pb^+ (μg kg^{-1} as TriML).

Table 6.11 *Certified mass fractions (dry matter) of trimethyllead in CRM 605 (μg kg^{-1} as TriML)*

Component	*Certified value/ μg kg^{-1} as TriML*	*Uncertainty/ μg kg^{-} as TriML*	p^a
Trimethyllead	7.9	1.2	7

[a] p = number of sets of results

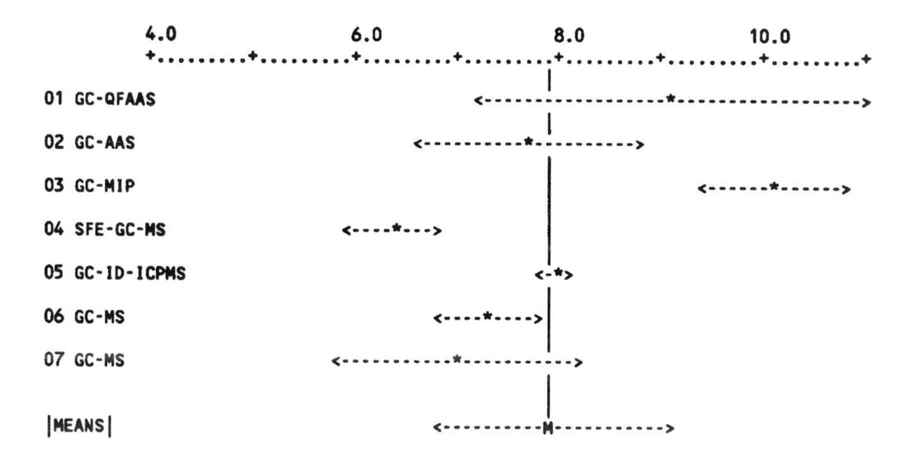

Figure 6.7 *Trimethyllead in urban dust in μg kg^{-1} as TML*

CHAPTER 7

Arsenic Speciation

7.1 Aim of the Project and Coordination

Justification

Arsenic is an ubiquitous element which occurs in the form of various chemical species in the environment. In biological tissues, the main species identified is arsenobetaine which is considered to be non-toxic and present at more than 90% in fish tissues but does not exceed 50% in mollucs [128]. Other species such as arsenocholine, tetramethylarsonium ion, trimethylarsenoxide, dimethylarsinic acid and arsenosugars have also been identified [129–132].

The Programme and Timetable

Several intercomparisons were organized from 1989 to 1995 to improve the state of the art of As speciation analysis [133]. The number of exercises needed to enable the certification of some arsenic species in tuna fish and solutions [134] illustrated the high degree of difficulty of this type of analysis. As described below, the interlaboratory exercises followed a stepwise approach consisting of six exercises of increasing difficulty, namely solutions of six pure arsenic species [arsenite As(III), arsenate As(V), monomethylarsonic (MMA) and dimethylarsinic (DMA) acids, arsenobetaine and arsenocholine], solutions containing a mixture of the six arsenic species, solutions containing the six arsenic species together with interfering cations and anions, fish and mussel raw extracts, fish and mussel cleaned extracts, and shark and mussel powders. The certification of total arsenic, dimethylarsinic acid and arsenobetaine in tuna fish (CRM 627) was completed in 1996, along with the certification of arsenobetaine in solution (CRM 626).

Coordination

The overall coordination of the interlaboratory studies and certification campaign was managed jointly by the Laboratoire de Chimie Analytique et Minérale of the University Louis Pasteur in Strasbourg (France) and the CNRS, Service Central d'Analyse in Vernaison (France). The synthesis of arsenobetaine and arsenocholine calibrants for the interlaboratory studies was carried out by the Laboratoire des Matériaux Organiques à Propriétés Spécifique in Vernaison (France) in collaboration with the coordinating institutes. Shark and mussel powders used in the intercomparisons were provided by the Environment Institute of the Joint Research Centre of Ispra (Italy), who also prepared the candidate reference material of tuna fish. Arsenobetaine calibrant for the certification was synthesized and characterized in the Laboratoire de Chimie Analytique et Minérale in Strasbourg (France).

Participating Laboratories

The following laboratories participated in all the interlaboratory studies: CNRS, Service Central d'Analyse, Vernaison (France); CNRS, Laboratoire de Chimie Analytique et Minérale, Strasbourg (France); IFREMER, Nantes (France); Institut Pasteur, Service Eaux-Environnement, Lille (France); Istituto Superiore di Sanità, Roma (Italy); National Food Agency, Søborg (Denmark); State Laboratory, Dublin (Ireland); Universidad de Barcelona, Departamento de Química Analítica, Barcelona (Spain); University of Plymouth, Department of Environmental Sciences, Plymouth (United Kingdom); University of Southampton, Department of Chemistry, Southampton (United Kingdom).

7.2 Techniques Used in Arsenic Speciation

Several methods have been developed for the determination of arsenic species, involving different extraction, derivatization, separation and detection steps [19]; these include performant hyphenated techniques based on liquid chromatography coupled to detectors such as ICP-MS [131,132,135] or ICP-AES [136], hydride generation in line with QFAAS [137], and UV degradation followed by ICP-AES detection [138]. The techniques which were selected in the certification campaign are described in detail in the following paragraphs.

Determination of As Species in Fish Tissue

UV Irradiation/Hydride Generation/ICP-AES

A subsample of 1 g was ultrasonically extracted with water/methanol (1/1 v/v) (3×20 mL); the extract was evaporated and diluted with water. Clean-up was carried out with C_{18} Sep-Pack cartridge. The extract was stored in a PTFE flask at 4 °C in the dark. Separation was by cation exchange LC (Hamilton PRP X 200) with gradient elution (HNO_3 and NH_4NO_3) for arsenobetaine and anion exchange LC (Hamilton PRP X 100) with gradient elution (NaH_2PO_4) for DMA. Hydride generation was carried out [using 1% $NaBH_4$ in NaOH/H_2SO_4 (1 mol L^{-1})] after UV irradiation (254 nm, 15 W for 2 min). Final detection was by ICP-AES.

UV Irradiation/Hydride Generation/QFAAS

2 g of sample were ultrasonically extracted with water/methanol (1/1 v/v) (3 x 20 mL); the extract was evaporated and diluted with water. Clean-up was performed with a silica column; however, it was shown that this should be avoided since it may lead to losses of arsenobetaine. The extract was stored in a glass flask at 4 °C in the dark. Separation was by anion exchange LC (Partisil SAX) in isocratic mode (NaH_2PO_4). Hydride generation (with 4% $NaBH_4$ in NaOH/HCl) after UV irradiation (4 W in the presence of $K_2S_2O_8$). Final detection was by QFAAS.

Gas Chromatography/Hydride Generation/QFAAS

2 g of sample were mixed with water (2 × 5 mL); no clean-up was performed. The extract was stored in a glass flask at 4 °C in the dark. Separation was by cryogenic trapping (in liquid N_2) in a U-tube packed with Chromosorb WAW DMCS (3% OV-101). Hydride generation was carried out with 2% $NaBH_4$ in $NaOH/H_2SO_4$. Final detection was by QFAAS.

Liquid Chromatography/Hydride Generation/ICP-AES

1 g of sample was ultrasonically extracted with water/methanol (1/1 v/v) (5 × 10 mL); the extract was evaporated, diluted with 6 mL water, filtered (0.45 μm filter) and stored in a PTFE flask at 4 °C in the dark. Separation was by anion exchange LC (Hamilton PRP X 100) with gradient elution (NaH_2PO_4/Na_2HPO_4). Hydride generation was carried out using 1% $NaBH_4$ in $NaOH/H_2SO_4$. Final detection was by ICP-AES (As betaine).

Liquid Chromatography/QFAAS

First method: 1 g of sample was extracted with water/methanol (1/1 v/v) (3 × 20 mL) by mechanical stirring; the extract was evaporated, diluted with 10 mL water, filtered (0.2 μm filter), passed on a C_{18} Sep-Pack cartridge and stored in a glass flask at 4 °C in the dark. Separation was by anion exchange LC (Hamilton PRP X 100) in isocratic mode (phosphate). Hydride generation was carried out (using 1% $NaBH_4$ in NaOH/HCl) followed by QFAAS detection.

Second method: 1 g sample was ultrasonically extracted with water/methanol 1/1 v/v (5 x 10 mL); the extract was evaporated, diluted with 6 mL water, filtered (0.45 μm filter), and stored in a PTFE flask at 4 °C in the dark. Separation was by anion exchange LC (Hamilton PRP X 100) with gradient elution (NaH_2PO_4/Na_2HPO_4). Hydride generation was carried out using 1% $NaBH_4$ in $NaOH/H_2SO_4$. Final detection was by QFAAS (DMA).

Liquid Chromatography/ICP-MS

First method: a subsample of 1 g was ultrasonically extracted with water/methanol (1/3 v/v) (5 × 20 mL); the extract was evaporated and diluted with

20 mL water. Clean-up was carried out by filtering on a Fluorisil Sep-Pack cartridge. The extract was stored in a brown glass flask at 4 °C in the dark. Separation was by ion-pair LC (Hamilton PRP 1) in isocratic mode (TBAP/Na_2HPO_4/MeOH). Final detection was ICP-MS of mass 75.

Second method: 1 g of sample was extracted with water/methanol (1/1 v/v) (5 × 10 mL) by mechanical stirring; the extract was evaporated, diluted with 25 mL water, cleaned-up by filtration (0.2 μm filter) and the extract was stored in a PTFE flask at 4 °C in the dark. Separation was by anion exchange LC (Dionex AS-7) with gradient elution (H_2O/$NaHCO_3$). Final detection was by ICP-MS of mass 75.

Third method: a sub-sample of 0.5 g was extracted by addition of 0.1 g trypsin and 15 mL ammonium bicarbonate (buffering at pH 8) by mechanical stirring at 37 °C for 4 h. The extract was stored in a glass flask at 4 °C in the dark. Separation was by anion exchange LC (2 × Benson AX 10) with gradient elution (K_2SO_4). Final detection was by ICP-MS of mass 75.

Fourth method: a 1 g sample was mixed with 15 mL water, followed by a microwave assisted extraction (20 W for 10 min). The extract was cleaned-up by filtration (0.45 μm filter), and stored in a PTFE flask at 4 °C in the dark. Separation was by anion exchange LC (Hamilton PRP X 100) with gradient elution [$(NH_4)_2HPO_4$/$(NH_4)H_2PO_4$ + MeCN]. Final detection was by ICP-MS of mass 75.

Fifth method: a subsample of 150 mg was ultrasonically extracted with a mixture of methanol/methane (5/2 v/v) (4 × 2.8 mL) and back-extracted in water; the extract was evaporated under N_2 stream and stored in a glass flask at 4 °C in the dark. Separation was by cation exchange LC (Chrompack Ionosphere) in isocratic mode (pyridinium formate). Final detection was by ICP-MS of mass 75.

7.3 Preparation of Pure Calibrants

One of the most difficult problems faced in the project was the lack of commercially available calibrants. Consequently, a set of calibrants was especially prepared for the purpose of the interlaboratory studies and certification. A full description of the synthesis is given elsewhere [126].

Arsenocholine was synthesized from arsenic trichloride, producing first trimethylarsine by addition of MeLi, followed by conversion of the trimethylarsine into arsenocholine by addition of bromoethanol. Arsenobetaine was prepared from trimethylarsine (synthesized from commercial arsenic trichloride).

The purity of the compounds was characterized by elemental analysis, molecular analysis (^1H NMR, mass spectrometry), thermodifferential and thermogravimetric analysis and separative techniques such as HPLC, GC and capillary zone electrophoresis [127]. This procedure provided highly pure products for use in the interlaboratory studies, *e.g.* arsenobetaine was shown to be more than 99.8% pure.

7.4 Interlaboratory Studies

Aqueous Solutions

Five mixtures containing different concentrations of arsenocholine, arsenobetaine, DMA, MMA, As(III) and As(V) were prepared in freshly boiled deionized water and distributed to the participants together with individual calibrant solutions. Twelve laboratories participated in this interlaboratory study on aqueous solutions, using LC-ICP-MS, GC-HAAS, LC-HAAS, LC-ICP, GC-HICP, CZE and LC-ETAAS.

The five solutions were found to be stable [except for As(III)] for four months if kept in the dark at +4 °C. Storage in the dark at +40 °C led to the formation of As(III) in some solutions. Arsenobetaine resulting from the degradation of arsenocholine was observed to occur significantly when solutions were stored at +20 °C in daylight but no trace of degradation was detected at +4 °C in the dark.

From this study, it was agreed that washing and preconditioning of the chromatographic column is a critical step. After a short period, the separation efficiency of the anion-exchange column can be seriously reduced, causing As(III) and arsenobetaine to coelute. It was suggested to wash the column at the beginning of each analytical cycle and after three consecutive runs [133].

The presence of chloride ions may interfere with the As signal in ICP-MS (M = 75 for As, as well as for the Ar^+Cl^- ion), but this interference is eliminated if chlorides are separated from As species by a pre-column or a properly selected analytical column [133].

Results of this interlaboratory study appeared satisfactory and it was decided to continue the evaluation with synthetic solutions before starting the exercises on real extracts. For a concentration of DMA of 5 μmol kg^{-1}, the mean of the mean values was very close to the target value [(5.05 $\pm$ 0.39) μmol kg^{-1}] and the coefficient of variation of the mean of means was only 7.7%.

Solutions Mimicking Fish and Soil Extracts

In a first interlaboratory study, two different solutions containing As species and interfering ions typically present in a soil (*e.g.* Mn^{2+}, Na^+, Al^{3+}, K^+, Fe^{2+}, Cu^{2+}, Zn^{2+}, Ca^{2+}, Cl^-, $PO_4{}^{3-}$) and in a fish extract (*e.g.* Na^+, K^+, Fe^{2+}, Cu^{2+}, Zn^{2+}, Ca^{2+}, Pb^{2+}, Cl^-, Br^-, $NO_3{}^-$, $PO_4{}^{3-}$) were prepared and distributed to the participants. As(V), DMA, MMA, arsenobetaine and arsenocholine were found stable in those solutions if they are kept at +4 °C in the dark. In the solution mimicking the soil extract, about 50% of As(III) was oxidized to As(V) after 45 days of storage at +4 °C. Storage should then be performed in the dark at +4 °C. Owing to the high risk of oxidation of As(III) to As(V) it was decided not to consider As(III) in the interlaboratory study.

Different types of columns were tested to detect possible effects of the separation column. Polymeric phases are preferred in comparison to silica since the low pH values (near 1) obtained for the tuna fish extracts may alter silica phases. Moreover, a clean-up procedure is needed to eliminate anions such as chlorides and phosphates which are present at high levels in marine organism extracts and which may affect the quality of separation. Two exercises on simulated extracts were needed to improve the accuracy and precision of the results before the organization of a third exercise on real fish and mussel extracts.

Fish and Mussel Cleaned Extracts

For this study, fish and mussel extracts were prepared using a water/methanol (1:1 v/v) extraction followed by clean-up with diethyl ether. Arsenobetaine, arsenocholine, DMA, MMA and As(V) in the extracts were stable at $-20\,^{\circ}C$ and +4 °C for at least 2 months.

The saturation of the column (ion exchange or ion pairing) due to the presence of high amounts of salts (especially in mussels) was avoided by dilution or use of a C_{18} pre-column. Chloride cartridges were tested but As(V) and DMA were partially and totally removed, respectively. Some discrepancies were attributed to calibration errors. The main observation reported for the mussel extract concerned the presence of two unknown compounds. It was suggested that the peaks might be due to arsenosugars, as already reported in the literature.

Mussel Raw Extract and Biological (Shark and Mussel) Tissues

Arsenocholine, arsenobetaine and DMA found in the samples were stable at +4 °C and +20 °C in the dark for at least 4 months.

The different purification methods tested on the mussel extracts were based on solvent extractions using chloroform, diethyl ether or petroleum ether, passage through a C_{18} cartridge, dilution or filtration. Lower values obtained for arsenobetaine were attributed to calibration errors or losses during the clean-up process. Higher values were certainly due to calibration or interferences (co-elution of an unknown compound with arsenobetaine).

Two major species (DMA and arsenobetaine) were detected in both biological tissues. Extraction efficiencies were very different from one laboratory to another (50 to 90%), which may be explained either by the determination of total arsenic content or by the exraction method. Therefore, it was recommended to check that the quantification of total arsenic was under quality control by using a reference material of similar composition (*e.g.* CRM 278 from BCR or DORM-1 from the NRCC).

Particular precautions were recommended to prevent contamination or losses during extraction, including careful cleaning of vessels. Evaporation to dryness has to be avoided (risks of degradation or volatilization). Recovery experiments have to be performed in order to ensure that no loss or transformation of the species occur during the extraction/clean-up process.

Conclusions

All the analytical steps (extraction, purification, separation, detection) needed for arsenic speciation in fish and mussel have been analysed in detail and optimized during the BCR project. The following recommendations were given:

- Vessels should be cleaned very carefully to avoid contamination. Solvent and reagent blanks should be made with each new batch of products and each analytical series of samples. Blanks and recoveries should be determined under the same circumstances as the analyte contents
- The procedure used for total As quantification should be validated with existing reference materials of the same type
- Prior to final determination, efficient purification should be applied to remove interfering salts or organic matter. Evaporation to dryness or at temperature above +40 °C should be avoided owing to risks of transformation or volatilization of the species
- Optimal conditions for HPLC are required to achieve a sufficient resolution of the peaks. Polymeric phases should be preferred to silica because of the low pH values encountered in the extracts
- Calibration solutions should be freshly prepared before use, preferably gravimetrically rather than volumetrically
- Extracts have to be analysed immediately after obtention or stored in the fridge with a small quantity of methanol to avoid bacterial growth

Certification

Preparation of the Candidate CRMs

The tuna fish material (CRM 627) was prepared by the Joint Research Centre, Environment Institute, of Ispra (Italy) whereas the arsenobetaine solution (CRM 626) was prepared by the Laboratoire de Chimie Analytique et Minérale in Strasbourg (France) [134].

Tuna Fish Material

The material was obtained from the city of Venice (fish removed from the fish market, owing to its too high content in mercury). The fish were caught in the Messina strait, kept frozen for approximately four weeks, dissected and the dorsal muscles were taken; these were minced and freeze-dried, ground in a zirconium dioxide mill, passed over a 125 μm sieve and the fraction >125 μm was discarded.

The fraction less than 125 μm was collected and homogenized in a special PVC mixing drum filled with dry argon. The homogenized powder was subsampled and tested for bulk homogeneity by XRF, choosing a number of minor and trace key elements. The material was filled in brown glass bottles with plastic inserts and screw caps. The bottles were primarily flushed with dry nitrogen and stabilized by irradiation (^{60}Co) to avoid microbiological decay.

The homogeneity was verified by repeated determinations of total As, arsenobetaine and DMA. The total As content was determined by HG-QFAAS after microwave assisted digestion, whereas DMA was determined by HPLC-ICP-MS. The within-bottle homogeneity was assessed by 10 determinations in each of two bottles, and the between-bottle homogeneity was evaluated by two determinations out of each 20 bottles; the method uncertainty was evaluated by five replicate determinations of a digest or extract solution. The within- and between-bottle *CV* ranged from 0.5 to 1.2% for total As, from 2.1 to 5% for arsenobetaine, and from 7.1 to 10.6% for DMA, which was comparable to the method uncertainty (respectively 2.6, 3.1 and 6.8%); therefore, no inhomogeneity was suspected at a level of 0.3 g for total As and 1 g for As species, and the material was considered to be suitable for use as a CRM.

The stability of arsenobetaine and DMA in tuna fish was tested over a period of 9 months at $-20\,^{\circ}C$, $+20\,^{\circ}C$ and $+40\,^{\circ}C$ by performing one determination on each of five bottles stored at different temperatures after 1, 3, 6 and 9 months. The reproducibility of the analytical method (same as the one used in the homogeneity study) was verified by determining a portion of raw extract (prepared at the beginning of the study and stored at $-20\,^{\circ}C$) at each occasion of analysis. In addition, the stability of the material and of the raw extract was also verified by qualitative control of the chromatograms; no unexpected peaks containing arsenic were detected. The results showed that no instability could be demonstrated for both arsenobetaine and DMA at $+20\,^{\circ}C$ and $+40\,^{\circ}C$ [134].

Arsenobetaine Solution

Arsenobetaine calibrant is not commercially available and had to be synthesized according to the following scheme:

$$AsMe + BrCH_2COOEt \longrightarrow Me_3As^{+}CH_2COOEt\ Br^{-}\ \mathbf{(1)}$$
$$\mathbf{(1)} + \}X^{+}OH^{-}(\text{anion exchange resin}) \longrightarrow Me_3As^{+}CH_2COO^{-} + EtOH + \}X^{+}Br^{-}$$

The solid collected at the end of the first step of the reaction was identified as $Me_3As^{+}CH_2COOEt$ by ^1H NMR in CD_3OD. No organic impurity was detected. The second step enabled a solid to be obtained, which was further purified using acetone, and dried under vacuum; this solid was stored under dry nitrogen to avoid water contamination. A yield of 87% was achieved.

The characterization of arsenobetaine was performed using element analysis, ^1H NMR, mass spectrometry, thermogravimetric/thermodifferential analysis and separative methods such as HPLC, GC and CZE. From all the results obtained it was concluded that the maximum amount of water present in the solid (if stored and handled under dry atmosphere) is 1% (w/w). In addition, arsenic impurities represent less than 0.15% (w/w); other impurities can be neglected. Therefore, the arsenobetaine purity of the calibrant obtained is >98.9%.

A stock solution was prepared by dissolving (accurate weighing) 29.367 g arsenobetaine in 2000.8 g double deionized water. Then 345.7 g of the solution were taken, poured into a 2 L bottle and completed to 2055.7 g with double deionized water (Milli-Q). This operation was repeated four times in order to obtain five stock solutions. Ten bottles of 2 L capacity were filled with 170.0 g of each stock solution and completed to 2000.0 g with double deionized water. Each bottle was used to fill penicillin-type flasks of 10 mL previously cleaned for at least one night and rinsed with double deionized water. This operation was carried out under a nitrogen flow.

Arsenic concentrations were determined in the five stock solutions as well as in the 10 final solutions using ICP-AES. An independent method (EDXRF) was used to confirm the values obtained for these latter solutions. Calibrations were performed using calibrants of well defined purity. The mean value of concentrations found in the five stock solution was (1016 ± 16) mg kg^{-1} which was in good agreement with the calculated value of (1017 ± 6) mg kg^{-1}. The mean As concentrations in the final solutions ranged between (431.6 ± 7.0) mg kg^{-1} and (435.0 ± 11.4) mg kg^{-1} [mean of the means for five series of 10 determinations: (431.7 ± 4.6) mg kg^{-1}], which agrees very well with the calculated value [(432.4 ± 2.6) mg kg^{-1}].

Evaluation of the Results

Each participating laboratory was requested to perform a minimum of five independent replicate determinations on at least two bottles and on no less than two separate days. Total arsenic as well as arsenobetaine and dimethylarsinic acid had to be quantified. Inorganic species [As(III) and As(V)], as well as arsenocholine and monomethylarsonic acid were not detected by the various methods described in Section 7.2.

Almost all the participants used an extraction with a water/methanol (1:1 v/v) mixture with or without ultrasonic assistance. It has been verified that this process does not induce arsenobetaine, DMA, MMA, As(V) or arsenocholine degradation. The optimization of the power of the ultrasonic bath is an important feature for achieving the best extraction yields; nevertheless, the bath must be refrigerated to avoid degradation of the compounds. Enzymatic digestion with trypsin is also possible, but the activity of the enzyme must be strictly controlled before use to guarantee the reproducibility of the method.

One laboratory used purification on a silica column, which is a source of possible losses and should be avoided. Filtration with C_{18} proved to be efficient and did not lead to losses of arsenobetaine, DMA, MMA or As(V).

In term of mass balances, two laboratories found a concentration of total arsenic in the powder lower than the sum of the various arsenic species detected. In one case, this was due to an incomplete digestion of the solid, and in the other case, to a calibration error in the determination of arsenobetaine. Some participants found traces of arsenocholine, MMA or As(V), but the amounts were too small to be quantified; consequently, only DMA, arsenobetaine and total As contents were certified.

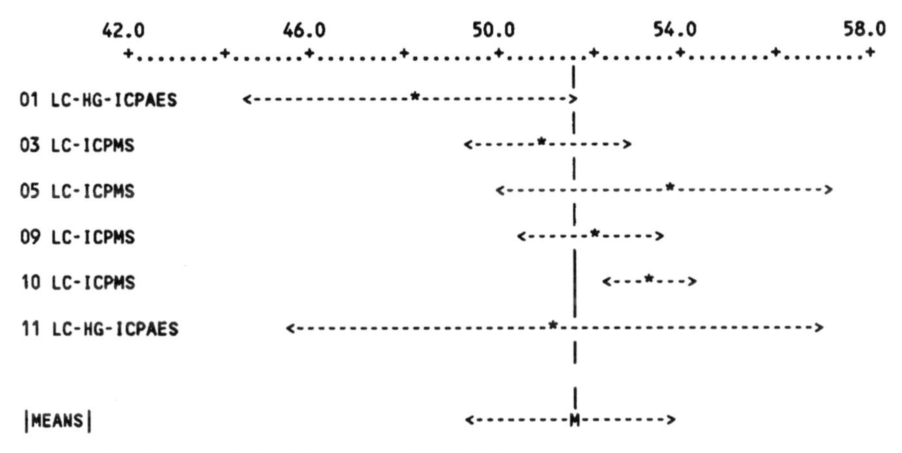

Figure 7.1 *Bar graph for Asbetaine in CRM 627*

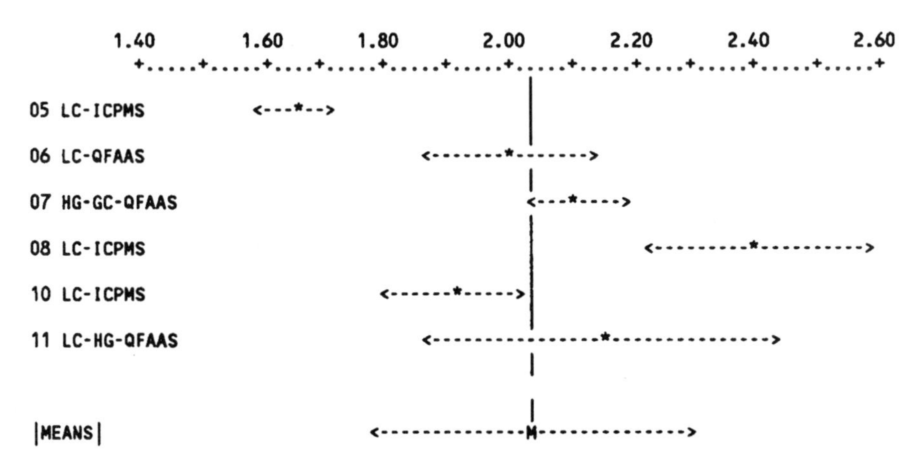

Figure 7.2 *Bar graph for DMA in CRM 627*

Certified values

The certified values of CRM 627 (tuna fish) and their uncertainties (half-width of the 95% confidence interval of the mean of laboratory means) were (51.5 ± 2.1) mmol kg^{-1} for arsenobetaine (6 sets of results); (2.04 ± 0.27) mmol kg^{-1} for dimethylarsinic acid (6 sets of results); and (4.8 ± 0.3) mg kg^{-1} for total As (9 sets of results). The certified value of arsenobetaine in CRM 628 (determined gravimetrically) is (5.77 ± 0.03) mmol kg^{-1}. The results are presented in the form of bar graphs in Figures 7.1 and 7.2.

CHAPTER 8

Selenium Speciation

8.1 Aim of the Project and Coordination

Justification

There is a growing interest in the determination of chemical forms of selenium in environmental matrices, owing to the different biological and toxic effects of the various species. Environmental studies dealing with selenium have mostly focused on the determination of inorganic species (selenite and selenate) and organic compounds such as dimethylselenium and dimethyldiselenium (released into the air from soils, lake sediments and sewage sludge), selenocysteine and selenomethionine, and trimethylselenium [141]. The main reasons that justify the need to speciate the different forms of selenium in the environment are ambiguous: species can either be considered as essential (*e.g.* selenite is added to foodstuffs to remedy selenium deficiency problems) or can be toxic when in excess (*e.g.* selenate).

The Programme and Timetable

The project was started in 1992 by a feasibility study on the optimal storage conditions for solutions containing selenate and selenite [142]. The verified stability of Se species in solutions enabled the organization of an interlaboratory study in 1993–94 [143], which was followed by a certification campaign in 1994–95 [144,145].

Coordination

The feasibility study was carried out at the Universidad Complutense, Departamento de Química Analítica, Madrid (Spain); further stability experiments were performed at the Université de Bordeaux I, Laboratoire de Photophysique et Photochimie Moléculaire (France). The preparation of the materials for the intercomparison and the certification campaign was carried out at the Universidad Complutense, Departamento de Química Analítica, Madrid (Spain) which also coordinated the exercises.

Participating Laboratories

The following laboratories participated in the interlaboratory study and/or certification: CISE, Milano (Italy); CNRS, Service Central d'Analyse, Vernaison (France); CSIC, Departamento de Química Ambiental, Barcelona (Spain); Institut Pasteur, Service Eaux et Environnement, Lille (France); EHICS, Laboratoire de Chimie Minérale et Analytique, Strasbourg (France); National Food Agency, Søborg (Denmark); Perkin-Elmer, Überlingen (Germany); School of Analytical Sciences, Dublin (Ireland); Technische Universität Wien, Institut für Analytische Chemie, Wien (Austria); Universidad Complutense, Departamento de Química Analítica, Madrid (Spain); Universidad de Córdoba, Departamento de Química Analítica, Córdoba (Spain); Università la Sapienza, Dipartimento di Chimica, Roma (Italy); Université de Bordeaux, Laboratoire de Photophysique et Photochimie Moléculaire, Talence (France); Université de Pau, Laboratoire de Chimie Analytique, Pau (France); Universiteit Antwerp, Dept. Scheikunde, Antwerp (Belgium); University of Plymouth, Department of Environmental Science, Plymouth (United Kingdom); University of Sheffield, Centre for Analytical Sciences, Sheffield (United Kingdom); University of Southampton, Department of Chemistry, Southampton (United Kingdom).

8.2 Techniques Used in Selenium Speciation

Techniques developed for the determination of selenite and selenate involve a succession of analytical steps (*e.g.* reduction, separation, detection) [22]. The methods described below were all used successfully in the interlaboratory study and certification campaign. For the determination of total selenium, the same methods as the ones described below were used, with the addition of a reduction step of selenate to selenite using HCl (6 mol L^{-1}), *e.g.* for 30 min at +70 to 80 °C or for 10 min in a microwave oven.

Determination of Selenite in Solution

Hydride Generation/AAS

First method: 0.1 mL of sample was mixed with 0.5% $NaBH_4$, followed by addition of HCl (6 mol L^{-1}) and preconcentration (10 mL to 1 mL), followed by HPLC separation (Hamilton PRP-X100). Calibration was by matrix matching, using Na_2SeO_3 calibrant. Final detection was by quartz furnace AAS.

Second method: a continuous flow of sample was mixed with a solution of 2% $NaBH_4$, which was followed by the addition of HCl (3 mol L^{-1}). Calibration was by matrix matching, using Na_2SeO_3 calibrant. Final detection was by quartz furnace AAS.

Third method: 0.2 mL of sample was mixed with 0.5% $NaBH_4$, followed by addition of HCl (2 mol L^{-1}). Calibration was by calibration graph, using Na_2SeO_3 calibrant. Final detection was by quartz furnace AAS.

Hydride Generation/AFS

First method: a continuous flow of sample was mixed with a solution of 1.5% $NaBH_4$, which was followed by the addition of HCl (6 mol L^{-1}). Calibration was by matrix matching, using Na_2SeO_3 calibrant. Final detection was by AFS.

Second method: 0.5 mL of sample was mixed with 0.2% $NaBH_4$, followed by addition of HCl (6 mol L^{-1}); Calibration was by calibration graph, using Na_2SeO_3 calibrant. Final detection was by AFS.

Hydride Generation/ICP-MS

0.2 mL of sample was mixed with 0.2% $NaBH_4$, followed by addition of HCl (2 mol L^{-1}). Calibration was by calibration graph, using Na_2SeO_3 calibrant. Final detection was by ICP-MS of ^{82}Se.

Ethylation/GC-MIP-AES

5 mL of sample were mixed with a $NaBEt_4$ solution to form diethyl selenide which was trapped on a fused silica column at $-150\,°C$; separation was by capillary gas chromatography (HP-1, 25 m length, 320 μm × 0.17 μm); detection was by MIP-AES. Calibration was by calibration graph, using Na_2SeO_3 calibrant.

Microcolumn Preconcentration/ETAAS

0.2–0.5 mL sample was preconcentrated by chelation with APDC followed by silica C_{18} microcolumn; elution of the Se(IV)–APDC complex with ethanol. Detection was by ETAAS. Calibration was by calibration graph, using a commercial solution of Se.

Microcolumn Preconcentration/ICP-MS

0.05 mL sample was preconcentrated on an alumina column; elution was with HNO_3 (0.016 mol L^{-1}). Detection was by ICP-MS of ^{82}Se. Calibration was by matrix-matching, using Na_2SeO_3 calibrant.

HPLC-ICP-MS

0.1 or 0.2 mL sample was injected on a HPLC column (*e.g.* Hamilton PRP-X100 or Merck Polyspher can-2) with anion exchange. Detection was by ICP-MS of ^{82}Se. Calibration was by calibration graph, using Na_2SeO_3 calibrant.

Determination of Selenate in Solution

As mentioned in the technical discussions, selenite was often determined by difference between total inorganic Se and selenate; this measurement was, however, not considered to be acceptable for the purpose of certification which should be based on direct measurements. Two different methods were used for direct selenate determination.

Hydride Generation/HPLC-AAS

0.1 mL sample was mixed with 0.5% $NaBH_4$, followed by HPLC separation (Hamilton PRP-X100), on-line microwave reduction of selenate to selenite in HCl (2 mol L^{-1}) and total determination as selenite by quartz furnace AAS. Calibration was by matrix-matching, using Na_2SeO_3 and Na_2SeO_4 calibrants.

HPLC-ICP-MS

0.1 or 0.2 mL sample was injected on a HPLC column (*e.g.* Hamilton PRP-X100 or Merck Polyspher can-2) with anion exchange. Detection was by ICP-MS of ^{82}Se. Calibration was by calibration graph, using Na_2SeO_4 calibrant.

8.3 Feasibility Study

Many problems occur in Se speciation analysis, owing to risks of adsorption on container walls, instability of species or contamination. Prior to conducting an interlaboratory project on this topic, it was hence decided to assess the stability of selenite and selenate according to various factors (effects of container materials, additives, temperature and light). The study focused on tests of the effects of physicochemical parameters on solutions stored in polyethylene and PTFE containers. Container volumes were 100 and 500 mL for polyethylene and 500 and 1000 mL for PTFE. Stock and initial working solutions were prepared in 1 and 5 L polyethylene containers previously cleaned with nitric acid (at pH 2) and rinsed with Milli-Q water. The stock solutions were prepared with sodium selenite and sodium selenate (purity >98%).

Eight initial working solutions with a total volume of 10 L were prepared. Two solutions with concentrations of 10 and 50 μg L^{-1} of each species, each of them at two different pH values (pH 2 by adding H_2SO_4 and pH 6 with no addition of extra reagent), with and without addition of chloride (100 mg L^{-1} as NaCl), were prepared in polyethylene containers. The homogeneity of the solutions was achieved by continuous pumping with a peristaltic pump and PTFE tubes for 5 h. Oxygen was removed from bottles by bubbling with N_2. These initial working solutions were placed in 1000 mL polyethylene and PTFE containers kept in the dark at −20, 20 and 40 °C and exposed to sunlight at 20 °C. Solutions were stored and total inorganic selenium and selenite were determined after 1 day, 2 weeks, and 1, 2, 6, 9 and 12 months. Determinations were carried out by hydride generation atomic absorption spectrometry after pre-reduction of selenate to selenite with HCl (6 mol L^{-1}); independent measurements were performed by neutron activation analysis for quality control. A full description of the analytical techniques used is described elsewhere [142], as well as the results of the study which are briefly summarized below.

The solutions were stable at $-20\,°C$ in all conditions tested (pH 2 and 6, with and without chloride, without addition of acid). At ambient temperature, samples stored at pH 2 in polyethylene containers showed instability of selenite after one month storage whereas selenate remained stable; this difference in behaviour was attributed to possible adsorption of selenite onto the container walls. The stability was better at pH 6 but instability of selenite was still observed after two months storage, whereas selenate remained stable. When a PTFE container was used, dramatic losses of selenite were observed at pH 6.

The effects of storage at 40 °C were studied in 100 mL vessels (instead of 500 mL as used in the other experiments). Surprisingly, the stability was found to be much better for both species in solutions stored at pH 2 and 6 in polyethylene containers (with and without addition of chloride). This was further attributed to the fact that, at 40 °C, bottles were opened only once, while at 20 °C, several control analysis were performed with the same bottle; thus, in the last case some dead air volume caused instability problems. Tests performed with samples stored in the dark and exposed to sunlight demonstrated that light had no significant effect on the stability of selenite and selenate for the period tested.

At this stage, the conclusions of the feasibility study were that no risk of selenium losses occur at the 10 and 50 μg L^{-1} levels over 12 months storage at $-20\,°C$. The stability of the species at both 20 and 40 °C depends upon the pH and the container type. Generally, both selenite and selenate stored in polyethylene containers at 40 °C were more stable than at room temperature, particularly at pH 6. The presence of chloride tended to stabilize both species.

On the basis of these preliminary conclusions, further experiments were carried out by spiking the solutions with increasing concentrations of Cl^- (up to 20 000 mg L^{-1}, simulating the salinity of seawater); this additional study was performed by preparing a series of selenite/selenate solutions containing approximately 10 and 50 μg L^{-1} total selenium at pH 6 and spiking them with Cl^- concentrations of 100, 500, 1000, 2000, 3000 and 5000 mg L^{-1}; a set of solutions was also prepared with seawater. Samples were maintained in polyethylene containers in the dark at ambient temperature for one year. The suitability of HCl addition at pH 2 for stabilizing Se species in comparison to H_2SO_4 was also tested; selenite/selenate solutions were prepared with a single addition of HCl (Cl = 350 mg L^{-1}) and stored in the dark in polyethylene containers at ambient temperature for one year. The results obtained in these new storage conditions showed that selenate was stable for one year in all conditions tested while selenite was unstable in HCl (0.01 mol L^{-1}) after two months storage; these results agreed with those obtained with H_2SO_4 in the feasibility study. The selenite stability improved with the addition of increasing Cl^- concentrations and this species was completely stabilized for 12 months as tested in seawater. It should be noted that the prepared samples were opened every two months (6 times) for analysis; however, previous studies have shown that the stability of these species decreases when the dead volume of the container increases. In order to check whether the stability of selenite would be improved in solutions containing 2000 mg L^{-1} of Cl^- stored in tightly closed containers, solutions containing 6 μg L^{-1} of selenite, 6 μg L^{-1} of selenate and 2000 mg L^{-1} of Cl^- were stored at room temperature and analyses were performed every week. Both Se species were found to be stable over 18 months when the containers were completely full and opened only once for analysis. However, when the same bottle was analysed at different times, selenite was only stable for two weeks (bottles being opened 5 times) whereas selenate was stable in all conditions tested.

The conclusions of the overall study were that a Cl^- concentration of 2000 mg L^{-1} or more is suitable to stabilize selenite but samples have to be opened only at the time of analysis to ensure complete stability. This recommendation was clearly stressed to participants in the first interlaboratory study (see below).

8.4 Interlaboratory Study

Two concentration levels of selenite and selenate solutions were prepared: a low concentration solution of 6 μg L^{-1} selenite + selenate (solution A) and a high concentration solution of 50 μg L^{-1} selenite + selenate (solution B). Sodium chloride (2000 mg L^{-1}) was added to stabilize the inorganic species (at pH 6).

Results of Selenite Determination

Sources of error detected in the technical discussion were mainly due to calibration errors or lack of quality control.

It was stressed by the participants using ICP-MS that polyatomic interferences from Cl^- can be removed to improve the accuracy of selenite determination; ways to do so are (i) to dilute the original sample after pre-concentration of the selenite species, (ii) to modify the plasma conditions by adding N_2 carrier to the plasma or (iii) to use anion-exchange chromatography to separate chloride from the selenium species. When diluting and pre-concentrating on a pre-column, care should be taken that selenite is quantitatively recovered. The use of silver nitrate to remove chloride is suspected to affect selenite.

The coefficient of variation (*CV*) between laboratories (raw data) was 23.7% for solution A; it decreased to 6.2% after removing outliers on technical grounds. The mean of laboratory means was (5.6 ± 0.4) μg L^{-1} which overlapped the expected value of 6.0 μg L^{-1}.

The *CV* between laboratories (raw data) was 17.5% for solution B; it decreased to 6.3% after removing outliers on technical grounds. The mean of laboratory means was (51.0 ± 3.2) μg L^{-1} which matched well with the expected value of 50.0 μg L^{-1}.

Results of Selenate Determination

The discussion focused on how to reduce selenite to selenate efficiently and calculate the recovery. Since a selenite calibrant was used to verify the efficiency of the reduction process, the laboratories were recommended to apply the same analytical procedure to both samples and calibrants. It was stressed that the HCl molarity and the reduction temperature should be strictly controlled. HCl molarities ranging between 4 and 6 mol L^{-1} and temperatures ranging between 60 and 100 °C (with different heating times) were found to be suitable to ensure complete reduction of selenate to selenite.

The determination of selenate by difference between total inorganic selenium content and selenite was questioned. Although this method is quite commonly used and acceptable for selenate determination, this does not correspond to a direct measurement of the species; hence, this method would not be accepted for certification and it was proposed that the results obtained by difference would be used as confirmative values of results obtained by techniques actually separating the Se species (*e.g.* HPLC-ICP-MS or HPLC-HGAAS).

For solution A, the *CV* between laboratories (raw data) was 31.9%; it decreased to 4.1% after removing outliers on technical grounds. The mean of laboratory means was (6.2 ± 0.3) μg L^{-1} which overlapped the expected value of 6.0 μg L^{-1}.

The *CV* between laboratories (raw data) was 29.4% for solution B and decreased to 7.0% after removing outliers on technical grounds. The mean of laboratory means was (50.3 ± 3.6) μg L^{-1} which overlapped the expected value of 50.0 μg L^{-1}

Additional Remarks on the Interlaboratory Study

It was recommended to use the term "inorganic Se(IV) and Se(VI)" (or selenite and selenate as used throughout this chapter) to make a clear distinction between these species and organic compounds. As stressed above, it was agreed that the method of selenate determination by difference would not be considered for certification; the selenate certification should be based on the results of techniques involving a separation of the species. Standard addition methods or matrix matching are recommended.

8.5 Tentative Certification

Preparation of the Candidate CRMs

Equipment and Cleaning Procedures

Two 250 L polyethylene tanks were used for the preparation and homogenization of the two candidate CRMs; they were cleaned thoroughly with deionized water (Milli-Q) and subsequently rinsed with the solution they would contain.

Polyethylene bottles were carefully cleaned with deionized water (Milli-Q);

Table 8.1 *Homogeneity study for CRMs 602 and 603*[a]

Component	*CRM 602*		*CRM 603*	
	CV/%[b]	*CV/%*[c]	*CV/%*[b]	*CV/%*[c]
Selenite	7.1 ± 1.6	3.8 ± 0.9	4.9 ± 1.1	5.3 ± 1.2
Selenate	1.1 ± 0.3	2.4 ± 0.6	6.5 ± 1.5	2.9 ± 0.7

[a] $(CV \pm U_{CV})\%$. Uncertainty on the *CV*'s: $U_{CV} \approx CV/\sqrt{2n}$.
[b] 10 replicate determinations on the content of one bottle.
[c] single determination in each of 10 different bottles.

each bottle was rinsed with the solution they would contain prior to final bottling. The bottle caps were treated in a similar way.

Homogenization and Bottling

The two candidate CRMs were prepared from deionized water (Milli-Q) to which the compounds of interest were added in the form of sodium salts (Na_2SeO_3 and Na_2SeO_4); the chloride ions added to stabilize selenium species (see Section 8.3) were spiked as NaCl.

Homogenization was carried out in the polyethylene tanks covered with a close fitting polyethylene lid. A centrifugal pump connected to the tank with polyethylene piping ensured constant recirculation of the solution. The pump had no metallic parts in contact with the water. The bottling was performed manually using pre-rinsed polymer tubes, avoiding any contact with metals. Each bottle was sealed and stored at ambient temperature.

Homogeneity Control

The between-bottle homogeneity was verified by analysing the content of 10 bottles of each solution taken at regular intervals during the bottling procedure. The method variability was assessed by 10 determinations on one bottle, assuming that the content of each bottle was homogeneous.

The analytical method used was HPLC-HGAAS. A preconcentration step was necessary in the case of the solution with low content (RM 602); this was achieved by retaining the selenium species on an alumina microcolumn conditioned in anionic form with nitric acid, and eluting the selenite and selenate species with ammonia solution.

The samples were analysed in the most repeatable way. Analyses of actual samples alternated with analyses of calibrant solutions, which enabled a correction to be made for a slight drift of sensitivity that occurred during the analytical procedure, thus increasing the obtained accuracy.

Table 8.1 gives the coefficient of variation (*CV*) of the method and the between-bottle *CV* for the two CRMs. The between-bottle *CV*'s are close to the *CV* of the method and, therefore, no inhomogeneities of the material were suspected. On the basis of the results, it was concluded that the homogeneity of the two materials would make them suitable for use as CRMs.

Table 8.2 *Stability study for RMs 602 and 603*

Species	*Time/months*	$R_t \pm U_t$	
		RM 602	*RM 603*
Selenite	1	1.00 ± 0.05	1.00 ± 0.08
	3	0.95 ± 0.06	0.99 ± 0.08
	6	0.98 ± 0.06	0.99 ± 0.08
	9	0.97 ± 0.04	1.01 ± 0.07
	12	0.98 ± 0.05	0.99 ± 0.08
Selenate	1	1.00 ± 0.05	1.00 ± 0.04
	3	0.98 ± 0.04	0.99 ± 0.04
	6	0.95 ± 0.04	1.02 ± 0.05
	9	0.99 ± 0.05	0.98 ± 0.05
	12	0.98 ± 0.05	1.00 ± 0.04

Stability Control

The stability of the materials was tested at +20 °C over a period of 12 months and selenite and selenate were determined at the beginning of the storage period and after 1, 3, 6, 9 and 12 months. Samples were analysed using the same procedures as for the homogeneity study. Selenite and selenate were each determined in quadruplicate (one replicate analysis in each of four bottles) at each occasion of analysis. The evaluation of the stability was based on the procedure described in Chapter 3, using the results of the homogeneity study (performed immediately after bottling) as reference for the samples analysed at the various occasions.

The results (Table 8.2) showed that no instability could be demonstrated over a 12 month period. They hence justified that the certification campaign be started.

Technical Evaluation

Results for Selenite

Calibration or calculation errors justified the rejection of four laboratory sets of results from the 23 sets submitted.

Problems were experienced with ICP-MS, owing to the matrix effects of Cl^-; while three of the four laboratories using ICP-MS could correct these effects, one set was rejected due to high (outlying) results.

Oxidation of selenite to selenate was suspected in two cases, in particular with the CSV technique, explaining the low results which were withdrawn.

High standard deviations were obtained with the CGC-MIP-AES and GC-FPD techniques; this was considered to be inherent to the GC technique and the results were, consequently, accepted.

Table 8.3 *Reference values for total inorganic selenium, selenite and selenate in RM 602*

Component	*Reference value and uncertainty*	*Target value (obtained gravimetrically)*	*p*
Total Se	13.5 ± 0.4		18
Selenite	5.9 ± 0.2	6.0 ± 0.1	14
Selenate	8.1 ± 0.3	8.0 ± 0.2	3

Table 8.4 *Reference values for total inorganic selenium, selenite and selenate in RM 603*

Component	*Reference value and uncertainty*	*Target value (obtained gravimetrically)*	*p*
Total Se	80.3 ± 1.4		16
Selenite	34.9 ± 0.8	35.0 ± 0.8	14
Selenate	45.1 ± 2.4	45.0 ± 1.0	3

Results for Total Inorganic Selenium

No particular problems were noticed for the determination of total Se in the two RMs 602 and 603.

Results for Selenate

As stressed in the interlaboratory study, the results obtained by determining selenate by difference between total selenium content and selenite would not be accepted for certification. Indeed, although this method is quite commonly used and acceptable for selenate determination, it does not correspond to a direct measurement of the species.

The reference value of selenate was hence based on the results of techniques actually separating the Se species (*e.g.* HPLC-ICP-MS or HPLC-HGAAS), with a gravimetric confirmation (target values). The values were also confirmed by alternative techniques determining selenate by difference; the mean of laboratory means were (7.7 ± 0.7) μg L^{-1} and (44.8 ± 4.4) μg L^{-1}.

The reference values (unweighted mean of *p* accepted sets of results) and their uncertainties (half-width of the 95% confidence intervals) are given in Tables 8.3 and 8.4. Bar graphs are depicted in Figures 8.1–8.6.

Additional Stability Checks

Problems of leakage were observed in some of the bottles containing the reference materials after the certification campaign had been concluded.

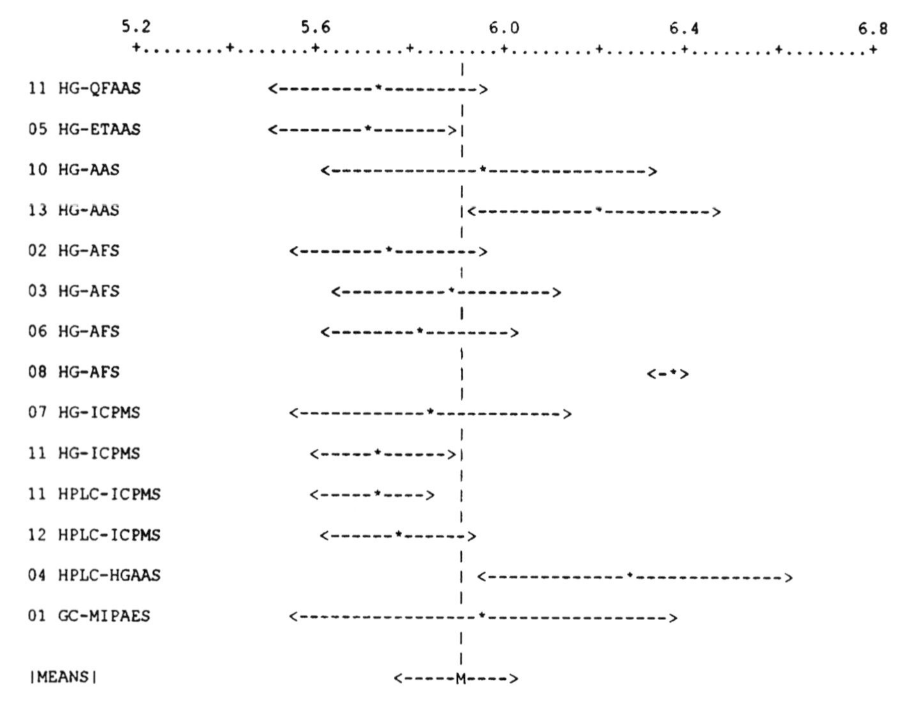

Figure 8.1 *Selenite in RM 602 in µg L^{-1}*

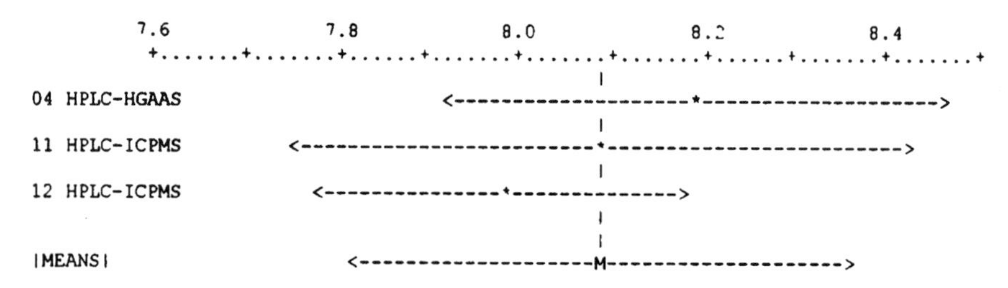

Figure 8.2 *Selenate in RM 602 in µg L^{-1}*

Consequently, it was decided to test other polypropylene bottles with tighter caps for the storage of the reference materials. This additional study was carried out 24 months after the initial stability study and led to the detection of instability problems of the Se species over a long-term period. A clear decrease in selenite content was observed in the new polypropylene bottles after 8 months storage, which was particularly acute for the low-concentration reference materials (Table 8.5), whereas selenate remained stable over the same period. On the basis of these results, if was found necessary to control the

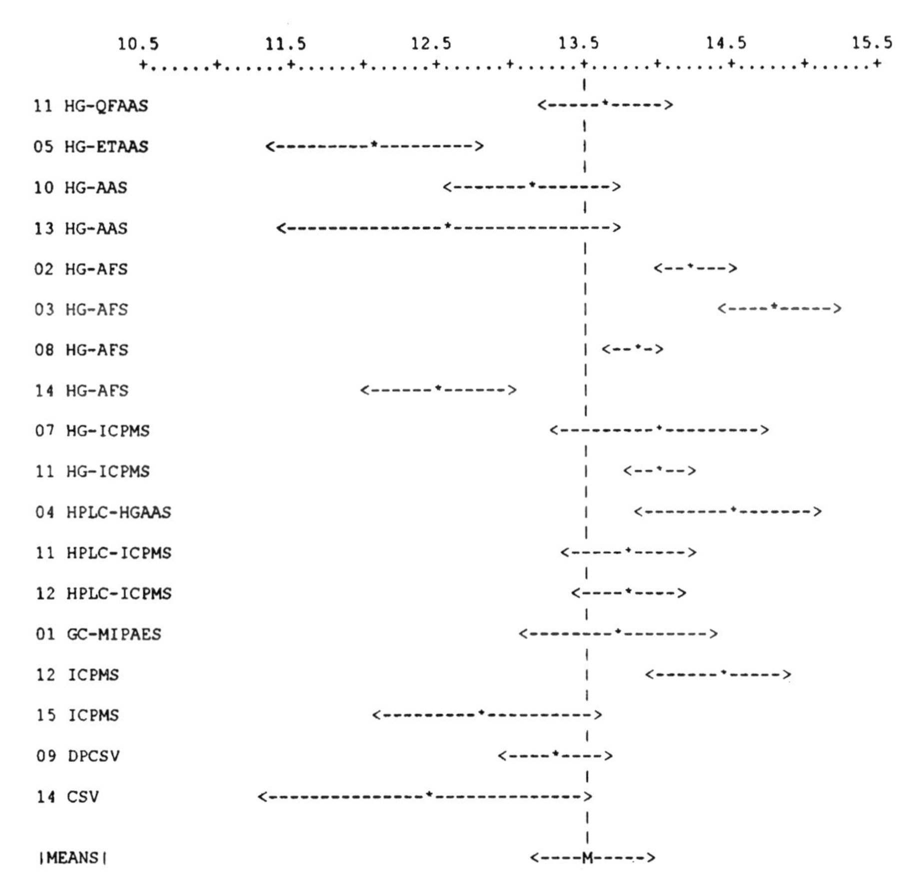

Figure 8.3 *Total inorganic Se in RM 602 in μg L^{-1}*

stability of the Se species in the materials stored in the original polypropylene bottles as well as in the stock solution stored in the polyethylene tank; the results are presented in Table 8.6, showing that both species are stable for 36 months at the two concentration levels in the polyethylene (60 L) tanks whereas a decrease of around 30% and 15% was observed for selenite in the two solutions stored in polypropylene bottles over the same period; selenate was found to slightly increase over the same duration but this change is not significant.

The reasons for this instability could be attributed to an adsorption process onto the container surface which was not observed in the 60 L tank, owing to a much smaller surface/volume ratio; in other terms, the ratio is 11 times smaller in the storage tanks in comparison to the 100 mL bottles, leading to a better stability of the species. Furthermore, the polyethylene material of the tank seems to be more suitable to achieve stability in comparison to polypropylene.

BAR-GRAPHS FOR LABORATORY MEANS AND 95% CI

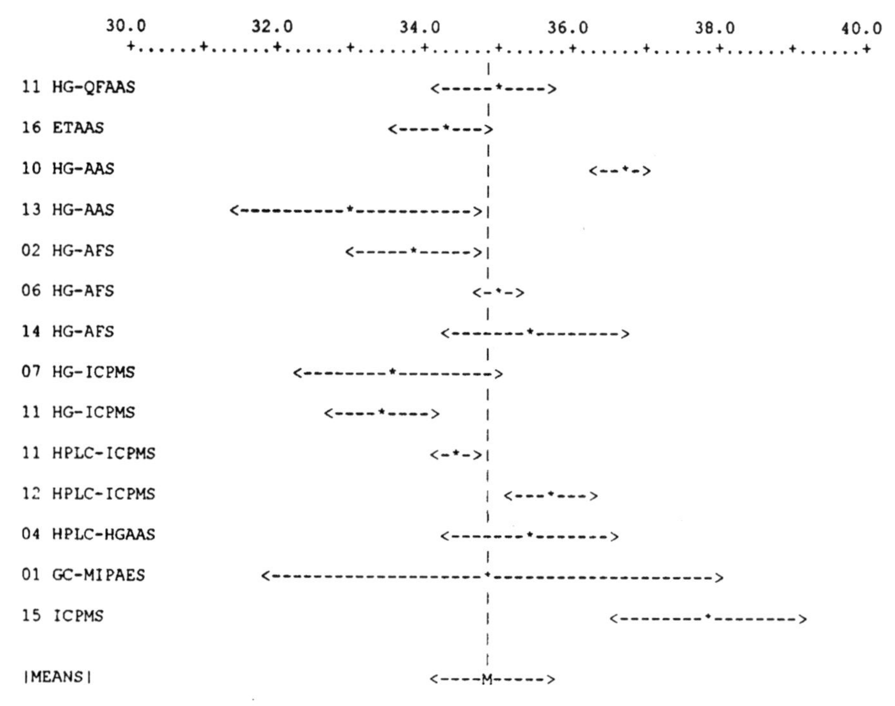

Figure 8.4 *Selenite in RM 603 in* $\mu g\ L^{-1}$

BAR-GRAPHS FOR LABORATORY MEANS AND 95% CI

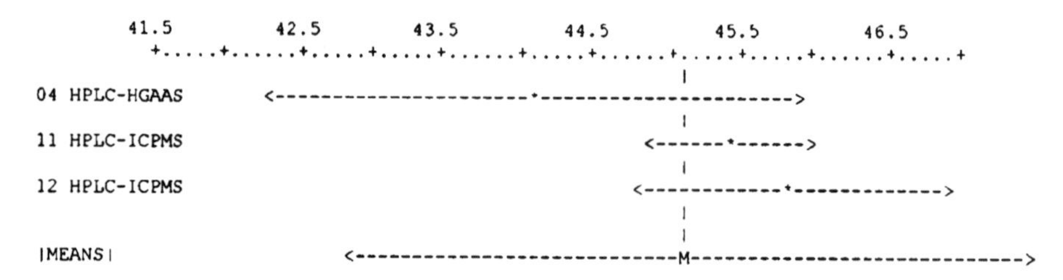

Figure 8.5 *Selenate in RM 603 in* $\mu g\ L^{-1}$

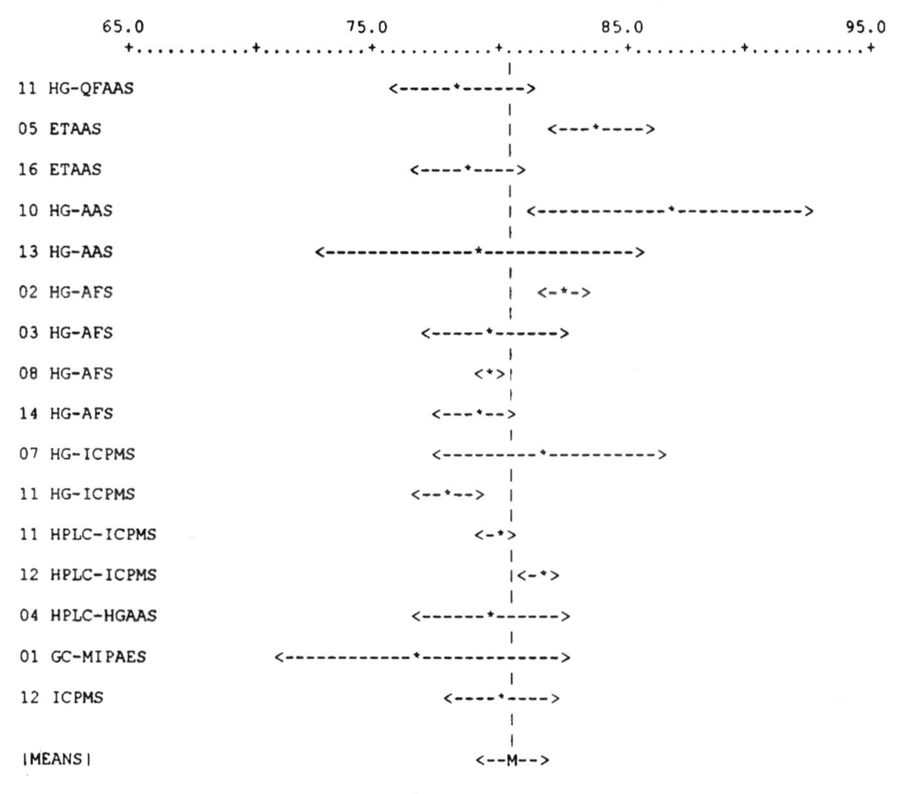

Figure 8.6 *Total inorganic Se in RM 603 in* $\mu g\ L^{-1}$

Table 8.5 *Se contents in the new polypropylene bottles*[a]

	RM 602		*RM 603*	
	Selenite	*Selenate*	*Selenite*	*Selenate*
Target values	6.00	8.00	35.0	45.0
Reference values	5.9 ± 0.4	8.1 ± 0.3	34.9 ± 0.8	45.1 ± 2.4
Measured value at bottling	5.9 ± 0.7	7.8 ± 0.8	36.0 ± 1.4	45.2 ± 3.3
1 month storage	5.9 ± 0.5	8.0 ± 0.2	34.8 ± 0.3	45.1 ± 2.3
4 months storage	5.6 ± 0.5	7.9 ± 0.6	35.8 ± 1.4	44.9 ± 2.1
8 months storage	2.5 ± 0.3	7.4 ± 0.5	33.1 ± 2.5	44.3 ± 4.3

[a] $\mu g\ L^{-1}$.

Table 8.6 *Se contents in the 60 L tanks and polypropylene bottles (36 months storage)*[a]

	RM 602		*RM 603*	
	Selenite	*Selenate*	*Selenite*	*Selenate*
Target values	6.00	8.00	35.0	45.0
Reference values	5.9 ± 0.4	8.1 ± 0.3	34.9 ± 0.8	45.1 ± 2.4
60 L tanks	5.7 ± 0.9	7.7 ± 0.7	36.2 ± 2.7	44.7 ± 4.9
Polypropylene bottles	4.2 ± 0.2	8.5 ± 0.9	29.6 ± 0.3	50.9 ± 4.3

[a] $\mu g\ L^{-1}$.

On the basis of these additional studies, these reference materials were not found to be suitable as CRMs [144,145].

Conclusions

The results of this tentative certification show that (1) the state-of-the-art for Se speciation is good enough for laboratories to obtain comparable data, (2) reference materials containing inorganic Se species can be stabilized and used over a 12-month period for the purpose of interlaboratory studies or routine quality control checks, (3) work remains to be done to find the optimal storage conditions for candidate CRMs to enable long-term stability and availability of the materials. An important aspect of this work is that reference materials can be prepared by laboratories for their own quality control, following the 'cooking recipe' given in this chapter; it is obvious that such reference solutions should be used with all necessary care to avoid instability problems and should certainly not be kept over a period longer than 12 months.

CHAPTER 9

Chromium Speciation

9.1 Aim of the Project and Coordination

Justification

The different toxicity and bioavailability of Cr(III) and Cr(VI) are a public health concern and therefore require strict control. Trivalent chromium is found to be essential for man, where it is involved in glucose, lipid and protein metabolism, whereas the deleterious effects to living organisms of Cr(VI) are well documented. Cr(VI) is also a potent carcinogenic agent for the respiratory tract, requiring continuous monitoring of occupational air. Hence monitoring of the separate species in drinking water, occupational exposure or environmental samples is necessary. Determination of the total Cr content does not provide sufficient information about possible health hazards.

Norms were issued by the European Union for Cr in drinking water and occupational air. The maximum allowable concentration for chromium in drinking water is 50 $\mu g\ L^{-1}$ according to the European Community Directive 80/778/EEC, L229/20, D48. Hexavalent chromium is such a potent carcinogenic agent for the respiratory tract that continuous monitoring is imposed, stated in Directive 90/3941/EEC on exposure to carcinogenic substances. In occupational health, the OEL (occupational exposure limits) for water soluble and certain water insoluble compounds in indoor air is limited to 0.5 mg m^{-3} for chromium, to 0.5 mg m^{-3} for Cr(III) and to 0.05 mg m^{-3} for Cr(VI), which reflects the different toxicity of both species.

The present state of the art of Cr speciation leaves much to be desired and requires the improvement of quality assessment. In order to meet the requirements of the norms, the reliability of the methods needs to be improved substantially. Appropriate reference materials certified for Cr(III) and Cr(VI) were, therefore, considered necessary for laboratories to check and improve their performance [146]. Two types of samples were considered:

- A lyophilized water sample containing Cr(VI) and Cr(III) as in tap water, CRM 544
- A filter loaded with welding dust as normally encountered on the personal filter monitors of stainless steel welders, CRM 545

The Programme and Timetable

The project was started by a series of feasibility studies carried out under bursary projects in 1989 [147]. It was followed by the organization of an interlaboratory study in 1992 [148] and a certification campaign which was conducted between 1994 and 1996 [149,150].

Coordination

The interlaboratory study and certification campaign were coordinated by the Laboratorium voor Analytische Scheikunde, Universiteit Gent (Belgium), which also performed the preparation of the lyophilized solutions; the filters loaded with welding dust were prepared by the Arbejdsmiljøinstituttet in Copenhagen (Denmark).

Participating Laboratories

The following laboratories participated in the interlaboratory study and certification campaign: Arbejdsmiljøinstituttet, Dept. of Chemistry, Copenhagen (Denmark); Bayer Antwerpen, Centraal Analytisch Labo (Germany); Berufsgenossenschaftliches Institut für Arbeitssicherheit, Sankt Augustin (Germany); Ciba-Geigy Ltd., Central Analytical Dept., Basle (Switzerland); Fondazione Clinica del Lavoro, Laboratorio di Igiene Industriale, Pavia (Italy); General Chemical State Laboratory, Athens (Greece); Health and Safety Executive, Sheffield (United Kingdom); Institut für Anorganische Chemie, Univ. Regensburg, Regensburg (Germany); Institut Pasteur, Service Eaux et Environnement, Lille (France); Institute of Occupational Health, Helsinki (Finland); Institute of Occupational Health, Oulu (Finland); Laboratorium voor Analytische Scheikunde, Universiteit Gent, Gent (Belgium); Oceanography Laboratory, University of Liverpool, Liverpool (United Kindgom); School of Science, Sheffield Hallam University, Sheffield (United Kingdom); University of Plymouth, Department of Environmental Sciences, Plymouth (United Kindgom); Vandkvalitetsinstituttet, Hørsholm (Denmark); VITO, Mol (Belgium).

9.2 Feasibility Studies

Preliminary studies investigating the feasibility of producing such reference materials, as well as their stability, were carried out [146,147]. The stability of both Cr species was investigated in different media, different pH and different container materials. By choosing a hydrogen carbonate buffer solution at pH 6.4 as the agent to prevent hydrolysis of Cr(III), a matrix very close to that of real waters was achieved [146]. Therefore, initially an aqueous buffered sample, kept under a CO_2 blanket in sealed quartz ampoules, was foreseen. This evolved into "lyophilized" samples to be reconstituted in the same buffer solution at the right pH. This step presumably ensures an indefinite shelf life. The optimal conditions for lyophilization, avoiding losses of material, and the reduction of Cr(VI) to Cr(III), the optimal sample volume and measures to avoid possible adsorption on the wall, *etc.*, were all investigated, and kept under strict control. The work is described in more detail elsewhere [146,147].

9.3 Techniques Used in Chromium Speciation

A range of techniques of proven performance have been used in this project. A new isotope dilution technique (IDMS) has been developed (funded by the BCR) in the frame of this project to serve as a reference method for the certification campaign (see description below) [151].

Determination of Cr Species in Lyophilized Solution

The techniques described below correspond to the determination of Cr(III) and Cr(VI) in lyophilized solution reconstituted with 20 mL HCO_3^-/H_2CO_3 buffer at pH 6.4, following the procedure described elsewhere [152].

Ion Chromatography with Spectrometric Detection

Ion chromatography was used with columns of Dionex Ionpac AG-7-AS 7 or CG5 [mobile phase: $(NH_4)_2SO_4$–NH_4OH]; post-column reaction with 1,5-diphenylcarbazide (DPC)–MeOH–H_2SO_4. The detection was UV-visible light spectroscopy with DPC as post-column reagent. An aqueous calibrant of $K_2Cr_2O_7$ was used for Cr(VI) determination and aqueous $Cr(NO_3)_3$ as calibrant for Cr(III).

Ion Chromatography with Chemiluminescence

Ion chromatography was applied with a column of Dionex AG4 A, eluting with KCl at pH 2.5. The effluent was reduced with SO_2 solution, followed by a post-column reaction with luminol reagent, orthoboric acid (0.1 mol L^{-1}) and H_2O_2 (0.01 mol L^{-1}) at pH 11.5. The detection was by chemiluminescence. An aqueous calibrant of $K_2Cr_2O_7$ was used for Cr(VI) determination, and an aqueous calibrant of $KCr(SO_4)_2$ for Cr(III) determination.

UV Digestion/DPCSV

Silica was added to the reconstituted sample to remove Cr(III), which was followed by addition of NaOAc buffer (0.25 mol L^{-1}, pH 8), UV digestion for 3 h, addition of ACSV reagents (mixture of 5 mol L^{-1} $NaNO_2$, 0.5 mol L^{-1} NaOAc, and 0.025 mol L^{-1} DPTA). The detection was by DPCSV. An aqueous calibrant of K_2CrO_4 was used.

Anion Exchange/ETAAS

First method: 50 mL H_2O, 5 mL phthalate buffer at pH 4 and 5 mL NaDDTC (2%) were added to 2 mL of sample, followed by an extraction into MIBK. The detection was by ETAAS. Calibration was carried out using a commercial solution of Cr checked by an independent method.

Second method: extraction was performed with liquid anion exchange solution (Amberlite La 2/MIBK). Detection was by ETAAS. Calibration was carried out using a commercial solution of Cr in HNO_3 checked against aqueous Na_2CrO_4 and aqueous $CrCl_3$, or with K_2CrO_4 for Cr(VI) and aqueous $Cr(NO_3)_3$ for Cr(III).

Anion Exchange/IDMS

This method has been developed in the frame of the BCR project as described in more detail elsewhere [152].

A known amount of ^{53}Cr(III)–^{53}Cr(VI) enriched spike was added to the reconstituted sample. Extraction was carried out with liquid anion exchange solution (Amberlite LA 2/MIBK), followed by back-extraction with 1.5 mL of ammonia solution (6%). The aqueous ammonia phase was washed with *n*-hexane, followed by electrodeposition on Pt wire in ammonia solution (6%) at a voltage of 2.45 V. The detection was by IDMS (thermal ionization of masses 52 and 53). The calibrants used were $Na_2Cr_2O_7$ for Cr(VI) and a commercial solution of $CrCl_3$ for Cr(III).

Microcolumn Preconcentration/ICP-MS

Cr(VI) was retained on an acidic microcolumn and eluted with NH_4OH (2 mol L^{-1}). Detection was by ICP-MS of mass 52. The calibrants used were K_2CrO_4 for Cr(VI) and aqueous $Cr(NO_3)_3$ checked against NIST SRM 1643c for Cr(III).

Determination of Cr(VI) in Welding Dust

The techniques described below correspond to the determination of Cr(VI) (and total leachable Cr) in welding dust loaded on a filter, following a procedure based on leaching with NaOH–Na_2CO_3 buffer and agitation and/or sonication [150].

Ion Chromatography with Spectrometric Detection

First method: the leaching was carried out with 10 mL NaOH (0.5 mol L^{-1})–Na_2CO_3 (0.3 mol L^{-1}) buffer and agitation in a heated ultrasonic bath (70 °C) for 40 min. The leachate was diluted with HCO_3^-/H_2CO_3 buffer and filtered over a 0.45 μm membrane filter. Separation was carried out by ion chromatography using columns of Dionex Ionpac AG-7-AS 7 [mobile phase:

$(NH_4)_2SO_4$–NH_4OH]; post-column reaction with 1,5-diphenylcarbazide (DPC)–MeOH–H_2SO_4. The detection was UV-visible light spectroscopy with DPC as post-column reagent. An aqueous calibrant of $K_2Cr_2O_7$ is used.

Second method: the filter was leached with 10 mL of pyridinedicarboxylic acid (0.02 mol L^{-1}), disodium hydrogen phosphate (0.02 mol L^{-1}), sodium iodide (0.1 mol L^{-1}), ammonium acetate (0.5 mol L^{-1}) and lithium hydroxide (0.028 mol L^{-1}). This was followed by filtration over a Millipore filter (0.45 μm). Separation was carried out by ion chromatography using columns of Dionex Ionpac CG5-CS5, eluting with the same mixture as used for the leaching. The detection was UV-visible light spectroscopy with DPC as post-column reagent. An aqueous calibrant of $K_2Cr_2O_7$ was used.

Third method: the leaching was carried out with 10 mL of NaOH–Na_2CO_3 buffer at pH 6.4, sonication for 5 min, and heating at 80 °C for 30 min. The leachate was diluted with $NaHCO_3$ at pH 6.4 and filtered over a HPLC filter. Separation was carried out by ion chromatography using columns of IonSpher A [eluent: gradient elution with Tris-HCl (0.2 mmol L^{-1}) at pH 8.6 and NaCl (0.5 mol L^{-1})]. Detection was by UV-visible light spectroscopy. An aqueous calibrant of $K_2Cr_2O_7$ was used.

Anion exchange/ETAAS

The leaching was carried out with 10 mL NaOH (0.5 mol L^{-1})–Na_2CO_3 (0.3 mol L^{-1}) buffer with agitation in a heated ultrasonic bath (50 °C or 70 °C) for 30 min. The leachate was diluted with HCO_3^-/H_2CO_3 at pH 6.4, followed by centrifugation. Extraction is performed with liquid anion exchange solution (Amberlite LA 2/MIBK). Detection was by ETAAS. Calibration was carried out using the NIST SRM 2109.

FAAS

The leaching was carried out with 2.5 mL of NaOH–Na_2CO_3 buffer for 2 h at room temperature with occasional shaking. The solution was filtered over a cellulose acetate filter (0.2 μm); the residue was washed with H_2O_2 and 5 mL HNO_3 (2 mol L^{-1}) were added, followed by a dilution with HNO_3. Detection was by FAAS, using $K_2Cr_2O_7$ as calibrant.

ICP-MS

The filter was leached with 5 mL of NaOH–Na_2CO_3 buffer solution; samples were heated to near boiling point for 45 min, and diluted with H_2O. Detection was by ICP-MS, using a commercial solution of Cr as calibrant (checked against NIST SRM 3112).

Anion Exchange/IDMS

The leaching was carried out with NaOH–Na_2CO_3 buffer with agitation in a heated ultrasonic bath (70 °C) for 70 min, followed by dilution with HCO_3^-/

H_2CO_3 at pH 6.4. A known amount of $^{53}Cr(III)$–$^{53}Cr(VI)$ enriched spike was added to the leachate. Extraction was carried out with liquid anion exchange solution (Amberlite LA 2/MIBK), followed by back-extraction with 1.5 mL of ammonia solution (6%). The aqueous ammonia phase was washed with *n*-hexane, followed by electrodeposition on Pt wire in ammonia solution (6%) at a voltage of 2.45 V. The detection was by IDMS (thermal ionization of masses 52 and 53). The calibrant used was $Na_2Cr_2O_7$.

First Interlaboratory Study

An intercomparison on a European scale, in which 24 laboratories participated, was organized in 1994. It was designed to identify pitfalls and sources of error and consequently to improve the skills of many laboratories to such a degree that certification of a reference material became feasible. The intercomparison would also prove the suitability of the materials produced. Two hundred bottles of two different water samples were prepared according to the procedures developed by Dyg *et al.* [146,147]. The "A" water sample was representative for tap or natural water, whereas the "B" sample was intended to simulate a filter leaching solution in which much higher Cr(VI) concentrations were encountered. The B solution should help to detect analytical problems that are not connected with the leaching of the filters from monitors of stainless steel welders. The concentration range and stabilizing matrix of both materials are indicated below:

Sample A: concentration range of Cr(III) and Cr(VI): 10–40 mg L^{-1}
in a HCO_3^-/H_2CO_3 buffer solution, pH 6.4, under a CO_2 blanket
Sample B: concentration of Cr(VI): 5–10 mg L^{-1}
in a CO_3^{2-}/HCO_3^- buffer solution, pH 9.6

The results of the intercomparison were scrutinized on scientific grounds during a meeting with all the participants. They are discussed below and described in more detail elsewhere [148]. A summary is given in Table 9.1.

Technical Evaluation

Solution A, Cr(VI)

High values were suspected to be due to the high blank value of the Amberlite LA-2 used for the separation of the Cr species or inadequate separation at this low concentration leading to interference by Cr(III). Conversely, low values for Cr(VI) were related to possible incomplete extraction, using Amberlite which was supported by a too high Cr(III) value. One laboratory experienced similar difficulties with home-made reference materials. A reasonable explanation could be that the Amberlite used was not acidic enough. All of the above mentioned results were excluded. The 15 remaining results were used to calculate the mean of mean values, (14.07 ± 1.12) μg L^{-1}, which was in very good agreement with the target value of 14.11 μg L^{-1}.

Table 9.1 *Results of the intercomparison for Cr speciation*

Cr species	*Mean of means and SD*	*Initial content made up*
Cr(VI) in A solution	14.07 ± 1.12 μg L^{-1}	14.11 μg L^{-1}
Cr(III) in A solution	22.24 ± 3.01 μg L^{-1}	25.14 μg L^{-1}
Total Cr in A solution	39.55 ± 2.88 μg L^{-1}	39.25 μg L^{-1}
Cr(VI) in B solution	5.64 ± 0.40 mg L^{-1}	5.644 mg L^{-1}
Total Cr in B solution	5.62 ± 0.26 mg L^{-1}	5.644 mg L^{-1}
Cr(VI) in filters	34.5 ± 1.5 mg per kg dust	

Solution A, Cr(III)

The overall picture revealed that most laboratories had difficulties in determining the Cr(III) content of the A solution. The reason was mainly due to a too high detection limit. Only a few laboratories hit the target value of 25.14 μg L^{-1}. Higher Cr(III) values could be explained by incomplete extraction of Cr(VI) with Amberlite as already mentioned above. One laboratory reported a contamination problem. Another laboratory submitted results for Cr(III) determined by ETAAS after its complexation with quinolin-8-ol and extraction into MIBK; they had problems with sputtering of the sample due to evaporation of CO_2 from the carbonate buffer when heating the sample for the complexation of Cr(III), which explained the rather low results. The mean of means calculated with the five results taken into consideration became (22.24 ± 3.01) μg L^{-1}, which was notably lower than the target value.

Solution B, Cr(VI)

The overall picture of the results was satisfying. Only two out of the 24 laboratories that sent in results had to be withdrawn. The mean of means, (5.64 ± 0.40) mg L^{-1} calculated from 22 results, was in perfect agreement with the target value of 5.644 mg L^{-1}.

The outcome of this intercomparison was promising enough to justify a certification exercise on a material identical to the A solution but with slightly different concentrations.

Filter

One of the parameters that might influence the results for the Cr(VI) content in welding dust is the leaching method. Leaching at low pH could cause reduction of Cr(VI) due to Fe(II) or organic substances leached as well. Leaching at high pH in the presence of oxygen, on the other hand, might lead to oxidation of Cr(III), yielding too high results for Cr(VI). Most of the laboratories used the NIOSH method (2% NaOH–3% Na_2CO_3) [153]. Laboratory 10 used a $NaHCO_3$ buffer at pH 6.4, laboratory 24 made use of a NaOAc buffer at pH 4, laboratory 25 used their pyridinecarboxylic acid buffer while laboratories 3 and 14 used H_2SO_4. The latter two, as well as laboratory 2, found Cr(VI) values in the filters that exceeded the instrumental neutron activation analysis results for total Cr [(39.3 ± 1.3) mg Cr per kg dust] obtained by the coordinating laboratory. The results submitted by laboratories 2, 3 and 14 were consequently omitted. The calculated mean of means resulting from 12 laboratories was (34.5 ± 1.5) mg Cr(VI) per kg dust.

The outcome of this intercomparison was promising enough to justify a certification exercise on identical filters with similar concentrations.

9.5 Certification of the Lyophilized Solution

Preparation of the Candidate Reference Material

A batch of 1100 vials containing a lyophilized solution, similar to the previous "A sample" was prepared. The preparation of the samples was carried out under class 100 clean room facilities. All laboratory ware (pipettes, volumetric flasks, *etc.*) was thoroughly cleaned by leaching in dilute acids.

1.009 g $Na_2CrO_4.4H_2O$ and 0.7067 g $CrCl_3.6H_2O$ (analytical grade) were weighed and dissolved in 100 mL Milli-Q water or 100 mL HCl (subboiled, 0.1 mol L^{-1}) respectively for the initial stock solutions. They were diluted 10 times, and 5 mL/10 mL of the Cr(VI)/Cr(III) stock solution respectively were transferred to a 5 L volumetric flask. 21 g $NaHCO_3$ (p.a. grade) and 124 mL HCl (subboiled, 1 mol L^{-1}) were then added to obtain the HCO_3^-/H_2CO_3 buffer solution, pH 6.4. The whole was made up to volume with Milli-Q water and kept under a CO_2 blanket.

2 mL of the solution were pipetted into cleaned brown-glass vials and weighed. The concentration of Cr(VI) and Cr(III) was calculated to be 22.42 µg L^{-1} and 27.58 µg L^{-1}, respectively, after reconstitution in 20 mL of buffer. The samples were deep frozen for at least 12 h prior to lyophilization. The lyophilization was done under carefully controlled conditions to avoid losses of material due to sputtering, reduction of Cr(VI) or adsorption on the wall. Owing to the limited volume of the freeze-drying equipment, the lyophilization had to be done in seven batches. The homogeneity test revealed no systematic differences between the different batches.

When the freeze drying was completed, the equipment was purged with pure N_2 and the bottles were sealed with Butyl stoppers with a Teflon coating, and an alumina cap. Purging with N_2 ensures the samples are stored under an inert atmosphere, thus preventing possible reduction of Cr(VI). The samples were further stored at 5 °C.

Homogeneity Control

To check the between-bottle homogeneity, 50 vials were randomly set aside after lyophilization. They were taken from the seven lyophilization batches.

The lyophilized samples were reconstituted with 20 mL of HCO_3^-/H_2CO_3 buffer, pH 6.4. The separation of Cr species took place immediately after

Table 9.2 *Comparison of the CV of measurements between bottles for CRM 544 with the CV of the method, together with their uncertainty*

Identification	*CV between vials*[a] $\pm U_{CV}$	*CV of method*[b] $\pm U_{CV}$
Cr(VI)	5.74 ± 0.58	2.98 ± 0.94
Cr(III)	3.30 ± 0.33	0.45 ± 0.14
Total Cr	3.20 ± 0.32	1.49 ± 0.47

[a] Single determination on the content of 50 vials.
[b] Five replicate determinations on one reconstituted sample.

reconstitution of the sample. The separation method was based on the cationic behaviour of Cr(III) and the anionic behaviour of Cr(VI). It relies on an ion exchange extraction using the liquid anion exchanger Amberlite LA-2. The liquid anion exchange solution (LAES) was obtained by stripping the Amberlite in HCl (6 mol L^{-1}) and diluting it in methyl isobutyl ketone (MIBK) (2:1:2). Cr(VI) was extracted completely into the organic phase while Cr(III) remained in the aqueous phase.

2 mL of the reconstituted samples and 1 mL of LAES were mixed on a whirling mixer for 1 min and centrifuged at 2500 rpm for 10 min. The two phases were separated and analysed for their Cr content by ETAAS (Perkin Elmer 4100 ZL). Cr(VI) was measured in the organic phase and Cr(III), after two-fold dilution, in the aqueous phase. For the determination of the total Cr content the reconstituted sample was diluted four times. The samples were measured against matrix-matched calibration curves.

The analyses were performed on different days to take into account the day-to-day variability of the technique used. Care was taken to perform the measurements in the most repeatable way. The methodological uncertainty was determined by five replicate analyses of one reconstituted sample.

The *CV*'s for the different species in the lyophilized solutions are presented together with their uncertainty in Table 9.2. An *F*-test was applied on the results. The test did not reveal any systematic difference at the 0.05 signifiance level between the *CV* of the results from the 50 bottles and the *CV* of the method for Cr(VI) and total Cr. For Cr(III), no difference could be observed at the 0.01 significance level. It could therefore be concluded that the samples were homogeneously distributed over the different vials.

Stability Control

The stability of the material was tested at +5 °C and +20 °C over a period of 12 months. The Cr(VI), Cr(III) and total Cr contents were determined at the beginning of the storage period, after 1 week, and after 1, 6 and 12 months. The samples were analysed using the same procedures as for the homogeneity study. Cr(III), Cr(VI) and total Cr were each determined in triplicate (one replicate analysis on each of three bottles stored at both temperatures) at each occasion of analysis. The evaluation was carried out, following the procedure described in Chapter 3 (using the samples stored at +5 °C as comparator for

Table 9.3 *Normalized results of the stability study for storage of CRM 544 at +5 °C and +20 °C*

Species	*Time*	*+20 °C/+5 °C*	
		Value of R_t	*U_t*
Cr(VI)	1 week	0.9809	0.0464
	1 month	0.9973	0.0117
	6 months	0.9995	0.0357
	12 months	0.9348	0.0433
Cr(III)	1 week	0.9878	0.0217
	1 month	0.9580	0.0779
	6 months	1.0202	0.0250
	12 months	1.0208	0.0809
Total Cr	1 week	0.9845	0.0214
	1 month	1.0117	0.0448
	6 months	0.9963	0.0187
	12 months	0.9752	0.0116

those stored at higher temperatures). As shown in Table 9.3, no instability could be detected for Cr(III) and Cr(VI) at any of the temperatures tested [149].

Technical Evaluation

One participant reported his findings when analysing the lyophilized solutions using a Dionex AG4 A column for the separation of the species, followed by detection with chemiluminescence [154]. According to his results, Cr(III) in the lyophilized samples is not present as a simple aquated Cr^{3+} ion (purple colour) but rather as a $[Cr(H_2O)_4Cl_2]^+$ ion (dark green colour), which is the main species in the commercially available analytical quality chromic chloride with empirical formula $CrCl_3.6H_2O$ [155]. When dissolved in water this species will convert only very slowly to Cr^{3+}. This does not influence the results when using non-specific ion chromatography columns (*e.g.* alumina), but might cause problems when using a more specific column producing multiple peaks or peaks with varying retention times. The very low results from one laboratory using chemiluminescence could be explained by the presence of different Cr(III)–Cl complexes instead of Cr(III) as such.

Some of the laboratories reconstituted the lyophilized samples in H_2O instead of the HCO_3^-/H_2CO_3 buffer solution. This does not influence the results as such, but enhances the risk for reduction of Cr(VI). Reconstitution in HCO_3^-/H_2CO_3 buffer is therefore recommended.

One laboratory reported results obtained after storing the reconstituted samples in the dark under a CO_2 blanket for a few weeks. The Cr(VI) results were much lower while the Cr(III) results were much higher than the ones obtained directly after reconstitution and they did not agree with the rest of the laboratories. Clearly a reduction of Cr(VI) had occurred in the reconstituted sample during storage. It is therefore recommended to start the separation of the species as soon as possible after reconstitution.

Table 9.4 *Certified values for Cr(III), Cr(VI) and total Cr in BCR CRM 544, expressed in* $\mu g\ L^{-1}$

Component	*Certified value*	*Uncertainty*	*p*
Cr(VI)	22.8	1.0	13
Cr(III)	26.8	1.0	9
Total Cr	49.4	0.9	14

Certified Values

The certified values (unweighted mean of *p* accepted sets of results) and their uncertainties (half-width of the 95% confidence intervals) are given in Table 9.4; the bar graphs are shown in Figures 9.1–9.3. The use of the material is fully described in the certification report [149].

BAR-GRAPHS FOR LABORATORY MEANS AND 95% CI

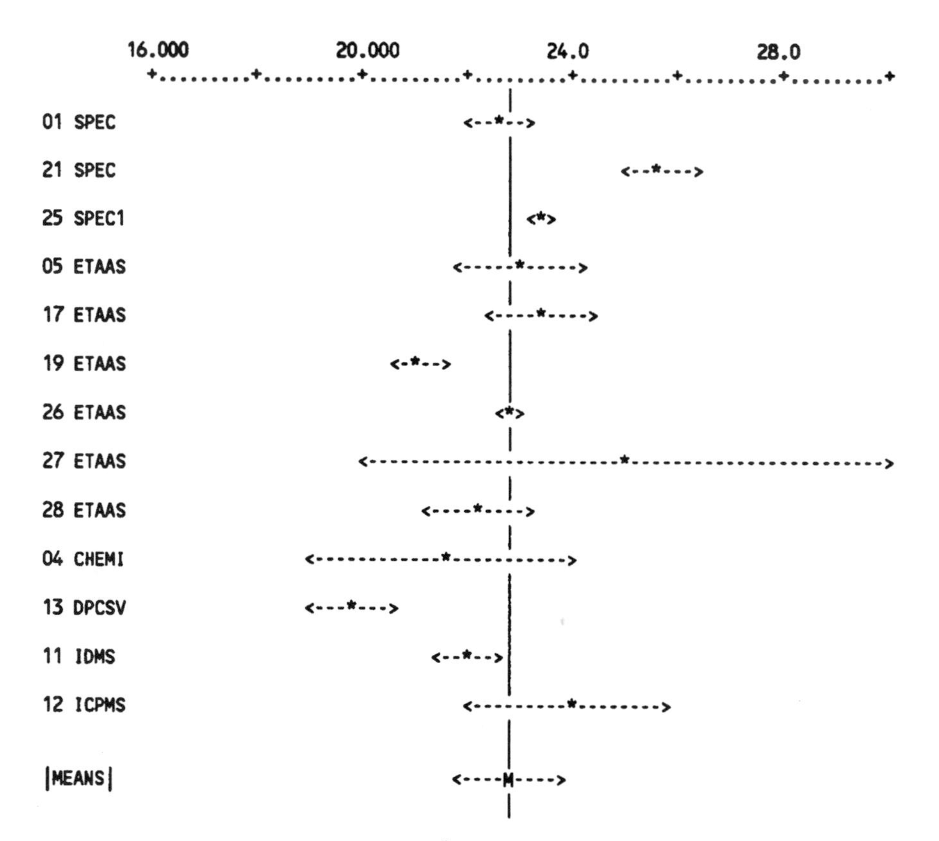

Figure 9.1 *Cr(VI) in CRM 544* ($\mu g\ L^{-1}$)

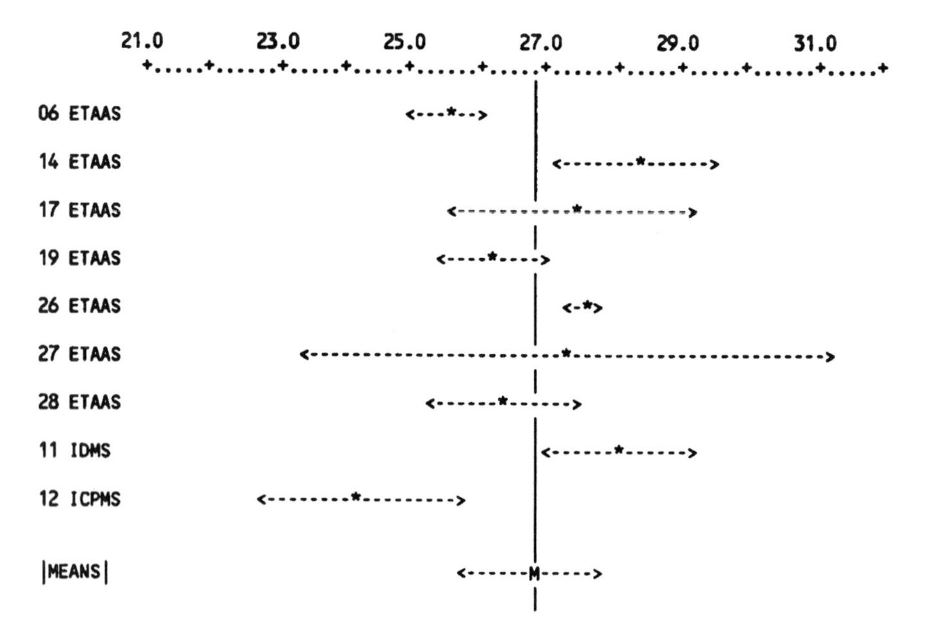

Figure 9.2 *Cr(III) in CRM 544 (µg L^{-1})*

9.6 Certification of Welding Dust

Preparation of the Candidate CRM

A batch of 1100 filters, loaded with welding dust containing approximately 100 µg of Cr(VI), was produced. The procedure described by Dyg *et al.* [146] was followed in general.

Prior to dust collection, the filters (borosilicate microfibre glass discs without resin binder) were acclimatized in an air-conditioned laboratory (22 °C and 50% relative humidity) for 24 h, and weighed carefully together with the two PTFE rings intended to avoid losses of glass fibre on the monitor when pressed together.

The monitors were assembled and mounted in the Sputnic air sampling unit. The Sputnic air sampler consists of a round vacuum chamber of stainless steel in which 100 critical orifices (Ø 0.4 mm) are placed. They are positioned in four rings containing 36, 29, 22 and 13 orifices, respectively. Two vacuum pumps secure a constant vacuum of 0.66 bar in the pressure chamber, which in turn secures a homogeneous airflow of 1.9 L min^{-1} through each orifice and filter. A pilot study demonstrated that the airflow through the inlet of the monitor mounted on the critical orifice was not disturbed nor changed either by different types of filters placed in the monitor or by the number of filters mounted on the manifold or critical orifice plugged by a rubber tube.

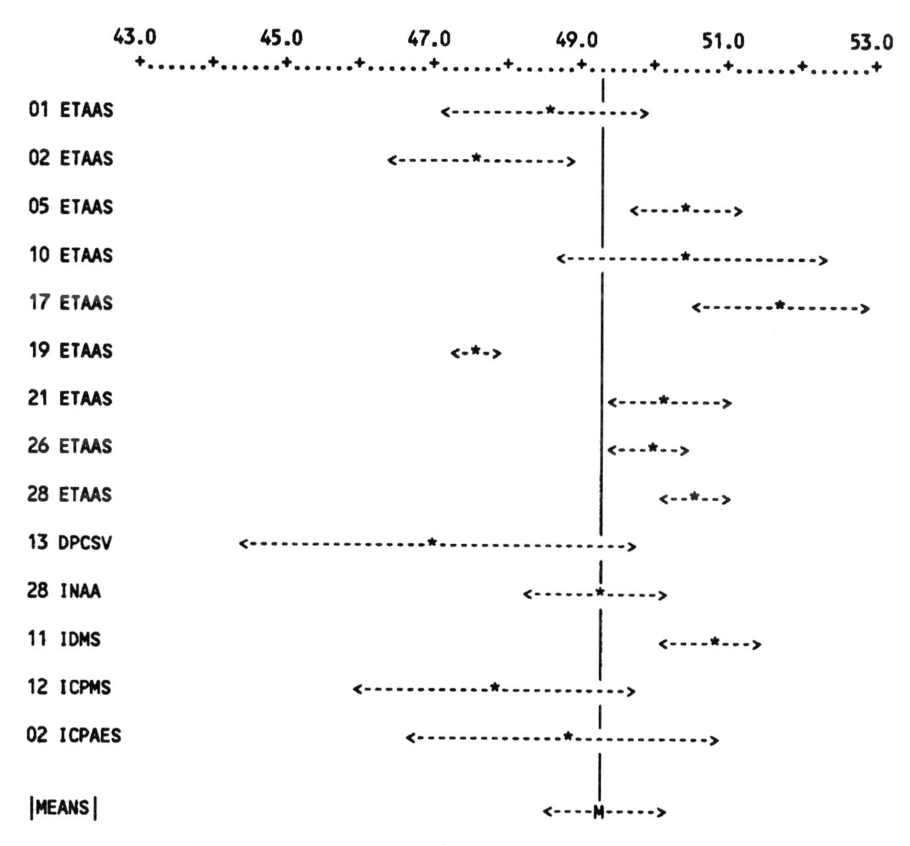

Figure 9.3 *Total Cr in CRM 544 ($\mu g\ L^{-1}$)*

Loading was done with welding fume dust originating from a manual metal arc computerized welding system, using an electrode which produces a welding fume dust containing approximately 4.3% Cr(VI). Fume analysis by the manufacturer showed the following composition: Fe 6.5%; Mn 3.1%; F 16.5%; Pb 0.06%; Cu 0.02%; Ni 0.66%; Cr(III) 1.2%; and Cr(VI) 4.3%. Loading was done in 12 batches of 100 filters each. All batches were made on the same day by the same operator to guarantee reproducibility. To avoid contamination, the welding was performed in a special fume box.

After charging, the Sputnic was flushed with Ar for 5 min while the pumps were still on. A second weighing of the filters took place following 24 h acclimatization in the air-conditioned laboratory. The monitors were re-assembled and flushed with N_2 to enure storage of the filters in an inert atmosphere. They were finally sealed with the end caps. The filters were stored in the dark at room temperature.

Table 9.5 *Comparison of the CV of measurements between filters with the CV of the method, and their uncertainty (CRM 545)*

Identification	*CV between filters*[a]	*CV of method*[b] $\pm U_{CV}$
Cr(VI)	4.98 ± 0.56	4.82 ± 1.52

[a] Single determination on the content of 40 filters.
[b] Five replicate determinations on one filter leachate.

Homogeneity Control

To check the between-filter homogeneity, 40 filters were randomly set aside after loading them with welding dust. They were taken from the different batches.

The filter and PTFE rings were placed in a 25 mL beaker with the dust-loaded side facing upwards. The filter was covered with 10 mL of an alkaline buffer (2% NaOH–3% Na_2CO_3) and a small conical PTFE ring was placed on top to secure the filter beneath the surface of the liquid. This assembly was covered with laboratory film and subjected to agitation in a heated (70 °C) ultrasonic bath for 30 min. Before separation and analysis the supernatant was diluted 1000 times with a HCO_3^-/H_2CO_3 buffer (pH 6.4) to reach the linear range in ETAAS.

Separation of the Cr species took place immediately after leaching of the sample. The separation method was based on the cationic behaviour of Cr(III) and the anionic behaviour of Cr(VI) species. It relied on ion exchange extraction using the liquid anion exchanger Amberlite LA-2. The liquid anion exchange solution (LAES) was obtained by stripping the Amberlite in HCl (6 mol L^{-1}) and diluting it in methyl isobutyl ketone (MIBK) (2:1:2). Cr(VI) is extracted completely into the organic phase while Cr(III) remains in the aqueous phase.

1 mL of the reconstituted samples and 1 mL of LAES were mixed on a whirling mixer for 1 min and centrifuged at 2500 rpm for 10 min. The two phases were separated and the organic phase was analysed for its Cr(VI) content by ETAAS.

To check the selectivity of the separation method, three blank filters were spiked with 100 μg Cr(III) and subjected to the analytical procedure. The Cr content in the aqueous fraction after extraction was always below the detection limit, indicating complete precipitation of Cr(III) in the initial alkaline leaching.

The analyses were done on different days to take into account the day-to-day variability of the technique used. Care was taken to perform the measurements in the most repeatable way. The methodological uncertainty was determined by five replicate analyses of one reconstituted sample.

The *CV* for Cr(VI) in the filters and the *CV* for the analytical method are presented together with their uncertainty in Table 9.5. An *F*-test was applied on the results. The test did not reveal any systematic difference at the 0.05

Table 9.6 *Normalized results of the stability study for storage at +5 °C and +20 °C*

Species	*Time*	*+20 °C/+5 °C*	
		Value of R_t	U_t
Cr(VI)	1 week	0.9497	0.0758
	1 month	1.0203	0.0614
	6 months	0.9525	0.0499
	12 months	0.9771	0.0264
Total soluble Cr	1 week	0.9951	0.0740
	1 month	1.0603	0.0740
	6 months	0.9651	0.0396
	12 months	–	–

significance level. It could therefore be concluded that Cr(VI) is distributed homogeneously over the different filters.

Stability Control

The stability of the content of Cr(VI) on the filters was tested at +5 °C and +20 °C over a period of 12 months, and the Cr(VI) and total soluble Cr contents were determined at the beginning of the storage period and after 1 week and after, 1, 6 and 12 months. Samples were analysed using the same procedure as for the homogeneity study. Cr(VI) and total soluble Cr were each determined in triplicate (one replicate analysis in each of three filters stored at different temperatures) at each occasion of analysis. The evaluation was similar to the one used for CRM 544, using a pooled leachate as reference for monitoring the stability.

The results are given in Table 9.6. On the basis of the results, it was concluded that no instability could be demonstrated.

Technical Discussion

Two laboratories using, respectively, SPEC and ETAAS as final determination, leached the filters with H_2SO_4; since leaching at low pH can cause reduction of Cr(VI) by Fe(II) or organic substances, those results were suspected and were withdrawn.

The total Cr content of the welding dust, including metallic Cr, Cr(III), *etc.*, was determined by instrumental neutron activation analysis (INAA) and was (49.57 ± 2.79) mg kg^{-1} dust.

Certified Values

The certified values (unweighted mean of *p* accepted sets of results) and their uncertainties (half-width of the 95% confidence intervals) are given in Table 9.7. The fact that the total leachable Cr content of the welding dust is

Table 9.7 *Certified mass fractions of Cr(VI) and total leachable Cr in CRM 544 (mg per kg welding dust)*

Component	*Certified value*	*Uncertainty*	*p*
Cr(VI)	40.16	0.56	11
Total soluble Cr	39.47	1.29	7

BAR-GRAPHS FOR LABORATORY MEANS AND 95% CI

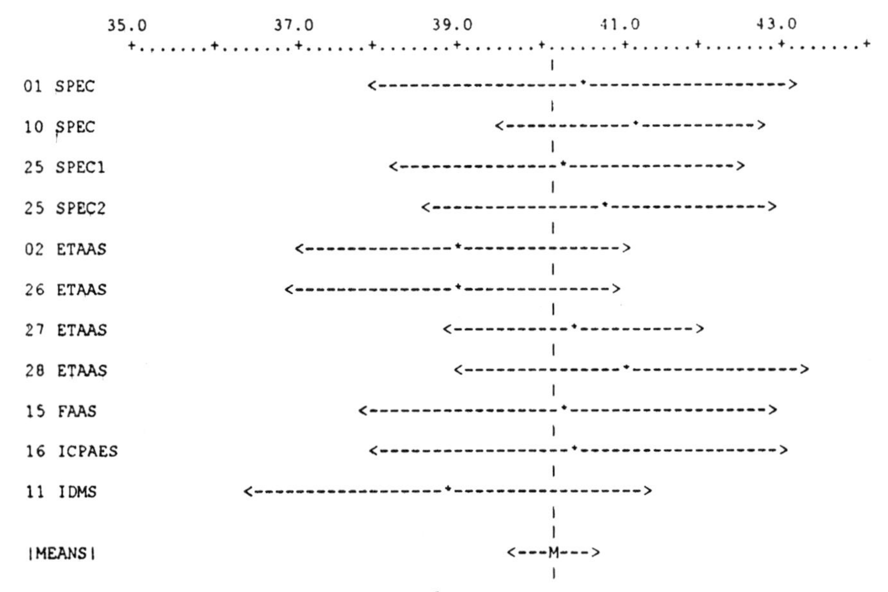

Figure 9.4 *Cr(VI) in CRM 545 (mg kg^{-1} dust)*

BAR-GRAPHS FOR LABORATORY MEANS AND 95% CI

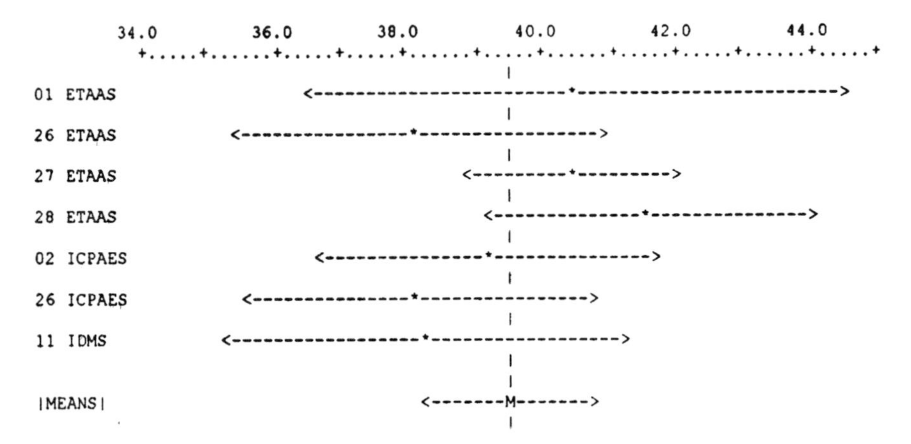

Figure 9.5 *Total leachable Cr in CRM 545 (mg kg^{-1} dust)*

somewhat lower than the Cr(VI) content is only due to the uncertainty on the results used for calculation of the certified values. The results are also shown in Figures 9.4 and 9.5.

Recommendations for the use of this CRM are given in the certification report [150].

CHAPTER 10

Aluminium Speciation

10.1 Aim of the Project and Coordination

Justification

The potential toxicity of aluminium has become a major medical and environmental issue. For a long time, aluminium was not considered to be toxic to human beings, but this attitude changed in the 1970's when aluminium was first associated with dialysis dementia syndrome [156,157]. Since then, comparatively high aluminium levels in body tissues have been implicated in various clinical disorders suffered by patients with chronic renal failure undergoing regular haemodialysis, where, for various reasons, water and dialysis fluids high in aluminium were used, *e.g.* epidemiological studies seem to demonstrate a positive link between the occurrence of Alzheimer's disease and mean aluminium levels in drinking water [158].

From an environmental standpoint, the recognition that "acid rain" mobilizes aluminium from poorly buffered soils into the aquatic environment has increased the awareness and concern about aluminium toxicity to aquatic organisms. The acidification of fresh water lakes and rivers in the USA, Canada and particularly the Scandinavian countries, and the subsequent rise in dissolved aluminium levels, has been linked to the decline in fish numbers and, in some cases, to the total elimination of entire fish populations [159,160]. In aquatic systems, it has been demonstrated that the $Al(OH)_2^+$ species seems to be most toxic to fish [159]. Other workers [161], using thermodynamic calculations in conjunction with fish toxicological experiments, have pointed to the $Al(OH)^{2+}$ species as also toxic to fish.

As with many elements, total aluminium determinations are of limited value for environmental toxicological purposes, owing to the different toxicology/bioavailability patterns related to the various Al species, which justifies Al speciation measurements.

This section is based on a project carried out under the auspices of the Measurements and Testing programme; it summarizes a report established by the University of Oviedo which describes the most frequently used methods and techniques for aluminium speciation analysis in water [156].

Programme and Coordination

The project was started in 1989 by a literature survey [156] and continued in 1990 by an interlaboratory study; it was coordinated by the Universidad de Oviedo, Departamento de Química Física y Analítica (Spain). The Universidad de Córdoba, Departamento de Química Analítica (Spain) and the University of Plymouth, Department of Environmental Sciences (United Kingdom) participated in the intercomparison.

10.2 Techniques Used in Aluminium Speciation

A number of fractionation procedures have been developed over the last 15 years in order to distinguish between the various aqueous forms of aluminium. These methods include dialysis, ion exchange (both batch and column), HPLC, F^- ion-selective electrode, NMR, species specific extractions, filtration and computational techniques. All of these procedures usually measure operationally defined aluminium fractions (*i.e.* groups of aluminium species are measured, rather than a single species, and the values obtained depend upon the precise procedure used for the analysis). Therefore, present speciation procedures may measure slightly different forms of aluminium depending on the conditions, and thus they should accordingly provide different results.

8-Hydroxyquinoline Extraction Procedures

Various methods have evolved based on the complexation of monomeric aluminium species with 8-hydroxyquinoline (oxine) and the rapid extraction of the formed $Al(oxinate)_3$ complex into organic solvents, *e.g.* chloroform [162], toluene [163] or methyl isobutyl ketone (MIBK) [164]; detection of the aluminium content was accomplished by electrothermal atomic absorption spectrometry.

Driscoll Methods

The so-called Driscoll methods are all based around ion-exchange separation of the monomeric inorganic aluminium from the monomeric organic aluminium. The column packing is Amberlite IR-120 cation exchange resin [159]. In essence, the Driscoll method directly measures three operationally defined fractions:

- Acid reactive aluminium (Al_r)
- Total monomeric aluminium (Al_{tm})
- Non-labile monomeric aluminium (Al_{nl})

The first published methods used a colorimetric 8-hydroxy-7-iodoquinoline

5-sulfonic acid (ferron)/ *o*-phenanthroline method for the complexation of the aluminium in the desired fractions [159,165]. The Al_r fraction was determined after acidification of the sample for 30 min before the final colorimetric determination. The Al_{tm} was determined through direct complexation with ferron, without acidification. The Al_{nl} (or monomeric organic) fraction was taken as the fraction determined by ferron without acidification, after the sample had been passed through the ion-exchange column. From these three experimentally obtained fractions, two more could be calculated: "acid soluble" aluminium is taken as the difference between Al_r and Al_{tm}; the "labile monomeric" aluminium fraction, Al_{lm}, considered the more toxic one, is taken as the difference between the Al_{tm} and Al_{nl} fractions. In this fractionation scheme the Al_{lm} fraction is thought to include free hydrated Al^{3+} along with its possible fluoride, sulfate and hydroxy complexes. Modifications of the Driscoll method include the use of the Barnes oxine/MIBK extraction [164], and probably the most popular variation in use at the present time, the technique using pyrocatechol violet (PCV) detection of the "Driscoll" aluminium fractions [166,167].

From the data available, there is no consensus as to the precise experimental conditions to be used. This is especially true where the analytical separation is concerned. The resin in the column has to be conditioned with a solution of NaCl having the same ionic strength and pH as the water samples to be analysed and this is one of the main drawbacks of the method because the solution of NaCl has to be passed through the column until the pH of the effluent does not change by more than ± 0.2 pH units. A full Driscoll/PCV speciation procedure including all the modifications has been published elsewhere [168].

Some authors claim that detection limits (DLs) of 5 mg L^{-1} are easily obtainable [167] using the batch PCV detection method. However, it has been reported that daily DLs range between 5 and 10 mg L^{-1} and that personnel who are highly acquainted with the method were needed to achieve DLs down to <10 mg L^{-1} [168]. This poses a problem when the non-labile or "monomeric organic fraction" of aluminium to be measured is very small. Errors for Al_{nl} can become high, and, because the Driscoll method is a subtraction method, the uncertainty of the calculated Al_{lm} fraction ($Al_{tm} - Al_{nl}$) can become unacceptable. To increase the precision, sample throughput and practicality of the Driscoll method, several flow injection analysis (FIA) systems have been designed both for the PCV method [169,170] and for detection with a fluorimetric determination of the 8-hydroxyquinoline-5-sulfonic acid (8-HQS) aluminium complex in a micellar medium (cetyltrimethylammonium bromide, CTAB) [168,171].

One important concern about the Driscoll method is which species are actually retained by the column: that is, the true chemical nature of the Al_{lm} fraction, considered the most toxic. Some authors have shown that the retention characteristics observed are sample flow dependent [172], and that each column needs to be individually characterized for aluminium retention.

Table 10.1 *Aluminium fractionation scheme according to the batch Chelex-100 method*[a]

Operational definition	*Functional definition*
Rapidly exchangeable (1 h reaction)	Monomeric and small polymeric cationically charged species
Moderately fast exchangeable (4 h reaction)	Mainly inorganic aluminium complexes
Slowly exchangeable (24 h reaction)	Mainly organically complexed forms of aluminium
Non-exchangeable	Strong alumino-organic complexes, colloidal or crystalline forms

Chelex-100 Based Methods

Several procedures using Chelex-100 (a styrene/divinylbenzene co-polymer with iminodiacetate functional groups) have also been developed. For batch systems, where conditioned resin is stirred with the water sample for up to 4 h, the "labile" aluminium fraction is calculated from the difference between the total aluminium concentration in the sample found before and after ion-exchange has taken place. The resin has to be very carefully preconditioned in a nitric acidified solution (pH 5.0) of $Ca(NO_3)_2$ and $MgSO_4$ (at concentrations normally found in natural waters) [173,174].

The main problem with using this type of technique for the determination of the toxic aluminium species in water is that it groups together the monomeric hydroxyaluminium, fluoroaluminium and low molecular weight (LMW) polymeric species in the same fraction (> 85% of these species exchange under 30 min [174]). Moreover, the operational definitions do not match adequately those of the more popular "Driscoll" methods (Table 10.1).

Further developments using Chelex-100 include a most recent, exhaustive, but rather complicated fractionation scheme, using a batch (unconditioned) Chelex-100 procedure combined with ultrafiltration [175]. Another recent and most interesting method, using a Chelex-100 column procedure, claims to be able to elute the AlF^{2+} species selectively from the other retained "labile" monomeric species using different strength eluents containing HCl [176]. However, the separation of the AlF^{2+} species from the other inorganic monomeric aluminium species was not quantitative.

Fluoride Electrode Methods

These are indirect methods based around computational procedures, using estimates of the monomeric inorganic aluminium fraction and "free" and "total" fluoride concentrations. Normally, "free" fluoride is determined by direct measurement with an ion-selective electrode. The "total" fluoride concentration is determined using the same ion-selective electrode technique, but after the addition of a total ionic strength buffer (TISAB). LaZerte [177] has reviewed the main fluoride electrode methods for aluminium speciation. When an estimate of the inorganic monomeric aluminium is available, "free" fluoride analytical data can be used to calculate Al^{3+} activities and the proportions of the other monomeric aluminium species. However, the concentrations of other possible binding ligands existing in the water have to be known as well. Driscoll [165] has described a method with fluoride electrodes which eliminates the necessity of direct measurement of the inorganic monomeric aluminium fraction. However, these methods should only be employed for water with a high aluminium content and a pH <5.5, as they need a large amount of inorganic monomeric aluminium to ensure that $AlF_x^{(3-x)+}$ complexes make up a large proportion of the total fluoride present in the sample. Finally, "free" fluoride measurements in natural water can be very slow and prone to error, particularly in the presence of other fluoride complexing cations (*e.g.* Fe^{3+}).

HPLC Methods

In principle, HPLC, with its more powerful separation abilities, should be a most useful speciation tool. Bertsch and Anderson [172] have described a system using a Dionex CG3 column with a NH_4Cl solution as the mobile phase at a pH range of 2.0–4.2. Aluminium was detected by post-column derivatization with 4,5-dihydroxybenzenedisulfonic acid (Tiron) at pH 6.3. They managed to separate the $Al(H_2O)_6{}^{3+}$ from two tentatively identified aluminium fluoride species (AlF^{2+} and $AlF_2{}^{+}$), and to quantify the monomeric aluminium fraction. They also reported some disadvantages regarding eluent ionic strengths and that the pH seriously influenced the observed speciation of aluminium in the presence of F^- and citrate ligands. Jones [178] has reported the use of a Dionex CG2 guard column maintained at 50 °C in a water bath for aluminium speciation. He used a potassium sulfate (pH 3.0) mobile phase with post-column derivatization of the aluminium with 8-HQS at pH 4.1. Fluorescence was measured at 512 nm (excitation at 360 nm) [179]. He was able to quantify the $Al(OH)_x^{(3-x)+}$ fraction with this system, excluding the fluoride and organically bound aluminium species.

The major point of these HPLC methods is that they separate the aluminium fluoride species from the other inorganic monomeric aluminium species (classic "Driscoll" methods do not).

^{27}Al NMR Studies

NMR techniques have been utilized by Bertsch *et al.* [180] for the determination of Al^{3+} concentrations and for other $Al(OH)_x^{(3-x)+}$ species. Problems with maintaining identical instrumental conditions, differences between acidified and unacidified solutions and low sensitivity are the main drawbacks. However, good agreement with values obtained by the HPLC technique for the uncoupled $Al(H_2O)_6{}^{3+}$ species were obtained [172]. The authors concluded that their method may be particularly useful for aluminium speciation in

Table 10.2 *Summary of some published intermethod comparison exercises for aluminium speciation*[a]

Techniques	*Comments*
Computational F^- *vs.* Oxine/dialysis	At pH < 5.5, $R^2 = 0.82$ for inorganic monomeric aluminium
Dialysis *vs.* Driscoll/PCV	For inorganic monomeric aluminium $R^2 = 0.99$. Dialysis underestimates PCV values by 6%.
Driscoll/Ferron *vs.* Driscoll/PCV	Al_r gave good and Al_{tm} and Al_{nl} gave satisfactory agreement. PCV method generally better
Driscoll/oxine/ MIBK extraction *vs.* Driscoll/PCV	Systematic differences between Al_{tm} and Al_{nl} fractions. However, good agreement for Al_{lm}, $R^2 = 0.99$.
Driscoll/PCV *vs.* three alternative methods	Al_r and Al_{lm} gave good agreement. Care has to be taken as one method (HPLC) did not include Al fluoride species in the Al_{lm} fraction

[a] Adapted from [156]

synthetic solutions, *i.e.* as utilized in kinetic and fish toxicological investigations. This has indeed been the case with reports of the use of ^{27}Al NMR to study the binding and stability of a variety of aluminium complexes such as acetate and oxalate as well as citrate [181,182].

10.3 Method Validation

Comparisons of different aluminium speciation methods are rather scarce. A selection of such exercises is given in Table 10.2. From these limited results it would seem that most of the aluminium speciation methods currently in use (particularly the Driscoll-based ones) uniformly assess the toxic aluminium fraction in acidic water. However, in all of these comparisons the compared methods were both performed in the same laboratory, and presumably by the same analysts. It is well known in inter-method comparison circles that this type of validation approach nearly always produces good results.

Some laboratories have tried either to standardize or validate different methodologies by interlaboratory comparison exercises [183]. However, little or no data from these programmes have found their way into the international scientific literature.

One interlaboratory method comparison exercise for aluminium speciation has been organized within the framework of the BCR project on Al speciation [184]. Results obtained using a standard Driscoll/PCV method for the determination of the "labile monomeric" aluminium fraction were tested and compared to three other aluminium speciation procedures. Each of the three participant laboratories carried out both the standard Driscoll method, used as

Table 10.3 *Results for a Interlaboratory method comparison exercise using a standardized batch Driscoll/PCV fractionation method against three other methods [184].*

	Batch Driscoll/PCV		*Alternative method*
	Al_r	Al_{lm}	Al_{lm}
Sample A			
Laboratory 1	58.1 ± 2.7	37.5 ± 2.5	46.7 ± 1.0
2	62.6 ± 2.6	46.7 ± 4.3	48.3 ± 5.0
3	57.1 ± 2.4	36.3 ± 4.2	52.1 ± 7.5
Sample B			
Laboratory 1	146 ± 5.4	84.4 ± 2.6	83.7 ± 6.1
2	102 ± 3.5	84.5 ± 2.9	62.1 ± 11[a]
3	113 ± 15	–	81.1 ± 9.4

[a] excludes Al fluoride species

a reference method, and its own method applied to the same water samples. For certain aluminium content in water (ranging from 100 to 200 mg L^{-1}), the standardized Driscoll/PCV method produced extremely good results (as summarized in Table 10.3), with pooled values for the Al_{lm} fraction having RSDs of 6–15%. It was concluded that for sufficiently stable aluminium levels in water the Driscoll/PCV method was portable, and could be used as the basis for a standardized aluminium speciation scheme. A programme to control each participating laboratory's ability to analyse trace levels of total aluminium in water samples was also recommended [184].

Briefly, although the results presented in Table 10.3 appear quite positive and promising, aluminium speciation data are still too scarce and procedures have yet to be rigorously tested for water samples in international interlaboratory method comparison exercises for adequate validation of the proposed aluminium speciation schemes and methods to be finally achieved.

10.4 Future Perspectives

Aluminium speciation in water samples has been shown to be virtually "operationally defined" at the present time. Final improvements in analytical techniques or in the quality control of such speciation determinations demand an answer to the basic question of what aluminium species do we need to analyse. A possible way forward would be standardization of the analysis for the aluminium species toxic to fish, *e.g.* $Al(OH)^{2+}$, as an indication of water quality. However, there is no evidence that other inorganic species, *e.g.* $Al(OH)_2{}^+$, $AlO_2{}^-$ or $AlCl^{2+}$, which quickly convert to each other, do not produce similar toxicity problems [185]. It is also wondered whether the analytical methods to distinguish and measure reliably only a given species, *e.g.* $Al(OH)^{2+}$, in real samples (water, biological fluids and tissues, *etc.*) are available at present. While these questions are not properly answered, perhaps it is better to use the term "speciation" in a not so rigorous sense, describing a group of species instead of only one [156]. The authors of the survey believe that "something is better than nothing", and therefore aluminium speciation analysis should be developed at least along two main paths: a) research on new bioanalytical techniques to answer the toxicity questions above, and b) refinement of the "operationally defined" present methods along with inter-laboratory method comparisons in order to obtain comparable results for the "toxic aluminium species" in water [156].

In order to pursue this latter objective effectively, at least two practical actions can be envisaged:

(i) Establishment of a standard method of analysis: this would offer many advantages. One of the most obvious would be in the comparison and validation of new speciation techniques (which could be checked against the standard method). Moreover, water quality, relative to aluminium toxicity to fish, could be internationalized by accepting the value of the Al_{lm} fraction provided by this standard method. However, it should be noted that more research is needed in order to establish finally such a "standard method". For example, if the Driscoll/PCV method was accepted as the standard, some problems such as the inclusion of the aluminium fluoride species in the toxic fraction would have to be looked at. Such a standard method of analysis would also have to include standardized sampling and sample storage protocols, orientated especially towards aluminium speciation analysis. Modest advances along these lines are probably the way forward, while basic progress is being made towards evaluation of individual species toxicity, their analytical separation and final reliable determination.

(ii) Availability of reference waters with certified aluminium fractions: several reference waters already exist with certified total elemental content and for a variety of matrices (*e.g.* artificial freshwater [186]). However, certified trace metal species content in such reference waters is virtually non-existent at the present time. In the case of aluminium, the availability of such an important quality control tool is hindered by some important problems in:

a) Certification: a clear definition of species or aluminium fractions on which to provide certification data is missing. In the case of aluminium it is obvious that at the present time the particular preconcentration/separation/detection methods chosen to provide the certification data will determine the aluminium species or fraction selected.

b) Formulation of unified sampling and sample storage procedures. These procedures or protocols are necessary if samples taken in the field are to be treated the same as any reference water in the laboratory, in order to preserve the integrity of the aluminium species present (also for long-term storage of possible reference materials). The question of filtration, or otherwise, and with what size filter, is a key point here. Maybe one approach would be the isolation of the desired species in the field (such techniques already exist for other elements [187]), mainly based around the retention of the desired species in mini-columns. The analysis of the retained element or species is carried out later in the laboratory. These procedures would solve the problems of preserving the integrity of the species during sample storage, as described in detail elsewhere [156].

c) Data on the main aluminium ligands present in any reference water. It is clear that such data, together with their concentrations, should be provided. This is specially important with regards to F^- as discussed earlier [156].

All of the problems outlined above indicate that the production of a "synthetic" reference water, with well defined total aluminium and Al_{lm} content, could prove most useful for aluminium toxicity control in natural waters before a certified natural water sample of well known and stable aluminium species content becomes available.

In conclusion, a lot of work remains to be done before aluminium speciation analysis in waters becomes a validated and recognized test, able to meet today's requirements. In particular, pre-normative research would be desirable to optimize and validate existing methodologies which could be proposed as international standards. Such necessary collaborative work should be undertaken at the international level through the organization of interlaboratory studies which, eventually, could serve as a good basis for the certficiation of reference materials containing Al species.

CHAPTER 11

Single and Sequential Extraction

11.1 Aim of the Project and Coordination

Introduction

As mentioned in Chapter 2, the eco-toxicity and mobility of metals in the environment depends strongly on their specific chemical forms or types of binding rather than the total element content. Consequently these have to be determined in order to assess the toxic effects and geochemical pathways. The determination of specific chemical species or binding forms is difficult and often hardly possible. Therefore, in practice, determinations of broader "operationally or functionally defined" forms or phases can be a reasonable compromise, *e.g.* "bioavailable" forms of trace elements can give sufficient information to arrive at a sound environmental policy. The increasing concern to assess the bioavailable metal fraction (and thus to estimate the related phyto-toxic effects) and the environmentally accessible trace metals upon disposal of sediment on to a soil (*e.g.* contamination of ground waters) is reflected by a considerable increase in the frequency of analysis over the last ten years. Single and sequential extraction schemes were designed in the 1980s in order to assess the different retention/release of metals in soil and sediment samples [188–192]. However, the lack of uniformity in the different procedures used did not allow the results to be compared worldwide nor the procedures to be validated. The results obtained are defined by the determination of extractable elements using a given procedure; therefore their significance is highly dependent on the extraction protocol performed. Moreover, the lack of suitable reference materials for this type of study did not enable the quality of the measurements to be controlled. Because of the many pitfalls likely to occur in the use of extraction protocols for soil analysis, the Measurements and Testing Programme has launched a project aimed at harmonizing measurements for extractable trace metal content in sediment and soil. This project followed a stepwise approach (through interlaboratory studies) of which the final aim was to certify sediment and soil reference materials for their extractable trace element content. Two soils (terra rossa and sewage sludge amended soils) were prepared and certified for their EDTA- and acetic acid-extractable trace element content, and one calcareous soil was certified for its EDTA and DTPA-extractable trace element content;

in addition, a sediment was certified using a three-step sequential extraction protocol.

The Programme and Timetable

The project was started in 1987 by a consultation of European experts examining the possibility of harmonizing single and/or sequential extraction schemes for soil and sediment analysis [60]. This inquiry was followed by the design of single extraction schemes (EDTA, acetic acid and ammonium acetate) and a sequential extraction protocol which were proposed to a group of *ca.* 30 laboratories for possible harmonization in 1989 [193]. Interlaboratory studies (two on soils and two on sediment) were carried out between 1989 and 1993 [194–196] and were followed by certification campaigns conducted between 1994 and 1996 [197–199].

Coordination

Feasibility Study

The feasibility study was coordinated by the Macaulay Land Use Research Institute in Aberdeen (United Kingdom); it involved the following laboratories: Technical University of Hamburg-Harburg (Germany); Institute for Soil Fertility, Haren (The Netherlands); Laboratoire Central des Ponts et Chaussées, Bouguenais (France); EC Joint Research Centre, Ispra (Italy).

Certification of Sewage Sludge-amended Soils

The certification of the sewage sludge-amended soils (CRMs 483 and 484) was coordinated jointly by the Strathclyde University in Glasgow (United Kingdom) and the Universidad de Barcelona (Spain). The preparation of the material was carried out at the EC Joint Research Centre of Ispra (Italy) and the homogeneity and stability were verified by the Macaulay Land Use Research Institute in Aberdeen (United Kindgom).

Certification of Calcareous Soil

The certification of the calcareous soil (CRM 600) was coordinated jointly by the Estación Experimental del Zaidin in Granada (Spain) and the Universidad de Barcelona (Spain). The preparation of the material was carried out at the EC Joint Research Centre of Ispra (Italy); the homogeneity and stabitility were verified by the Institut National de Recherche Agronomique in Villenave d'Ornon (France).

Certification of Sediment

The collection and preparation of the sediment reference material (CRM 601) was carried out by the EC Joint Research Centre in Ispra (Italy) and the Universidad de Barcelona (Spain).

Participating Laboratories

The following laboratories participated in the interlaboratory study and certification campaigns on soils: Agriculture and Food Development Authority, Wexford (Ireland); Agricultural Research Centre, Jokioinen (Finland); Bundesanstalt für Materialforschung und Prüfung, Berlin (Germany); Institut National d'Agronomie, Laboratoire de Chimie Analytique, Paris (France); Kemiteknik, Teknologisk Institut, Taastrup (Denmark); Institute for Soil Fertility, Haren (The Netherlands); Institut National de Recherche Agronomique, Station d'Agronomie, Villenave d'Ornon (France); Estación Experimental de Zaidin, CSIC, Granada (Spain); Estação Agronómica Nacional, Oeiras (Portugal); Laboratoire Central des Ponts et Chaussées Bouguenais (France); Landbouw Universiteit, Wageningen (The Netherlands); Leiter Labor, Arbeit- und Umweltschutz, Nordenham (Germany); Station Fédérale de Recherches en Chimie Agricole, Liebefeld-Bern (Switzerland); The Macaulay Land Use Research Institute, Aberdeen (United Kingdom); Universidad de Barcelona, Departamento de Química Analítica, Barcelona (Spain); Università di Bari, Istituto di Chimica Agraria, Bari (Italy); Università di Udine, Istituto de Produzione Vegetale e Tecnologia Agraria, Udine (Italy); Universiteit Gent (Belgium); Universidade Nova, Lisbon (Portugal); University of Strathclyde, Pure and Applied Chemistry, Glasgow (United Kingdom); University of Reading, Department of Soil Science (United Kingdom).

The following laboratories participated in the sediment interlaboratory studies and certification campaign: Chalmers University of Technology, Göteborg (Sweden); The Macauley Land Use Research Institute, Aberdeen (United Kingdom); Estación Experimental de Zaidin, CSIC, Granada (Spain); Institut National d'Agronomie, Paris (France); Institut National de Recherche Agronomique, Villenave d'Ornon (France); Universidad de Sevilla (Spain); Technical University of Hamburg (Germany); Institut de Géologie du Bassin d'Aquitaine, Talence (France); Laboratoire Central des Ponts et Chaussées, Bouguenais (France); Bundesanstalt für Materialforschung und Prüfung, Berlin (Germany); Agricultural and Food Development Authority, Wexford (Ireland); Vrije Universiteit, Brussels (Belgium); EC Joint Research Centre, Ispra (Italy); Nova Universidade de Lisboa (Portugal); Agricultural Research Centre, Jokoinen (Finland); Institut National de Recherche Agronomique, Arras (France); Strathclyde University, Glasgow (United Kingdom); University of Gent (Belgium); Estação Agronómica Nacional, Oeiras (Portugal); University of Reading, Department of Soil Science (United Kingdom).

11.2 Feasibility Study

Some 35 years ago, the Macaulay Institute for Soil Research in Aberdeen was approached by the NIST, then the National Bureau of Standards, as to the feasibility of preparing a reference soil certified for its extractable content as distinct from its total content. At the time, the difficulties of preparation, stability and analysis were considered too great. The collaborative approach adopted by the BCR could overcome these difficulties. An initial study of the literature and a consultation of European experts were carried out, the results of which are summarized elsewhere [60,193].

Temporal Stability of Extractable Contents

One of the major hurdles to be surmounted, before the preparation of a reference material certified for extractable trace metal content could be contemplated, was the question of temporal stability. This problem has been studied by Salomons and Scheltens [200] who repeated, in 1987, the sequential extraction analysis of freshwater sediments first carried out in 1975. While those authors were somewhat disappointed in that significant changes did occur in the measured extractable content after 12 years, it is likely that the temporal stability was sufficient for the sediments to be characterized on each occasion in such a way that decisions on the management or use of the sediments would be unchanged.

A test of the stability of the extractable contents of a single soil was carried out for EDTA, acetic acid, ammonium acetate and calcium chloride extracts. With the exception of chromium, the EDTA extractable contents were shown to be stable over 3 year intervals within about 10% for the elements Cd, Cu, Fe, Mn, Ni, Pb and Zn [193]. For acetic acid extracts, one-year changes for Cr, Fe, Mn and Zn were poorer than for EDTA, but for Cd, Cu, Ni and Pb the extracts were reasonably stable. For ammonium acetate and calcium chloride extracts, however, the results, apart from Cd and Ni in ammonium acetate, were useless. These failures in ammonium acetate and calcium chloride were probably related to the fact that the solution concentrations measured were too low for reliable determination by the FAAS and ICP-AES methods used.

Preliminary Trial

Four laboratories took part in a preliminary trial of three different sequential extraction schemes with seven sediments and one soil supplied by the EC Joint Research Centre of Ispra. The sequential extraction methods were:

- The modified Tessier [188] procedure of Förstner [201]
- The short method of Salomons and Förstner [202]
- The Meguellati method [203]

The results of this preliminary trial are described in detail elsewhere [60]. The fractional contents obtained by the different procedures were in good enough agreement for the sediments to be sufficiently well chacterized for decisions to be made on the management or use of the sediments. There were serious failures in detail and it could be concluded that, while the results were promising, improvements were necessary if reference material certification was to be attempted. Furthermore, any future study should be based on a well specified procedure, which was designed as described in Section 11.4.

11.3 Sequential Extraction for Sediment Analysis

Design of the Sequential Extraction Procedure

An initial study of the literature on the "speciation" of metals in soils and sediments by chemical extraction, and a consultation with European experts, was carried out by A. Ure on behalf of the BCR [60]. The outcome of this study was discussed in a meeting of *ca.* 40 representatives from leading European laboratories in the field of soil and sediment analysis and a programme was adopted involving (i) the design of single and sequential extraction procedures for the analysis of soil and sediment, respectively, (ii) interlaboratory trials on these extraction schemes and (iii) the preparation of reference material certified for their extractable trace metals. Therefore, a three-step extraction procedure was designed based on acetic acid extraction (step 1), hydroxyammonium chloride extraction (step 2) and hydrogen peroxide attack/ammonium acetate extraction (step 3).

This scheme (given in the Annex to this chapter) was tested in a first interlaboratory trial on Cd, Cr, Cu, Ni, Pb and Zn [196] and the results showed that, while promising, improvements were necessary prior to attempting the certification of a reference material.

Interlaboratory Trials

Sediment Samples Used in the Interlaboratory Studies

The sediment sample used in the first interlaboratory trial was collected in Yrseke, The Netherlands, with grab samplers, wet sieved at 2 mm and air dried at room temperature (*ca.* 20 °C). The air-dried material was then ground, sieved at 90 µm and homogenized before bottling. The bulk homogeneity of the sediment was tested by X-ray fluorescence (XRF) by determining the major components.

The sediment sample used in the second interlaboratory trial was collected in the River Besòs, Spain. The material was sampled with a grab, air-dried, then sieved at 63 µm, homogenized and bottled. Homogeneity and stability studies of extractable trace metals were carried out and the material was found to be homogeneous and stable enough to be used in the intercomparison exercise [196].

Technniques Used in the Intercomparisons

The techniques used to determine metal concentration in extracts were generally FAAS (flame atomic absorption spectrometry) or ETAAS (electrothermal atomic absorption spectrometry). ICP-OES (inductively coupled plasma optical emission spectrometry) and ICP-MS (inductively coupled plasma mass spectrometry) were also used by some laboratories.

Table 11.1

	First round-robin			*Second round-robin*		
Steps	*Mean*	*SD*	*CV*	*Mean*	*SD*	*CV*
Cadmium						
1	7.18	0.81	11.3	0.18	0.02	8.5
2	3.41	0.63	18.5	0.08	0.02	16.9
3	1.03	0.20	20.1	1.02	0.17	16.6
Chromium						
1	1.36	0.20	14.7	1.28	0.41	32.0
2	3.29	1.07	32.5	1.21	0.31	25.6
3	76.3	10.4	13.6	866	126	14.5
Copper						
1	3.69	0.76	20.6	0.23	0.08	34.8
2	3.13	1.96	20.1	0.53	0.33	62.3
3	63.4	13.2	20.8	90.1	7.6	8.4
Nickel						
1	9.76	4.36	44.7	13.0	2.08	16.0
2	5.79	1.54	26.6	1.80	0.23	12.7
3	10.2	3.32	32.5	15.2	2.31	5.2
Lead						
1	5.06	2.50	49.4	0.30	0.17	56.7
2	11.0	8.87	80.6	0.14	0.09	64.3
3	6.93	4.78	69.0	47.7	7.61	5.9
Zinc						
1	262	35.1	13.4	93.9	16.9	18.0
2	140	34.2	24.4	79.7	16.9	21.2
3	89.7	9.14	10.2	676	44	6.5

Technical Evaluation

The results of the two interlaboratory studies are summarized in Table 11.1.

Copper. A high spread of results was observed in both steps 1 and 2 for copper, which was suspected to be due to the final determination rather than to the extraction procedure itself. Ca interferences were suspected to occur at the emission line used in ICP (324.8 nm) by one laboratory; however, two other laboratories used the same emission line and did not observe such interferences. The University of Barcelona tested the likelihood for such interferences at the emission line of 324.754 nm and the presence of a high background signal was observed, which was confirmed by the addition of Ca to the calibrant solutions. The copper content was actually considered to be close to the determination limits for a number of laboratories, and the accepted sets of results (6 ETAAS, 1 ICP and 1 ICP-MS) were too low to draw any firm conclusions on the state of the art of steps 1 and 2 in this material.

In the case of step 3, the effects of shaking type and speed or room temperature on the spread of results were thoroughly discussed. Changes due to temperature effects were not noticeable. However, it was found more difficult to standardize the shaking procedures. Six laboratories used an end-over-end shaker whereas the other laboratories used a horizontal shaker. No effect of shaker type was suspected in this case; however, effects of shaking speed were suspected as lower results were generally obtained at a speed of less than 40 rpm, whereas higher results corresponded to speeds of up to 150 rpm. The need for maintaining the sediment in suspension during shaking was found to be highly critical. The proposal to use a glass ball during shaking was not accepted as it would lead to possible grinding effects. From these results, a study of the influence of shaking type and speed was carried out, which demonstrated that the copper extracted in this step using an end-over-end shaker, operated at 30 rpm, was 20% higher than when using a horizontal shaker operating a 130 rpm. The range of data (*CV* between laboratories of *ca.* 8%) was considered to be good for this kind of measurement (16 sets of results accepted of which 5 were ETAAS, 6 FAAS, 4 ICP and 1 ICP-MS).

The low levels of copper clearly represented a major problem in the first two extraction steps in comparison with the first exercise (Table 11.1). In the case of step 3, the results improved drastically, which led to supposition that the extraction procedure itself was not responsible for the large spread observed before, but that the sources of error were rather due to the technique of final determination.

Lead. Ca interferences were again suspected to affect the signal when ICP-AES was used. This was verified by the addition of Ca to the calibrant solution, which yielded similar interferences. Here again the very low Pb content extracted in steps 1 and 2 in the second intercomparison hampered firm conclusions to be drawn. In the case of step 2, a very high discrepancy was noted between the ETAAS results and the FAAS results (one order of magnitude). However, all the FAAS results were actually recognized to be below the limits of determination and were consequently withdrawn. When ETAAS was used as the technique of final determination, the use of the matrix modifier was highly recommended. As for copper, the small number of accepted sets of data (5 laboratories of which 3 were ETAAS, 1 ICP and 1 ICP-MS) were found to be insufficient to draw any firm conclusions.

In the case of step 3, the shaking speed was again suspected to have some effects on the spread of results, *i.e.* that lower speeds could lead to lower results. The results ranged from 34.6 to 60.0 mg kg^{-1} (coefficient of variation of *ca.* 16%) and covered a good range of techniques (5 ETAAS, 6 FAAS, 4 ICP and 1 ICP-MS). The participants stressed once more that it is of paramount importance to verify that the sediment is continually in suspension during the extraction. It was agreed to add the following recommendation in the protocol: **"The sediment should be continually in suspension during the extraction. If this is not verified, the shaking speed should be adapted in order to ensure a continuous suspension of the mixture".** Step 3 was considered to be critical in the presence of high amounts of organic matter (which is the case of the sample used in the second exercise) as the incomplete destruction of organic matter, as well the difficulty in oxidizing sulfide, may be the source of a high uncertainty which could explain the spread of results.

In some cases, it was difficult to obtain a "cake" after centrifugation and, consequently, fine particles were still present in solution, which created problems in ICP analysis (nebulizer clogging). In these circumstances a filtration step was necessary. Filtration was not recommended to be included in the protocol; however, when ICP is used, the participants strongly recommended this step for ICP users. The protocol was hence modified by adding the following sentence: **"After centrifugation, a filtration step (0.45 μm) is recommended for ICP determinations"**. Later studies performed by the University of Barcelona showed the occurrence of interference due to high content of Al and the presence of a high background signal due to Ca. The effects of interferences can be taken into account at the calibration step by adding both elements to the calibrant. Violent reactions could be observed with hydrogen peroxide with this sediment. The participants recommended special precautions to be taken in the handling of this reagent (*i.e.* slow addition).

As observed for copper, the low Pb contents extracted in steps 1 and 2 were likely to be too close to the limits of determination for the different techniques. The results in step 3 indicated that a reasonable agreement was found among the laboratories (Table 11.1).

Nickel. The influence of the size of the extraction vessel was noted by one laboratory: the bigger the extraction tube, the higher the results. This was, however, not observed by other laboratories.

When using AAS techniques, different interferences were observed and the use of a background corrector and matrix modifiers (phosphate or palladium) were strongly recommended. When possible, the modifier should be dried before introducing the sample to avoid carbide formation with nickel. For this element the effects of type/speed of shaking, although noticeable, were less pronounced than for Cu and Pb.

The selected results in step 1 ranged from 10.85 to 17.33 mg kg^{-1}, with a sufficient range of techniques (2 ETAAS, 6 FAAS, 4 ICP and 1 ICP-MS), which was found to be a reasonable picture for the present state of the art and the Ni content involved.

In the case of step 2, contamination problems were throught to be the major causes of the discrepancies observed. After doubtful results were withdrawn, the remaining data ranged from 1.65 to 2.28 mg kg^{-1} (6 ETAAS, 1 FAAS, 3 ICP and 1 ICP-MS).

In the case of step 3, two laboratories withdrew their results due to contamination problems and dilution error, respectively. The remaining results ranged from 12.53 to 19.47 mg kg^{-1} (*CV* between laboratories of *ca.* 15% with the following range of techniques: 3 ETAAS, 6 FAAS, 5 ICP and 1 ICP-MS), which was found to be a reasonable figure for the present state of the art.

The results of Ni determination in this exercise demonstrated a considerable inprovement in comparison with the first interlaboratory exercise (Table 11.1).

Both the levels of Ni content and the range of techniques would allow certification to be contemplated at a later stage.

Cadmium. In the first two steps, FAAS was not considered to be a suitable technique for the determination of the low Cd content and the corresponding sets were consequently withdrawn. Some ETAAS results were below determination limits and were therefore rejected. Conclusions could hardly be drawn on the basis of the three remaining results, except that the extractable Cd content was too low for the purpose of testing the extraction scheme for both steps 1 and 2.

In the case of step 3, contamination problems were again experienced. FAAS results were thought to be too close to the determination limits and were recommended to be used as indicative values only. A systematic difference was noted between ETAAS and ICP results. Measurements were suspected of being affected by iron interferences and calibration by standard addition was strongly recommended for ETAAS. This recommendation was actually included in the protocol as follows: **"When ETAAS is the final method of element determination, the method of standard additions is strongly recommended for calibration"**. In this step, the effect of the type of shaker was again noticed, the results obtained being higher when an end-over-end shaker was used. The obtained *CV* between laboratories [*ca.* 17% with a range of techniques covering ETAAS (9 sets), FAAS (5 sets), ICP (3 sets) and ICP-MS (1 set)] was considered to reflect the present state of the art.

Chromium. Some results in step 1 were rejected as they were below the limits of determination. Other sets of results suffered from contamination problems (high results with high standard deviations) and were consequently withdrawn. The remaining sets in this step ranged from 0.72 to 2.00 mg kg^{-1} (3 ETAAS and 4 ICP). Improvement was considered to be necessary before contemplating certification.

In the case of step 2, influences of the shaker type were again suspected, which was confirmed by studies performed by the University of Barcelona.

Lower results (factor 2) were observed for ICP in comparison to ETAAS. It was assumed that interferences were taken into account in the calibration in the ICP measurement. However, when standard additions were used or Ca was added to the calibrants, ICP results agreed with those of ETAAS, which demonstrated the need to perform a matrix matching calibration or a standard addition procedure. The scatter of results observed in step 2 (0.73 to 1.66 mg kg^{-1}) was found to be high but not surprising considering the low Cr content in the extract.

In the case of step 3, insufficient extraction was found to be responsible in one case for low results and the set of data was withdrawn. It was agreed that the use of nitrous oxide in FAAS was not suitable for eliminating Ca and Fe interferences, which could explain the low results obtained by one laboratory. Air–acetylene (with oxine as releaser) should perhaps be used instead. This last point was confirmed by investigations performed by the University of Barcelona [190]. However, one participant used nitrous oxide and did not experience any problem. It was difficult to draw any firm conclusion at this stage. The remaining results in step 3 ranged from 682 to 1005 mg kg^{-1} [*CV* between laboratories of *ca.* 12% with a range of techniques covering ETAAS (4 sets), FAAS (5 sets), and ICP (4 sets)]. This figure was considered to represent well the present state of the art.

The results obtained for this metal did not show any improvement on the first exercise (Table 11.1), but improvement will be necessary before certification can be contemplated.

Zinc. In the case of step 1, one laboratory experienced problems due to very high and variable blanks, which could be the reason for high results. The accepted results ranged from 72.7 to 125.8 mg kg^{-1}, which was considered to reflect the state of the art (the following techniques were used: 11 FAAS and 5 ICP).

In the case of step 2, typical errors were due to calibration and contamination. Insufficient extraction was also detected in one case. The range of the remaining values was from 61.5 to 99.5 mg kg^{-1}. As for step 1, the participants considered this figure to represent the state of the art (techniques used were: 1 ETAAS, 8 FAAS, 4 ICP and 1 ICP-MS).

In the case of step 3, similar problems as in step 2 were found. The accepted results ranged from 596 to 735 mg kg^{-1} with a *CV* between laboratories below 10% (with techniques covering 1 ETAAS, 9 FAAS, 4 ICP and 1 ICP-MS), which was considered to be acceptable for possible certification. As observed for the other metals, the type of shaking could be an important cause of the spread of results. The agreement between laboratories was found to be similar or slightly better than in the first interlaboratory exercise. However, the range of techniques should be enlarged in order to avoid a possible "method effect".

Conclusions

The second interlaboratory exercise on extractable trace metals in sediment showed a consequent improvement in comparison with the results of the first exercise. Furthermore, these collaborative efforts allowed the sequential extraction procedure to be slightly improved by minor amendments. The general noticeable trend was the important decrease in the number of total and accepted sets of values for concentrations in the extract below 10 µg L^{-1}, which illustrated the difficulties experienced by a number of laboratories in the determination of such concentration levels in these matrices.

11.4 Certification of Sediment

Preparation of the Candidate Reference Material

The material was collected in March 1994 from different sampling sites of Lake Maggiore (Italy). Sampling operations were performed using grab

Table 11.2 *Bulk homogeneity testing by XRF (CRM 601)*[a]

CVs	*Si*	*Al*	*Ca*	*K*	*Mg*	*Ti*	*S*	*P*	*Pb*	*Fe*	*Zn*	*Cu*	*Ni*	*Cr*	*Mn*
CV drum	0.81	0.43	0.67	0.39	0.30	0.49	1.31	0.99	0.65	0.40	0.81	0.71	1.11	1.36	0.61
CV bottle	0.33	0.32	0.23	0.14	0.26	0.26	0.51	0.28	0.64	0.13	0.46	0.42	1.26	0.22	0.15

[a] CV drum: 10 determinations on samples taken randomly in the drum;
CV bottle: 10 determinations on the content of one bottle.

collectors. The wet sediment was passed through a 2 mm sieve in order to remove stones and other materials extraneous to sediments.

The sieved sediment was placed on stainless-steel free trays and exposed at ambient air temperature for drying, turning the lumps from time to time to accelerate the drying process.

The dry sediment with a mean moisture content of 3.5% (calculated by drying a separate portion of sediment at 105 °C for 3–4 h) was passed through a tungsten carbide-bladed hammermill and sieved to pass apertures of 90 μm. The <90 μm fraction was collected in a PVC mixing drum (filled with dry argon) and homogenized for 14 days at about 48 rpm.

Ten sub-samples were randomly taken in the drum and analysed by X-ray fluorescence spectrometry in order to assess the bulk homogeneity; the *CV*'s obtained were compared with the *CV* calculated from ten replicate measurements performed in one bottle (Table 11.2). The results obtained indicated that the bulk homogeneity was satisfactory (*CV* generally lower than 1%) and the bottling operation was, therefore, carried out. The bottling procedure was performed manually: after an additional period of mixing of 2 days, a first series of 20 bottles was filled and immediately closed with screw caps and plastic inserts. Several series of 20 bottles were hence filled, alternating with re-mixing of the powder (2 minutes). The bottles produced were stored at ambient temperature.

Homogeneity Control

The extractants were prepared as laid out in the Annex (Section 11.7). All precautions were taken to avoid contamination during the extraction procedures. The trace element contents (Cd, Cr, Cu, Ni, Pb and Zn) in the extracts were determined by flame atomic absorption spectrometry (FAAS) or electrothermal atomic absorption spectrometry with Zeeman background correction (ZETAAS).

For the homogeneity study, the six elements were determined in the candidate CRM by analysing 10 subsamples taken from one bottle (within-bottle homogeneity test) and one subsample in each of 20 different bottles selected during the bottling procedure (between-bottle homogeneity test).

The *CV*'s and the total uncertainty U_{CV} for the extractable trace element contents between (CV_B) and within (CV_W) bottles are given in Table 11.3.

Table 11.3 *Homogeneity tests*

	Cd	*Cr*	*Cu*	*Ni*	*Pb*	*Zn*
1st step						
CV_B	1.8 ± 0.3	5.2 ± 0.8	1.1 ± 0.2	6.7 ± 1.1	3.6 ± 0.6	0.6 ± 0.1
CV_W	1.5 ± 0.3	5.2 ± 1.2	1.2 ± 0.3	6.1 ± 1.4	3.9 ± 0.9	0.2 ± 0.1
2nd step						
CV_B	3.3 ± 0.5	21.2 ± 3.4	26.1 ± 4.1	12.2 ± 1.9	16.7 ± 2.7	4.2 ± 0.7
CV_W	5.2 ± 1.2	15.4 ± 3.4	7.9 ± 1.2	6.1 ± 1.4	7.4 ± 1.7	2.3 ± 0.5
3rd step						
CV_B	6.0 ± 0.9	4.0 ± 0.7	5.1 ± 0.8	7.9 ± 1.3	2.7 ± 0.4	4.8 ± 0.8
CV_W	5.3 ± 1.2	4.9 ± 1.1	6.4 ± 1.4	7.6 ± 1.7	4.4 ± 1.0	6.1 ± 1.4

Table 11.4 *Stability tests: first step*

Element	*Time*/month	$R_T \pm U_T$ (20 °C)	$R_T \pm U_T$ (40 °C)
Cd	1	0.97 ± 0.01	0.97 ± 0.02
	3	1.02 ± 0.01	1.03 ± 0.02
	6	1.00 ± 0.01	0.98 ± 0.03
	12	0.93 ± 0.03	0.89 ± 0.01
Cr	1	1.02 ± 0.09	1.07 ± 0.11
	3	1.41 ± 0.18	1.41 ± 0.20
	6	1.03 ± 0.06	1.12 ± 0.05
	12	1.06 ± 0.09	1.01 ± 0.12
Cu	1	1.06 ± 0.01	1.10 ± 0.01
	3	1.10 ± 0.03	1.25 ± 0.03
	6	1.12 ± 0.02	1.39 ± 0.03
	12	1.20 ± 0.09	1.39 ± 0.02
Ni	1	1.00 ± 0.11	0.94 ± 0.07
	3	1.01 ± 0.18	0.97 ± 0.15
	6	1.15 ± 0.21	0.90 ± 0.13
	12	0.93 ± 0.11	0.92 ± 0.10
Pb	1	1.03 ± 0.03	0.98 ± 0.06
	3	1.16 ± 0.08	1.15 ± 0.07
	6	0.97 ± 0.04	1.24 ± 0.04
	12	1.01 ± 0.03	1.08 ± 0.07
Zn	1	0.97 ± 0.01	0.98 ± 0.01
	3	1.05 ± 0.01	1.06 ± 0.01
	6	1.02 ± 0.01	1.03 ± 0.05
	12	1.01 ± 0.01	0.93 ± 0.02

For most of the extractable metal contents, the overlap was within the total uncertainty U_T of the CV (an approximation of the uncertainty U_{CV} of the CV is calculated as follows: $U_{CV} \approx CV/\sqrt{2n}$). Differences between the within-bottle and between-bottle CVs observed for step 2 were considered to be an analytical artefact rather than an indication of inhomogeneity which would have been reflected in the spread of results submitted in the certification. The material was then considered to be homogeneous for the stated level of intake (1 g).

Table 11.5 *Stability tests: second step*

Element	*Time*/month	$R_T \pm U_T$ (20 °C)	$R_T \pm U_T$ (40 °C)
Cd	1	1.00 ± 0.02	0.98 ± 0.02
	3	1.00 ± 0.03	0.95 ± 0.03
	6	0.99 ± 0.02	0.97 ± 0.02
	12	0.94 ± 0.03	1.07 ± 0.03
Cr	1	0.89 ± 0.07	0.88 ± 0.11
	3	0.80 ± 0.12	0.76 ± 0.12
	6	0.92 ± 0.05	0.70 ± 0.09
	12	0.97 ± 0.08	0.92 ± 0.22
Cu	1	0.88 ± 0.08	0.92 ± 0.16
	3	0.92 ± 0.11	0.87 ± 0.20
	6	0.98 ± 0.08	0.84 ± 0.07
	12	1.03 ± 0.14	0.98 ± 0.42
Ni	1	0.93 ± 0.09	0.87 ± 0.04
	3	0.98 ± 0.10	0.87 ± 0.10
	6	1.04 ± 0.08	0.73 ± 0.05
	12	0.93 ± 0.15	1.24 ± 0.14
Pb	1	1.02 ± 0.03	0.97 ± 0.06
	3	0.74 ± 0.09	0.67 ± 0.09
	6	0.87 ± 0.05	0.59 ± 0.04
	12	0.98 ± 0.13	0.86 ± 0.28
Zn	1	0.98 ± 0.01	0.97 ± 0.03
	3	0.95 ± 0.02	0.91 ± 0.02
	6	0.97 ± 0.03	0.79 ± 0.02
	12	1.00 ± 0.04	1.20 ± 0.05

Stability Control

The stability of the extractable trace element contents was tested at −20, +20 and +40 °C and the extractable contents of Cd, Cr, Cu, Ni, Pb and Zn were determined (in five replicates) after 1, 3, 6 and 12 months. The detection techniques used were the same as in the homogeneity study. The evaluation of the stability was based on the procedure described in Chapter 3, using the samples stored at −20 °C as reference for the samples stored at +20 and +40 °C, respectively. The results are given in Tables 11.4–11.6; in most cases, the uncertainty in the method could account for the deviations observed. Some risks of instability were, however, suspected at 40 °C owing to possible changes in the extractability of some elements (*e.g.* Cu and Pb); these changes induced by the high storage temperature could be related to changes in the status of the organic matter or in the crystallographic compounds of Fe or Mn. Hence, it is recommended to avoid storage at temperatures above 20 °C.

Technical Evaluation

At the technical meeting, it was recalled that strict observance of the extraction protocol would be a criterion for considering the results for discussion. The participants recommended that a tolerance of ±30% be included in the

Table 11.6 *Stability tests: third step*

Element	*Time*/month	$R_T \pm U_T$ (20 °C)	$R_T \pm U_T$ (40 °C)
Cd	1	0.88 ± 0.07	0.91 ± 0.02
	3	1.06 ± 0.06	1.03 ± 0.07
	6	1.02 ± 0.05	0.92 ± 0.04
	12	1.01 ± 0.07	1.00 ± 0.07
Cr	1	0.97 ± 0.04	1.02 ± 0.04
	3	0.95 ± 0.04	0.93 ± 0.05
	6	1.02 ± 0.03	0.96 ± 0.03
	12	1.00 ± 0.04	1.18 ± 0.04
Cu	1	1.05 ± 0.05	1.02 ± 0.04
	3	0.93 ± 0.04	0.91 ± 0.06
	6	1.02 ± 0.04	0.95 ± 0.03
	12	0.98 ± 0.03	1.03 ± 0.03
Ni	1	1.03 ± 0.08	1.15 ± 0.08
	3	1.04 ± 0.07	1.03 ± 0.11
	6	1.09 ± 0.06	1.00 ± 0.02
	12	0.99 ± 0.08	0.81 ± 0.06
Pb	1	0.95 ± 0.06	1.12 ± 0.07
	3	1.04 ± 0.05	1.02 ± 0.05
	6	1.01 ± 0.04	1.20 ± 0.06
	12	1.03 ± 0.09	0.96 ± 0.07
Zn	1	1.01 ± 0.06	1.09 ± 0.06
	3	1.06 ± 0.08	1.03 ± 0.09
	6	1.05 ± 0.06	0.94 ± 0.03
	12	1.00 ± 0.06	0.75 ± 0.05

extraction protocol for the shaker speed. The scrutiny showed unacceptable spreads of results in some cases (*e.g.* extractable Cr, second step); in the case where the spread was too high, it was decided not to give any indicative value which could mislead the user.

Cadmium was certified for the three steps; only one ETAAS set was rejected owing to very high standard deviation (*CV* of *ca.* 30%).

Chromium could only be certified for the first extraction step. A high spread was observed at the second step, *i.e.* a "camel distribution" was obtained with results ranging from 0.3 mg kg^{-1} to more than 2.9 mg kg^{-1} (*CV* of 66%). The third step demonstrated similar variability, with results ranging from 10 mg kg^{-1} to 31 mg kg^{-1} (*CV* of 29%).

The results of the extractable Cu were acceptable in the first step. However, certification was not recommended owing to a suspicion of instability. The data obtained at the second and third steps were too dispersed for being acceptable for certification.

The extractable Ni contents were certified at all steps, whereas only the first step was certified for Pb extractable contents (high spreads hampering certification for steps 2 and 3). A narrow distribution of results was obtained for Zn at the first step (*CV* of 1.9%) and the second step (*CV* of 6.0%), whereas the large spread observed at step 3 hampered certification.

Table 11.7 *Certified contents of extractable contents of Cd, Cr, Cu, Ni, Pb and Zn*

	Certified value/ mg kg^{-1}	*Uncertainty*/ mg kg^{-1}	p^a
First step			
Cd	4.14	0.23	11
Cr	0.36	0.04	12
Cu	8.32	0.46	9
Ni	8.01	0.73	10
Pb	2.68	0.35	11
Zn	264	5	12
Second step			
Cd	3.08	0.17	10
Ni	6.05	1.09	11
Pb	33.1	10.0	9
Zn	182	11	12
Third step			
Cd	1.83	0.20	11
Ni	8.55	1.04	9
Pb	109	13	12

[a] p = number of data sets.

Certified Values

The certified values (unweighted mean of p accepted sets of results) and their uncertainties (half-width of the 95% confidence intervals) are given in Table 11.7 as mass fractions of the respective extracts obtained at the first, second and third steps (based on dry mass) in mg kg^{-1} [191]. Examples of bar graphs (Cd contents in the three steps) are shown in Figures 11.1–11.3.

While this CRM was considered of high added value, improvements were deemed necessary for achieving a better comparability of data, using the three-step sequential extraction step. A further project has, therefore, started very recently to improve this scheme (without major modifications) and produce other CRMs (sediment and possibly soil).

11.5 Single Extractions for Sludge-amended Soils

Interlaboratory Study

A sewage sludge-amended soil from Great Billings Sewage farm, Northampton (by courtesy of Anglian Water, UK) was collected, air-dried, sieved and some 12 kg of the <2 mm soil homogenized and bottled in 100 g lots for the interlaboratory study. The large range of particle sizes in the <2 mm soil material normally used for extraction necessitates the use of large (5–20 g) sub-samples to ensure representative sampling of the bulk material for extraction and analysis. This posed the not trivial problem of homogenization of the bulk material and the determination of the minimum sample size required for

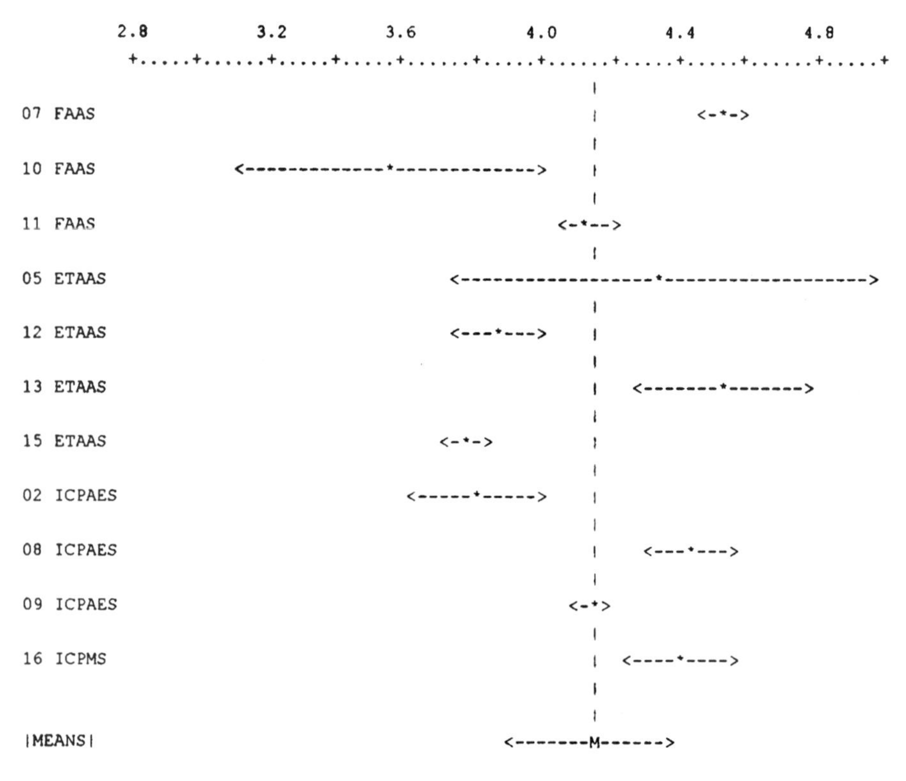

Figure 11.1 *Cadmium in CRM 601: first step in mg kg^{-1}*

each extraction and analysis and the verification that each sample bottle was statistically representative of the whole. Homogenization of the bulk soil was achieved by rolling it in a polyethylene bag. The whole material was sub-sampled into 100 bottles, each containing approximately 100 g of soil, by coning and quartering. It was determined that a 5 g sub-sample, taken from each bottle by coning and quartering the whole contents of the bottle, was representative of the whole material. The within- and bottle-homogeneity of the material was found to be acceptable for the EDTA- and ammonium acetate-extractable trace metal contents [193].

Since both the temperature of extraction and the vigour of the mechanical shaking could be expected to affect the amount of metal extracted, limits for both of these parameters were prescribed in the detailed protocol for extraction and analysis issued to the participating laboratories along with the solid sample [193]. As a check on each laboratory's calibration, reference solutions containing the trace metals of concern (Cd, Cr, Cu, Ni, Pb and Zn) were prepared. Three types of single extractant were used: EDTA (0.05 mol L^{-1}), acetic acid (0.43 mol L^{-1}) and ammonium acetate (1 mol L^{-1}).

The overall mean of the individual laboratory mean extractable contents for

BAR-GRAPHS FOR LABORATORY MEANS AND 95% CI

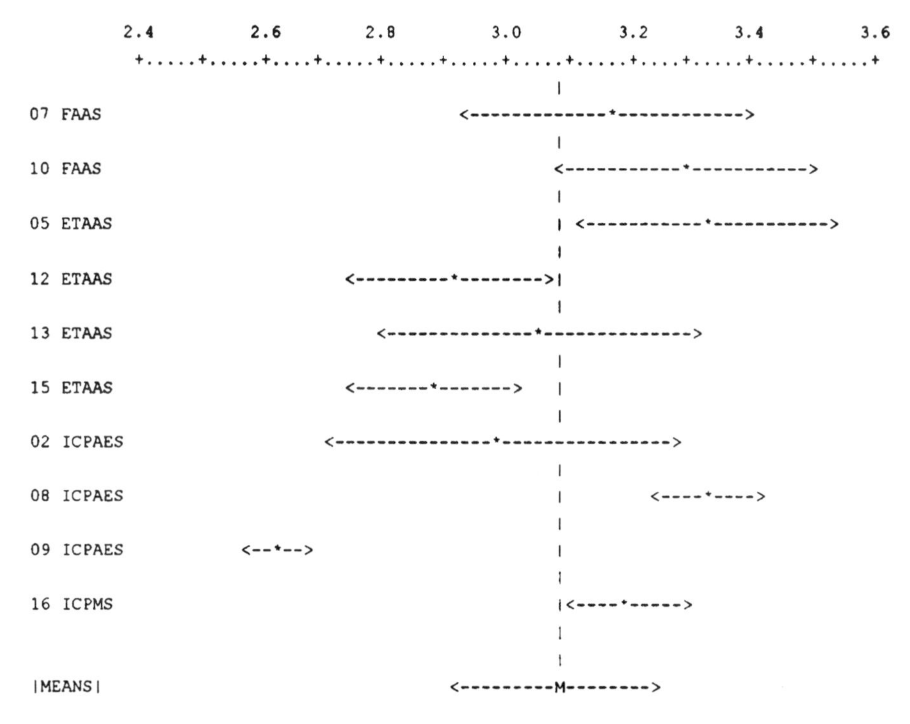

Figure 11.2 *Cadmium in CRM 601: second step in mg* kg^{-1}

the six elements in the three extractants are shown in Table 11.8. It was concluded that, with the exception of Cr, the *CV*'s for the contents in the EDTA extracts were acceptably low (*ca.* 10%). While for the acetic acid extracts the *CV*'s for Ni and Pb were a little higher than for the EDTA extracts, it was considered that certification of acetic acid extractable trace metal contents would, like for EDTA, be feasible. For ammonium acetate extraction, only Cd showed acceptable *CV* and it was concluded that certification of ammonium acetate extractable contents was not yet feasible in such, or similar, reference material and would require considerable improvement in the methodology. This failure was likely to be due to the low element contents extracted by the reagent. It was a notable feature of this trial that calibration by the individual laboratories was frequently in error, often by 10%.

Preparation of the Candidate Reference Materials

Sewage Sludge-amended Soil

The sewage sludge-amended soil was collected from Great Billings Sewage farm (Northampton, UK). Some 300 kg of field-moist soil was collected by multiple sampling to a depth of 10 cm and bulked into polyethylene bags for transport

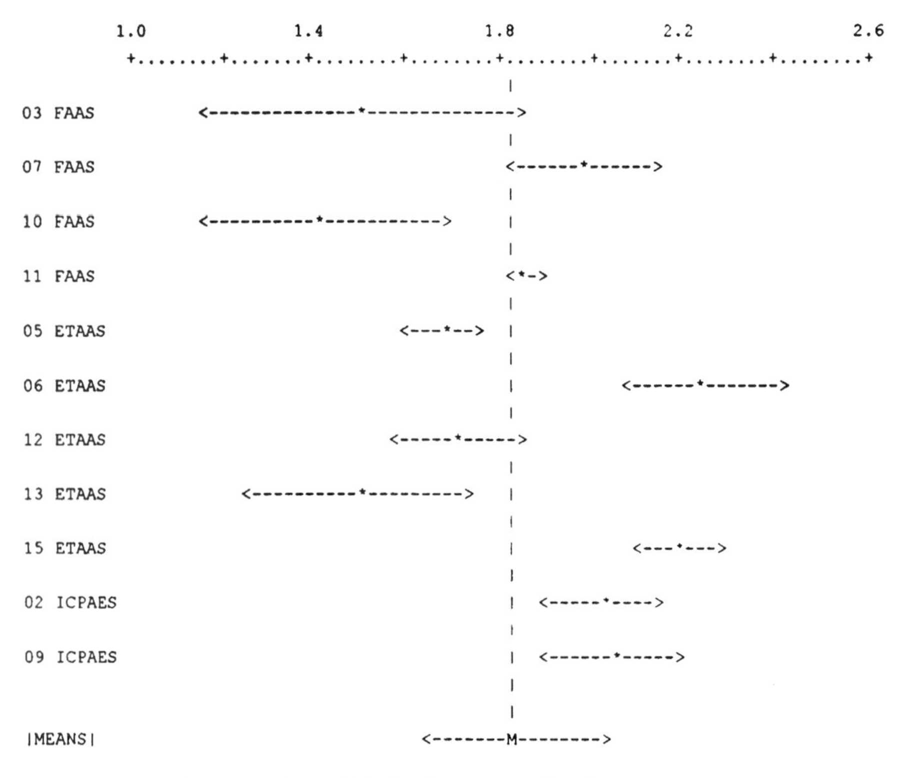

Figure 11.3 *Cadmium in CRM 601: third step in mg kg^{-1}*

Table 11.8 *Coefficients of variation[a] of the mean of laboratory means of EDTA and acetic acid extractable contents as obtained in the first interlaboratory study*

	Cd	*Cr*	*Cu*	*Ni*	*Pb*	*Zn*
EDTA extracts	9.2	25.9	8.5	13.0	11.8	5.8
Acetic acid extracts	7.5	8.2	10.0	18.1	24.6	6.9
Ammonium acetate extracts	10.9	40.6	23.4	22.5	26.8	22.5

[a] *CV*/%. Adapted from [193].

to the Macaulay Land Use Research Institute (Aberdeen, UK). The whole soil was air-dried at 30 °C for 3 weeks on paper-lined aluminium trays. The dried material was then gently rolled with a wooden roller to break up large aggregates, sieved through a 2 mm round-hole sieve and stored in tightly-sealed polyethylene bags. The soil sample was thoroughly mixed and homogenized by rolling on a clean polyethylene sheet for 3 days with occasional mixing by hand. The whole sample was then gently poured on to a clean polyethelene sheet, mixed, and coned and quartered by hand. The initial sample, nominally 150 kg of air-dry (<2 mm) soil, was split by coning and quartering, bulking opposite quarters to form the half samples, and setting one half sample aside. The remaining half sample was again coned and quartered. The coning and quartering procedure continued (six cycles) until the half-sample weight was approximately 2 kg. From opposite quarters of this half-sample, 20 subsamples were taken alternately by nylon spatula into pre-cleaned brown glass bottles (capped by polyethylene screwcaps), each containing approximately 70 g. A total of 1280 bottles were obtained and 128 bottles (2 from each final half-sample) were set aside for homogeneity and stability testing.

Terra Rossa Soil

The sampling of the terra rossa soil was carried out in a farm plot amended with sewage sludge from a water treatment centre located in North-East Catalonia, Spain. Samples were taken from an area of about 250 m^2 with a small shovel to a depth of about 10 cm and sifted on-site by hand through a 0.5 cm nylon sieve into polyethylene bags. The samples were taken to the water treatment centre and again sieved through a 20 cm diameter nylon sieve with a mesh size of 2 mm into polyethylene bags for transport to the analytical chemistry laboratory of the University of Barcelona. The soil was then spread over a polyethylene sheet and air-dried at 30 °C for one week to a final water content of 1.5%. The air-dried soil was packed into a 100 litre polyethylene container, tighly sealed and dispatched to the Environment Institute of the Joint Research Centre of Ispra (Italy) for homogenization and bottling.

The air-dried (<2 mm) soil sample was transferred in total (91 kg) to a mixing drum filled with dry argon and placed on roll-bed capable of handling 100 kg samples. The homogenization of this soil, with its large spread of particle sizes, from just below 2 mm down to fractions of a micrometer, required particular care. This procedure was, therefore, carried out by mixing in the drum for over 4 weeks. Ten subsamples were taken from the centre of the drum for a preliminary check of the homogeneity by analysis using X-ray fluorescence spectrometry. Pellets were prepared from each of the 10 bottles and a number of major elements, each of which represented a particular solid phase, *e.g.* quartz, clay minerals, mica or iron oxyhydrates, were determined by XRF. A number of trace elements were also determined. These preliminary analyses gave confidence that the material had been thoroughly homogenized. Following these tests, the soil was further homogenized by mixing in the drum for 3 days. To prevent segregation of fine particles, 10 samples were taken from the centre of the drum immediately upon stopping the rotation of the mixing drum. The samples were placed in 10 pre-cleaned brown glass bottles, each containing a minimum of 70 g of soil, and stoppered. The drum was again rotated for a further two minutes and a further 10 samples were sub-sampled in the same way into bottles. The sub-sampling and bottling operation was continued until 1000 bottles of soil were obtained. The residual soil material, amounting to about 10% of the whole, was discarded, as earlier experience

Table 11.9 *Homogeneity tests*

	Cd	*Cr*	*Cu*	*Ni*	*Pb*	*Zn*
Sewage sludge-amended soil						
EDTA						
S_B	5.7 ± 1.3	7.0 ± 1.6	5.8 ± 1.3	5.7 ± 1.3	5.3 ± 1.2	5.6 ± 1.3
S_W	5.2 ± 1.6	8.5 ± 2.7	5.4 ± 1.7	5.2 ± 1.6	5.1 ± 1.6	5.1 ± 1.6
Acetic acid						
S_B	1.9 ± 0.4	1.4 ± 0.3	1.0 ± 0.2	2.8 ± 0.6	11.0 ± 2.5	2.6 ± 0.6
S_W	1.6 ± 0.5	0.8 ± 0.3	1.0 ± 0.3	2.8 ± 0.9	14.0 ± 4.4	2.8 ± 0.9
Terra rossa soil						
EDTA						
S_B	5.0 ± 1.1	11.0 ± 2.5	5.7 ± 1.3	4.3 ± 1.0	4.8 ± 1.1	5.4 ± 1.2
S_W	7.1 ± 2.2	6.4 ± 2.0	8.0 ± 2.5	8.7 ± 2.8	8.2 ± 2.6	8.2 ± 2.6
Acetic acid						
S_B	5.3 ± 1.2	10.2 ± 2.3	3.4 ± 0.8	7.2 ± 1.6	–	6.8 ± 1.5
S_W	12.9 ± 4.1	10.7 ± 3.4	4.5 ± 1.4	8.3 ± 2.6	4.5 ± 1.4	6.8 ± 2.2

suggested that this might be less homogeneous. One hundred bottles, selected sequentially over the whole bottling procedure, were sent to the Macauley Land Use Research Institute for homogeneity and stability testing.

Homogeneity Control

The extractants (0.05 mol L^{-1} EDTA and 0.43 mol L^{-1} acetic acid) were prepared as laid out in the Annex (Section 11.7). All precautions were taken to avoid contamination during the extraction procedures. The trace element contents (Cd, Cr, Cu, Ni, Pb and Zn) in the extracts were determined by inductively coupled plasma atomic emission spectrometry (ICP-AES). Calibrant solutions were prepared from stock solutions (1000 mg L^{-1}) of the individual elements in 0.5 mol L^{-1} HNO_3.

For the homogeneity study, the six elements were determined in each candidate CRM by analysing 10 subsamples taken from one bottle (within-bottle homogeneity test, S_W) and one subsample in each of 20 different bottles selected during the bottling procedure (between-bottle homogeneity test, S_B). The *CV*'s and the total uncertainty U_{CV} for the extractable trace element contents between (S_B) and within (S_W) bottles are given in Table 11.9.

In the case of the sewage sludge-amended soil, little analytical difficulty was experienced, as illustrated by the good agreement obtained between the within-bottle and between-bottle CVs. For the terra rossa soil, however, the lower extractable contents were closer to the detection limits and consequently poorer analytical precision was observed, in particular for Cr (EDTA extractable contents), Cd and Pb (acetic acid extractable contents).

An *F*-test has been used to test for significant difference between the within-bottle and between-bottle test results. With the exception of two results for chromium, there was no significant difference between the within-bottle and between-bottle test results.

Table 11.10 *Stability tests – EDTA: sewage sludge-amended soil*

Element	*Time*/month	$R_T \pm U_T$ (20 °C)	$R_T \pm U_T$ (40 °C)
Cd	1	1.04 ± 0.05	1.05 ± 0.04
	3	0.98 ± 0.05	0.94 ± 0.04
	6	0.98 ± 0.04	0.98 ± 0.05
	12	0.99 ± 0.04	1.00 ± 0.06
Cr	1	1.03 ± 0.06	1.00 ± 0.09
	3	1.06 ± 0.06	0.95 ± 0.09
	6	0.94 ± 0.07	0.92 ± 0.09
	12	1.01 ± 0.08	0.96 ± 0.09
Cu	1	1.02 ± 0.05	1.01 ± 0.04
	3	0.99 ± 0.07	0.95 ± 0.05
	6	0.98 ± 0.04	1.01 ± 0.04
	12	0.99 ± 0.04	1.01 ± 0.08
Ni	1	1.03 ± 0.06	1.03 ± 0.05
	3	0.98 ± 0.07	0.93 ± 0.05
	6	0.98 ± 0.04	0.98 ± 0.06
	12	0.98 ± 0.04	0.99 ± 0.07
Pb	1	1.04 ± 0.05	1.03 ± 0.05
	3	0.99 ± 0.06	0.93 ± 0.04
	6	0.97 ± 0.04	0.97 ± 0.06
	12	0.99 ± 0.04	1.02 ± 0.10
Zn	1	1.02 ± 0.05	1.01 ± 0.04
	3	0.98 ± 0.06	0.94 ± 0.05
	6	0.98 ± 0.05	0.97 ± 0.06
	12	0.98 ± 0.04	0.98 ± 0.07

On the basis of these results, the materials were considered to be homogeneous at a level of 5 g (as specified in the extraction protocols).

Stability Control

The stability of the extractable trace element contents was tested at −20, +20 and +40 °C during a period of 12 months and the EDTA and acetic acid extractable contents of Cd, Cr, Cu, Ni, Pb and Zn were determined (in five replicates) after 1, 3, 6 and 12 months. The procedures used were the same as in the homogeneity study, with the exception of Pb for which ETAAS was used as the final determination. The evaluation of the stability was based on the procedure described in Chapter 3, using the samples stored at −20 °C as reference for the samples stored at +20 and +40 °C, respectively.

The results are shown in Tables 11.10–11.13. No instability could be demonstrated and hence the materials were suitable as candidate CRMs.

Technical Discussion

At the technical meeting, it was recalled that strict observance of the extraction protocols would be a criterion for considering the results for discussion. An example of a possible source of discrepancy occurring as the result of non-

Table 11.11 *Stability tests – acetic acid: sewage sludge-amended soil*

Element	*Time*/month	$R_T \pm U_T$ (20 °C)	$R_T \pm U_T$ (40 °C)
Cd	1	1.00 ± 0.04	1.00 ± 0.03
	3	1.00 ± 0.02	1.02 ± 0.03
	6	0.98 ± 0.02	1.02 ± 0.03
	12	0.96 ± 0.03	1.02 ± 0.03
Cr	1	1.00 ± 0.01	1.02 ± 0.02
	3	1.02 ± 0.02	1.08 ± 0.02
	6	0.98 ± 0.02	1.08 ± 0.03
	12	0.98 ± 0.03	1.16 ± 0.03
Cu	1	1.01 ± 0.02	1.02 ± 0.01
	3	1.01 ± 0.02	1.06 ± 0.02
	6	1.00 ± 0.01	1.06 ± 0.02
	12	1.00 ± 0.03	1.09 ± 0.03
Ni	1	1.02 ± 0.03	1.00 ± 0.03
	3	1.00 ± 0.04	1.03 ± 0.04
	6	0.95 ± 0.03	1.00 ± 0.03
	12	1.03 ± 0.05	1.02 ± 0.06
Pb	1	1.00 ± 0.26	1.02 ± 0.25
	3	0.92 ± 0.07	0.87 ± 0.08
	6	1.06 ± 0.25	1.10 ± 0.33
	12	1.15 ± 0.25	1.21 ± 0.25
Zn	1	1.02 ± 0.04	1.01 ± 0.09
	3	1.07 ± 0.03	1.04 ± 0.04
	6	0.97 ± 0.04	1.00 ± 0.04
	12	0.97 ± 0.05	1.01 ± 0.06

Table 11.12 *Stability tests – EDTA: terra rossa soil*

Element	*Time*/month	$R_T \pm U_T$ (20 °C)	$R_T \pm U_T$ (40 °C)
Cd	1	0.97 ± 0.04	1.00 ± 0.07
	3	0.98 ± 0.09	1.00 ± 0.10
	6	1.08 ± 0.11	1.03 ± 0.09
	12	1.00 ± 0.07	0.98 ± 0.07
Cr	1	1.03 ± 0.06	1.02 ± 0.10
	3	1.02 ± 0.07	0.99 ± 0.04
	6	1.01 ± 0.06	1.03 ± 0.07
	12	1.04 ± 0.05	0.99 ± 0.07
Cu	1	0.97 ± 0.04	1.02 ± 0.05
	3	1.00 ± 0.07	1.03 ± 0.12
	6	1.08 ± 0.11	0.99 ± 0.12
	12	1.00 ± 0.07	0.96 ± 0.05
Ni	1	0.96 ± 0.04	0.96 ± 0.07
	3	0.93 ± 0.09	0.93 ± 0.11
	6	1.05 ± 0.07	0.92 ± 0.07
	12	0.99 ± 0.07	0.90 ± 0.07
Pb	1	0.96 ± 0.03	0.98 ± 0.04
	3	0.98 ± 0.08	1.00 ± 0.11
	6	0.99 ± 0.17	0.92 ± 0.16
	12	1.03 ± 0.05	0.98 ± 0.06
Zn	1	0.97 ± 0.05	1.00 ± 0.05
	3	0.9 ± 0.07	1.00 ± 0.10
	6	1.08 ± 0.12	1.01 ± 0.11
	12	1.02 ± 0.07	0.98 ± 0.07

Table 11.13 *Stability tests – acetic acid: terra rossa soil*

Element	*Time*/month	$R_T \pm U_T$ *(20 °C)*	$R_T \pm U_T$ *(40 °C)*
Cd	1	1.00 ± 0.11	0.92 ± 0.08
	3	1.00 ± 0.09	1.00 ± 0.12
	6	1.00 ± 0.24	1.09 ± 0.22
	12	0.92 ± 0.11	0.92 ± 0.10
Cr	1	1.00 ± 0.16	1.00 ± 0.27
	3	1.09 ± 0.17	1.00 ± 0.16
	6	1.00 ± 0.14	1.09 ± 0.22
	12	0.89 ± 0.25	0.89 ± 0.23
Cu	1	1.00 ± 0.05	1.05 ± 0.05
	3	0.98 ± 0.04	1.00 ± 0.03
	6	0.83 ± 0.31	0.88 ± 0.34
	12	0.95 ± 0.08	0.98 ± 0.09
Ni	1	1.00 ± 0.05	0.98 ± 0.09
	3	1.03 ± 0.10	1.00 ± 0.08
	6	1.00 ± 0.16	1.05 ± 0.13
	12	0.87 ± 0.07	0.93 ± 0.12
Pb	1	0.94 ± 0.14	0.94 ± 0.14
	3	0.89 ± 0.17	1.04 ± 0.19
	6	1.03 ± 0.05	1.00 ± 0.05
	12	0.97 ± 0.08	1.00 ± 0.14
Zn	1	1.04 ± 0.07	1.02 ± 0.07
	3	1.05 ± 0.11	0.99 ± 0.06
	6	0.98 ± 0.10	1.05 ± 0.08
	12	0.87 ± 0.16	0.86 ± 0.17

adherence to the protocols was given by one laboratory which used a reciprocating shaker instead of the recommended end-over-end shaker and obtained systematically low results; the repetition of the analyses clearly showed that the error was due to this fact (Table 11.14). On the basis of this remark, the results of three laboratories were systematically rejected.

The shaker speed was also considered to be an important parameter since it represents one factor (along with the shaker type) that affects the maintenance of the samples in suspension during the extraction. The protocol stipulates a speed of 30 rpm but speeds ranging from 20 to 40 rpm were considered to be acceptable; exceptionally, results at 14 rpm provided by one laboratory were accepted since the soil samples were found to remain in suspension in this particular case.

The uses of filtration, centrifugation or centrifugation followed by filtration were all found acceptable.

For EDTA extracts, good agreement between the results of laboratories using high-speed reciprocating shaking and the results of the bulk of the participants (using end-over-end shaking) could be observed; as stressed before, these results could, however, not be retained for certification since the extraction protocol was not strictly followed. For chromium in CRM 483, it was considered that FAAS using an air/acetylene flame without a releasing agent was not acceptable. Two laboratories, a using nitrous oxide/acetylene

Table 11.14 *Differences in results obtained using a reciprocating and end-over-end shaker*

Shaker	*Cd*	*Cr*	*Cu*	*Ni*	*Pb*	*Zn*
CRM 483						
EDTA						
Reciprocating	19.9±1.0	18.2±1.9	165±6	24.3±1.3	149±12	478±15
End-over-end	24.6±1.3	27.5±1.0	218±12	30.4±1.6	232±10	610±29
Acetic acid						
Reciprocating	7.4±0.4	5.5±0.6	17.6±1.0	9.9±0.6	2.37±0.16	257±3
End-over-end	17.5±0.3	17.2±0.2	31.7±0.6	22.2±0.8	1.47±0.16	572±22
CRM 484						
EDTA						
Reciprocating	0.48±0.03	0.46±0.03	61.0±1.3	0.92±0.06	25.7±1.3	104±3
End-over-end	0.47±0.01	0.62±0.04	85.9±2.3	1.29±0.03	41.2±0.9	146±3
Acetic acid						
Reciprocating	0.22±0.03	0.21±0.02	18.8±1.8	0.93±0.14	2.46±0.11	96±12
End-over-end	0.44±0.03	0.35±0.04	30.8±1.0	1.58±0.21	1.85±0.14	172±11

Table 11.15 *Certified contents of EDTA extractable contents of Cd, Cr, Cu, Ni, Pb and Zn in CRM 483 (sewage sludge-amended soil)*

	Certified value/ mg kg^{-1}	*Uncertainty/* mg kg^{-1}	p[a]
Cd	24.3	1.3	18
Cr	28.6	2.6	9
Cu	215	11	17
Ni	28.7	1.7	17
Pb	229	8	17
Zn	612	19	17

[a] p = number of data sets.

flame, were on the high side and in the absence of detailed information on this flame in this matrix, these results were rejected; in the case of CRM 484, a high spread of results for chromium could not allow this element to be certified.

No particular problems were experienced with acetic acid, except for lead for which some results were too close to the detection limits. As observed for the EDTA extracts in CRM 484, the wide dispersion of chromium results could not allow certification; the results are given as an indicative value.

Certified Values

The certified values (unweighted mean of p accepted sets of results) and their uncertainties (half-width of the 95% confidence intervals) are given in Tables 11.15–11.18 as mass fractions of the respective extracts (based on dry mass) in

Table 11.16 *Certified contents of acetic acid extractable contents of Cd, Cr, Cu, Ni, Pb and Zn in CRM 483 (sewage sludge-amended soil)*

	Certified value/ mg kg^{-1}	*Uncertainty*/ mg kg^{-1}	*p*[a]
Cd	18.3	0.6	18
Cr	18.7	1.0	17
Cu	33.5	1.6	18
Ni	25.8	1.0	15
Pb	2.10	0.25	12
Zn	620	24	18

[a] *p* = number of data sets.

Table 11.17 *Certified contents of EDTA extractable contents of Cd, Cu, Ni, Pb and Zn in CRM 484 (terra rossa soil)*

	Certified value/ mg kg^{-1}	*Uncertainty*/ mg kg^{-1}	*p*
Cd	0.51	0.03	14
Cu	88.1	3.8	17
Ni	1.39	0.11	15
Pb	47.9	2.6	18
Zn	152	7	17

Table 11.18 *Certified contents of acetic acid extractable contents of Cd, Cu, Ni, Pb and Zn in CRM 484 (terra rossa soil)*

	Certified value/ mg kg^{-1}	*Uncertainty*/ mg kg^{-1}	*p*
Cd	0.48	0.04	13
Cu	33.9	1.4	18
Ni	1.69	0.15	16
Pb	1.17	0.16	11
Zn	193	7	17

mg kg^{-1}. Examples of bar graphs (Zn in both CRMs) are given in Figures 11.4–7.

Indicative Values

During the course of the certification, some laboratories carried out other types of extraction procedures which are described in the Annex (Section 11.7), *i.e.* using calcium chloride, ammonium nitrate or sodium nitrate. These results are given in Tables 11.19–21 together with the standard deviation obtained,

BAR-GRAPHS FOR LABORATORY MEANS AND 95% CI

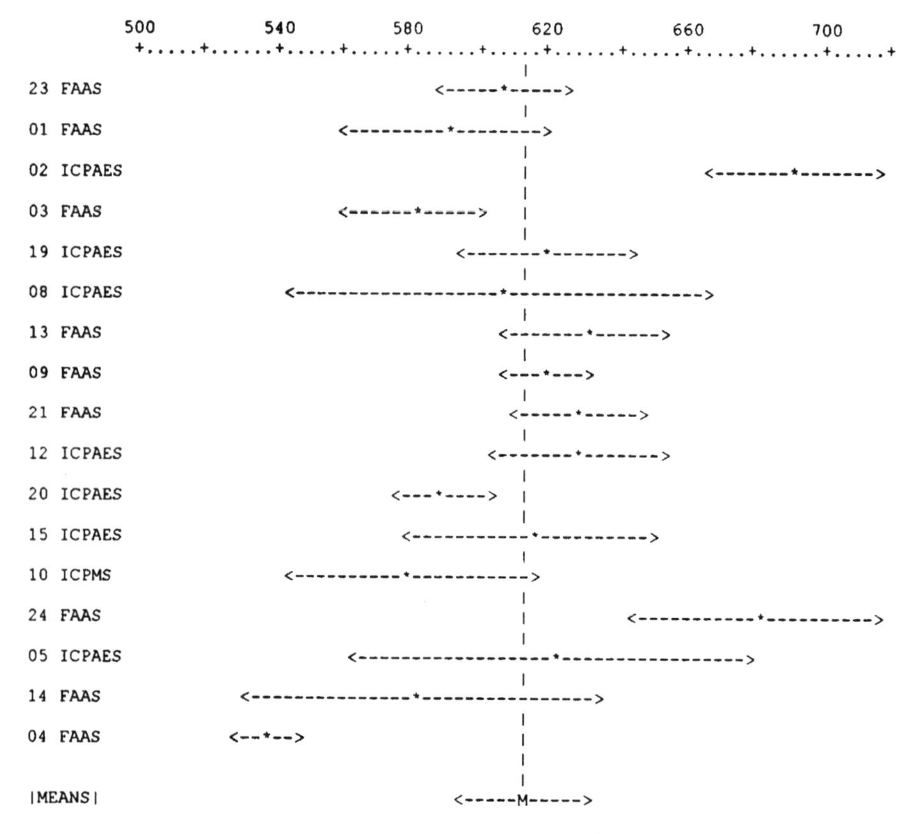

Figure 11.4 *EDTA extractable content of zinc in mg kg^{-1} (CRM 483)*

the method(s) used and the number of sets of results. It is emphasized that the values of this table are not certified.

11.6 Single Extractions for Calcareous Soil

Interlaboratory Study

Selected Reference Material

A sewage contaminated calcareous soil was selected from the bank of reference materials of the Environment Institute of the Joint Research Centre of Ispra (Italy) in order to present both the characteristics of a calcareous soil ($CaCO_3$ content of 228 g kg^{-1}) and with heavy metal content well above the determination limits of the currently used flame atomic absorption spectrometric method. The soil sample was composed of 15.4% sand (2 mm to 50 mm), 9.3% coarse silt (50 to 20 μm), 34.0% fine silt (20 to 2 μm) and 41.3% clay (<2 μm).

BAR-GRAPHS FOR LABORATORY MEANS AND 95% CI

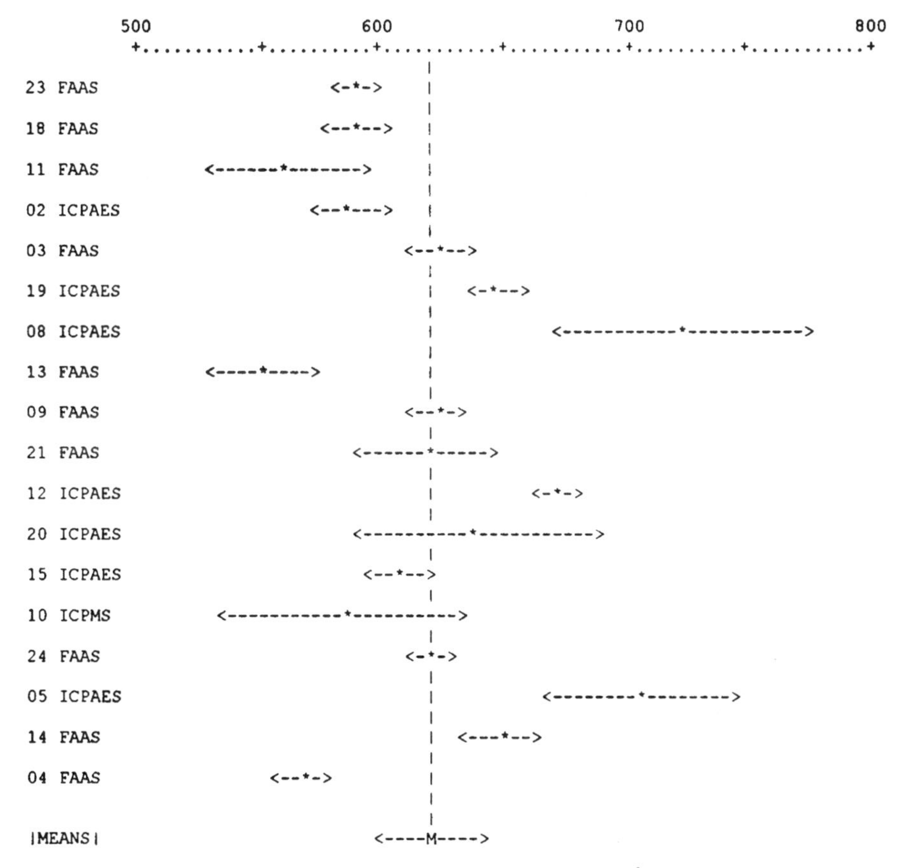

Figure 11.5 *Acetic acid extractable content of zinc in mg kg^{-1} (CRM 483)*

This material was sieved to 2 mm, homogenized, sterilized and bottled in dark brown bottles for homogeneity and stability studies, which showed that the material was sufficiently homogeneous and stable to be used in an interlaboratory study.

Selection of Extractants

Three different extractants were discussed for the feasibility study on calcareous soil, namely EDTA, DTPA and mixed acid ammonium acetate/EDTA.

EDTA (ethylenediaminetetraacetic acid) extracts of soils tend in general to correlate well with plant contents, in particular with the plant-available fraction for Cd, Cu, Ni, Pb and Zn [208–210]. EDTA (0.05 mol L^{-1}) at pH 7 was used in the certification of the two soils mentioned above [197]. This test is assumed to extract both carbonate-bound and organically-bound fractions of metals and was hence considered to be suitable for calcareous soil analysis.

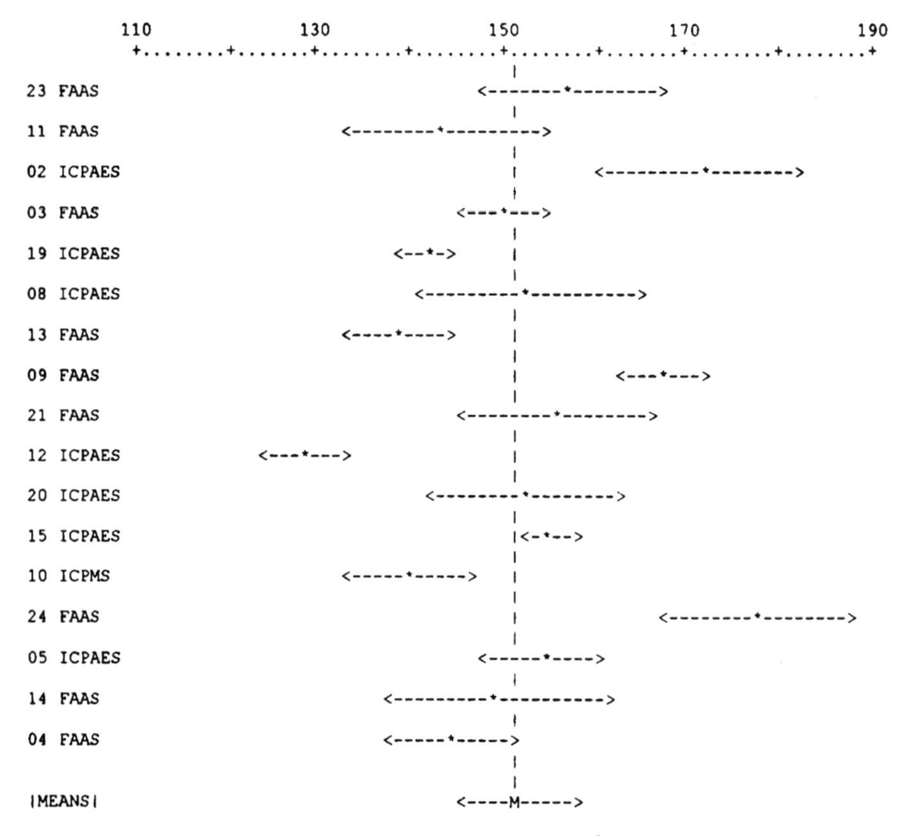

Figure 11.6 *EDTA extractable content of zinc in mg kg^{-1} (CRM 484)*

Mixed acid ammonium acetate/EDTA reagent introduced by Lakanen and Ervio [211] was found to provide good correlation with wheat and maize contents for Cu and Zn [212]. This method was discarded as there is some evidence that EDTA at pH 5.5 can precipitate Cr, Pb and Zn, as was observed in polluted sediments [213]. Moreover, the benefits of supplementing the acid ammonium acetate did not seem worth the more complicated procedure.

DTPA (diethylenetriaminepentaacetic acid) extracts have been extensively studied, particularly as a Zn diagnostic procedure for calcareous soils [210,214,215]. DTPA (0.005 mol L^{-1}) at pH 7.3 was thought to be a possibility for the determination of extractable Cd, Cu, Fe and Mn, but would be less suitable for Cr and Ni. This procedure was stressed to be more complicated than the EDTA one and was recognized to be often misused [201]; in addition, DTPA extracts less than EDTA which might lead to sensitivity problems. This method was, however, retained for the feasibility study owing to its high degree of acceptance. EDTA was the method of preference as it was extensively tested in previous studies and will be soon adopted as an ISO Standard.

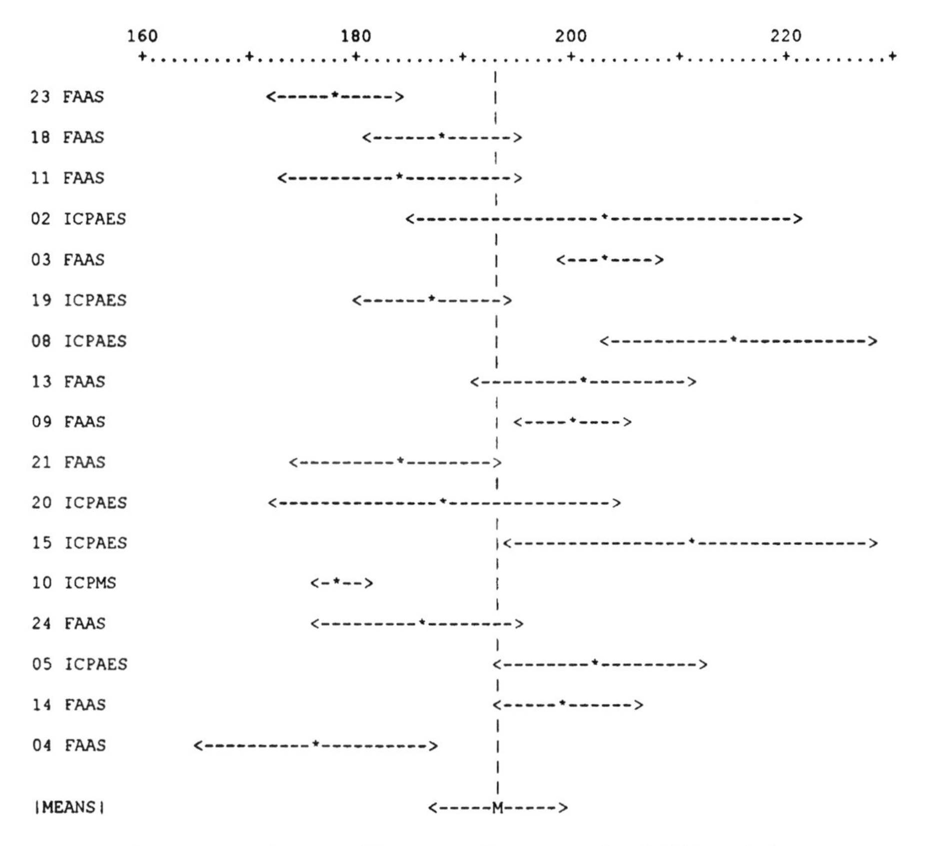

Figure 11.7 *Acetic acid extractable content of zinc in mg kg^{-1} (CRM 484)*

Table 11.19 *Indicative values: calcium chloride extractable contents*

Element/CRM	*Mean*	*SD*	p^a	*Techniques used*
CRM 483				
Cd	0.45	0.05	10	FAAS, ETAAS, ICP-AES
Cr	0.35	0.09	9	FAAS, ETAAS, ICP-AES
Cu	1.2	0.4	11	FAAS, ETAAS, ICP-AES
Ni	1.4	0.2	10	FAAS, ETAAS, ICP-AES
Pb	<0.06	–	8	FAAS, ETAAS, ICP-AES
Zn	8.3	0.7	9	FAAS, ETAAS, ICP-AES
CRM 484				
Cd	<0.08	–	9	FAAS, ETAAS, ICP-AES
Cr	<0.09	–	7	FAAS, ETAAS, ICP-AES
Cu	0.67	0.29	10	FAAS, ETAAS, ICP-AES
Ni	<0.05	–	9	FAAS, ETAAS, ICP-AES
Pb	<0.06	–	7	FAAS, ETAAS, ICP-AES
Zn	0.31	0.17	7	FAAS, ETAAS, ICP-AES

[a] p = sets of results (each of 5 replicates).

Table 11.20 *Indicative values: sodium nitrate extractable contents*

Element/CRM	*Mean*	*SD*	*p*[a]	*Techniques used*
CRM 483				
Cd	0.08	0.03	6	FAAS, ETAAS, ICP-AES
Cr	0.30	0.07	4	FAAS, ETAAS, ICP-AES
Cu	0.89	0.22	6	FAAS, ETAAS, ICP-AES
Ni	0.65	0.07	5	FAAS, ETAAS, ICP-AES
Pb	<0.03	–	4	FAAS, ETAAS, ICP-AES
Zn	2.7	0.8	5	FAAS, ETAAS, ICP-AES
CRM 484				
Cd	<0.05	–	7	FAAS, ETAAS, ICP-AES
Cr	<0.03	–	6	FAAS, ETAAS, ICP-AES
Cu	0.48	0.15	8	FAAS, ETAAS, ICP-AES
Ni	0.023	0.005	6	FAAS, ETAAS, ICP-AES
Pb	<0.06	–	7	FAAS, ETAAS, ICP-AES
Zn	0.09	0.04	6	FAAS, ETAAS, ICP-AES

[a] *p* = sets of results (each of 5 replicates).

Table 11.21 *Indicative values: ammonium nitrate extractable contents*

Element/CRM	*Mean*	*SD*	*p*[a]	*Techniques used*
CRM 483				
Cd	0.26	0.05	9	FAAS, ETAAS, ICP-AES
Cr	0.27	0.10	8	FAAS, ETAAS, ICP-AES
Cu	1.2	0.3	9	FAAS, ETAAS, ICP-AES
Ni	1.1	0.3	9	FAAS, ETAAS, ICP-AES
Pb	0.020	0.013	4	FAAS, ETAAS, ICP-AES
Zn	6.5	0.9	8	FAAS, ETAAS, ICP-AES
CRM 484				
Cd	0.003	0.002	7	FAAS, ETAAS, ICP-AES
Cr	<0.06	–	7	FAAS, ETAAS, ICP-AES
Cu	1.1	0.4	10	FAAS, ETAAS, ICP-AES
Ni	0.033	0.017	6	FAAS, ETAAS, ICP-AES
Pb	<0.06	–	7	FAAS, ETAAS, ICP-AES
Zn	0.17	0.05	9	FAAS, ETAAS, ICP-AES

[a] *p* = sets of results (each of 5 replicates).

Results and discussion

The results of the interlaboratory study were discussed in a technical meeting held in Brussels on 1 June 1994. The most critical steps discussed were the type and speed of shaking, filtration and centrifugation. As a general remark, it was stressed that the use of standard additions for calibration was a prerequisite for electrothermal atomic absorption spectrometry. Small standard deviations obtained by the laboratories (in particular for EDTA-extractable Cd) led to an apparent spread of results. It was not clear how these low variances could be obtained by performing five independant extractions as was requested in the instructions for participation.

Table 11.22 *Coefficients of variation[a] of the mean of laboratory means of EDTA and DTPA extractable trace metal contents. These data correspond to the results before technical scrutiny*

	Cd	*Cr*	*Cu*	*Ni*	*Pb*	*Zn*
EDTA extracts	29	30	24	28	21	27
DTPA extracts	26	79	24	39	35	48

[a] *CV*/%.

One laboratory used a horizontal shaker (whereas an end-over-end shaker is required in the extraction protocol) and obtained systematically low results. Its sets of data were rejected for all the elements.

Another laboratory observed a loss of Cd in the extract after an equilibrium time of 48 h which was, however, not confirmed by the other participants. This laboratory applied a second filtration step with a membrane filter (not included in the protocol), which could explain the losses observed. Losses due to precipitation were considered to be unlikely since the stability of the extracts over a period of 72 h was verified, providing that they were stored at 4 °C. These examples demonstrated how essential it was that the extraction protocols be strictly followed, since a slight variation may lead to an incomparability of data. The protocol should state that the analysis be carried out with the samples "as received" and that the moisture content in the soil intake be corrected for by determining the loss of mass on drying a separate soil sample.

The DTPA extraction was criticized owing to the high mass/volume ratio, which limited the volume of extract collected after centrifugation. This procedure was not considered to be applicable to Cr determination.

Table 11.22 gives the coefficients of variation between laboratories obtained before the technical scrutiny (*i.e.* including all sets of data). As shown in this table, the overall variability in the results of EDTA and DTPA extractable contents of Cd and Cu was comparable. However, the variability observed for the DTPA extractable contents of Ni, Pb and Zn appeared significantly higher in comparison with the EDTA extractable contents. Considering the variability of the raw set of data, it was stressed that a significant improvement would be needed to allow the possible certification of a calcareous soil for EDTA and DTPA extractable trace metals at this stage. As an example, the *CV*'s obtained in an interlaboratory study on EDTA-extractable trace metals in sewage sludge-amended soil ranged from 6 to 13% (for Cd, Cu, Ni, Pb and Zn) to 26% for Cr, which corresponded to the results of data selected after technical scrutiny. It is to be noted, however, that the *CV*'s of selected data in the present exercise would have ranged between 5–8% (Cu, Pb, Zn) and 15–27% (Cd, Cr, Ni) for the EDTA extraction, and between 6 and 11% (all elements except Cr) for the DTPA extraction, which would give good perspectives for attempting certification.

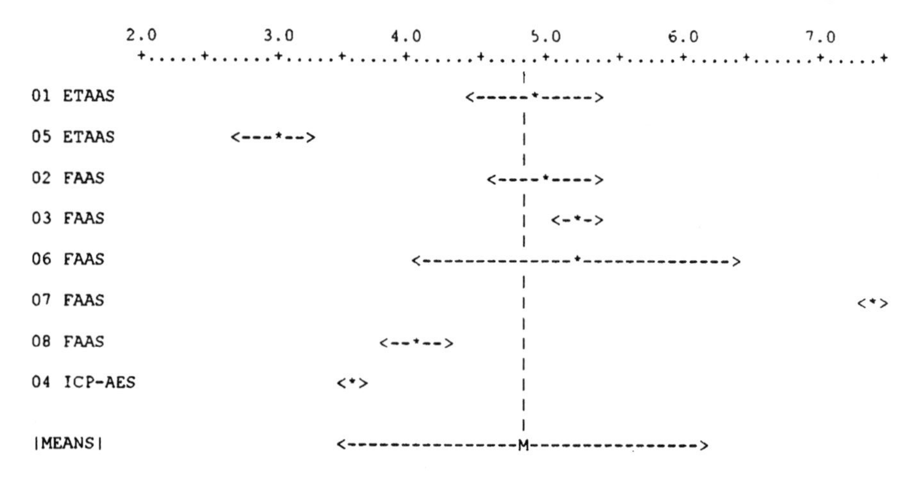

Figure 11.8 *Bar graph for laboratory means and standard deviation of EDTA extractable Ni*

It was commented that a linear relationship could be established between the EDTA and DTPA results. As a theoretical relationship actually exists between EDTA and DTPA extraction efficiency, it could be possible to detect systematic errors by establishing this relationship for each soil analysed and comparing the actual results obtained; this approach could be used at a later stage for certification.

On the basis of the results obtained in this exercise, the choice of EDTA and DTPA for certification was discussed. Whereas EDTA was widely accepted, the choice of DTPA was more criticized because of its operational difficulties; the wide use of this latter would, however, justify the use of this extraction in the certification campaign, provided that its limitations in comparison with EDTA were clearly identified.

The following recommendations were expressed for certification:

- Use of standard additions for electrothermal atomic absorption spectrometry
- Preparation of calibrant solutions in the same extractant to avoid possible problems of surface tension and other interference effects
- Verification of the purity of the extractant and calibrant used
- Possible amendment of the DTPA protocol prior to certification
- Selection of a broader range of final detection techniques

Preparation of the Candidate Reference Material

The material was collected at San Pellegrino Parmense (Italy) in February 1994, following a prospective study of various sites in Italy which aimed at identifying a material with reasonably high calcium carbonate content. Stones

Table 11.23 *Elemental composition of CRM 600 (total concentrations)*

	Major components /g kg^{-1}		Trace elements mg kg^{-1}
SiO_2	340	Cr	97
Al_2O_3	101	Cu	175
CaO	170	Mn	985
K_2O	12	Ni	232
Fe_2O_3	32	Pb	52
MgO	26	Zn	401
TiO_2	5		
S	11		
P_2O_5	48		

and large plant litter were removed prior to sieving at 2 mm mesh. The fraction less than 2 mm was collected in stainless-steel trays in which the material was disposed in thin layers of a few cm thickness to dry at ambient temperature. The material was sieved again after drying to remove lumps which were formed during the drying process. The residual moisture content at this stage was found to be 3.8% (measured by taking a separate portion of 1 g dried at 105 °C until constant mass was attained). The sieved material was transferred to a PVC mixing drum filled with dry argon, and was homogenized for 12 days at about 48 rpm. The final material was manually bottled in brown glass bottles; the bottling procedure was carried out by filling 10 bottles, closing the drum and mixing the material again for 2 minutes before bottling 10 further bottles, and so on until only some centimeters of soil remained in the drum (which were discarded). All bottles were closed with an insert and a screw cap and stored at ambient temperature, 1050 bottles each containing about 70 g were produced.

Soil Characterization

One bottle was selected and the elemental composition was determined by X-ray fluorescence spectrometry. Table 11.23 lists the contents of some major and trace elements (it is emphasized that these values are not certified and are given for information only).

Homogeneity Control

The extractants (0.05 mol L^{-1} EDTA and 0.005 mol L^{-1} DTPA) were prepared as laid out in the Annex (Section 11.7). All precautions were taken to avoid contamination during the extraction procedures. The trace element contents (Cd, Cr, Cu, Ni, Pb and Zn) in the extracts were determined by flame atomic absorption spectrometry or electrothermal atomic absorption spectrometry with Zeeman background correction.

Table 11.24 *Homogeneity tests*

	Cd	*Cr*	*Cu*	*Ni*	*Pb*	*Zn*
EDTA						
S_B	1.9 ± 0.4	10.8 ± 2.4	2.4 ± 0.5	2.3 ± 0.5	7.3 ± 1.6	3.1 ± 0.7
S_W	2.3 ± 0.5	6.9 ± 1.5	2.4 ± 0.5	2.5 ± 0.6	4.5 ± 1.0	3.1 ± 0.7
DTPA						
S_B	12.9 ± 2.9	5.6 ± 1.2	1.7 ± 0.4	1.4 ± 0.3	1.8 ± 0.4	2.3 ± 0.5
S_W	1.8 ± 0.4	4.0 ± 0.9	1.3 ± 0.3	1.3 ± 0.3	2.0 ± 0.4	1.4 ± 0.3

For the homogeneity study, the six elements were determined in the candidate CRM by analyzing 10 sub-samples taken from one bottle (within-bottle homogeneity test, S_W) and one sub-sample in each of 20 different bottles selected during the bottling procedure (between-bottle homogeneity test); in the case of the DTPA procedure, the within-bottle homogeneity test was carried out on two bottles instead of one, owing to the amount of sample intake required (10 g). The *CV*'s and the total uncertainty U_{CV} for the extractable trace element contents between (S_B) and within (S_W) bottles, are given in Table 11.24.

For most of the extractable metal contents, the overlap was within the total uncertainty U_T of the *CV* (an approximation of the uncertainty U_{CV} of the *CV* is calculated as follows: $U_{CV} \approx CV/\sqrt{2n}$). Differences between the within-bottle and between-bottle *CV*'s observed, *e.g.* for Cd or Cr, was considered to be rather an analytical artefact than an indication of inhomogeneity. The material was then considered to be homogeneous for the stated level of intake (5 and 10 g for EDTA and DTPA, respectively).

Stability Control

The stability of the extractable trace element contents was tested at −20, +20 and +40 °C during a period of 12 months and the EDTA and DTPA extractable contents of Cd, Cr, Cu, Ni, Pb and Zn were determined (in five replicates) after 1, 3, 6 and 12 months. The procedures used were the same as in the homogeneity study. The stability was assessed as described before, *i.e.* using the samples stored at −20 °C as reference for the samples stored at +20 and +40 °C.

The results showed that the stability was in most cases suited for the material to be used as a CRM (Tables 11.25 and 11.26). Some risks of instability were, however, suspected at 40 °C owing to possible changes in the extractability of some elements (*e.g.* Cu and Zn); these changes, induced by the high storage temperature, could be related to changes in the status of the organic matter or in the crystallographic compounds of Fe or Mn. Hence, it is recommended to avoid storage at temperatures above 20 °C.

Table 11.25 *Stability tests – EDTA extractable contents in CRM 600*

Element	*Time*/month	$R_T \pm U_T$ (20 °C)	$R_T \pm U_T$ (40 °C)
Cd	1	0.96 ± 0.05	1.00 ± 0.06
	3	1.01 ± 0.03	1.06 ± 0.01
	6	1.11 ± 0.09	1.14 ± 0.10
	12	1.11 ± 0.06	1.15 ± 0.09
Cr	1	0.94 ± 0.12	1.05 ± 0.16
	3	0.92 ± 0.09	0.97 ± 0.04
	6	1.06 ± 0.08	1.04 ± 0.09
	12	1.02 ± 0.06	1.16 ± 0.19
Cu	1	1.03 ± 0.03	0.89 ± 0.03
	3	1.12 ± 0.03	1.44 ± 0.03
	6	1.23 ± 0.10	1.70 ± 0.14
	12	1.20 ± 0.06	2.07 ± 0.17
Ni	1	0.80 ± 0.04	0.70 ± 0.03
	3	0.97 ± 0.06	0.96 ± 0.05
	6	1.10 ± 0.04	0.90 ± 0.05
	12	0.87 ± 0.03	0.88 ± 0.05
Pb	1	1.00 ± 0.04	0.86 ± 0.02
	3	1.00 ± 0.07	0.92 ± 0.06
	6	1.13 ± 0.08	1.00 ± 0.09
	12	0.92 ± 0.04	0.94 ± 0.08
Zn	1	0.97 ± 0.06	0.85 ± 0.05
	3	0.97 ± 0.03	0.93 ± 0.02
	6	1.00 ± 0.03	0.86 ± 0.02
	12	0.96 ± 0.05	0.94 ± 0.08

Table 11.26 *Stability tests – DTPA extractable contents in CRM 600*

Element	*Time*/month	$R_T \pm U_T$ (20 °C)	$R_T \pm U_T$ (40 °C)
Cd	1	1.03 ± 0.03	0.93 ± 0.03
	3	0.99 ± 0.01	0.87 ± 0.02
	6	0.99 ± 0.05	0.69 ± 0.04
	12	1.01 ± 0.06	1.01 ± 0.15
Cr	1	0.94 ± 0.06	0.88 ± 0.06
	3	0.85 ± 0.11	0.65 ± 0.06
	6	1.00 ± 0.08	0.93 ± 0.07
	12	0.95 ± 0.17	1.00 ± 0.36
Cu	1	1.07 ± 0.05	0.97 ± 0.09
	3	1.05 ± 0.02	1.15 ± 0.09
	6	1.11 ± 0.03	1.25 ± 0.11
	12	1.21 ± 0.15	1.89 ± 0.32
Ni	1	1.01 ± 0.04	0.97 ± 0.03
	3	0.98 ± 0.02	0.84 ± 0.02
	6	0.95 ± 0.05	0.81 ± 0.03
	12	0.96 ± 0.06	0.87 ± 0.08
Pb	1	1.00 ± 0.04	0.83 ± 0.04
	3	0.88 ± 0.02	0.65 ± 0.03
	6	0.88 ± 0.04	0.62 ± 0.04
	12	0.81 ± 0.04	0.82 ± 0.15
Zn	1	1.03 ± 0.04	0.84 ± 0.04
	3	0.92 ± 0.02	0.68 ± 0.03
	6	0.89 ± 0.10	0.58 ± 0.06
	12	0.87 ± 0.05	0.73 ± 0.14

Table 11.27 *Certified contents of EDTA extractable contents of Cd, Cr, Cu, Ni, Pb and Zn*

	Certified value/ mg kg^{-1}	*Uncertainty*/ mg kg^{-1}	p^a
Cd	2.68	0.09	18
Cr	0.205	0.022	9
Cu	57.3	2.5	17
Ni	4.52	0.25	17
Pb	59.7	1.8	17
Zn	383	12	17

[a] p = number of data sets.

Table 11.28 *Certified contents of DTPA contents of Cd, Cr, Cu, Ni, Pb and Zn*

	Certified value/ mg kg^{-1}	*Uncertainty*/ mg kg^{-1}	p^a
Cd	1.34	0.04	18
Cr	0.014	0.003	17
Cu	32.3	1.0	18
Ni	3.31	0.13	15
Pb	15.0	0.5	12
Zn	142	6	18

[a] p = number of data sets.

Technical Evaluation

As mentioned for the other similar projects, it was recalled that strict observance of the extraction protocols would be a criterion for considering the results for discussion. The participants recommended that a tolerance of ±30% be included in the extraction protocols for the shaker speed. Most of the errors detected were due to calibration errors rather than to the application of the extraction procedures.

In relation to the choice of EDTA or DTPA as extraction procedure for this type of analysis, there was a general consensus to prefer EDTA because this procedure is more easy to apply (better soil to solution ratio for EDTA). In addition, it was stressed that EDTA and DTPA extractions are closely correlated, which renders questionable the use of both extraction procedures at the same time. It was assumed that EDTA extraction enables a complete extraction to be achieved and mimics the mobility of trace metals from soils; DTPA is widely used in the USA and is applied to predict plant uptake. The choice of the extractant is, therefore, closely related to the objective of the study.

BAR-GRAPHS FOR LABORATORY MEANS AND 95% CI

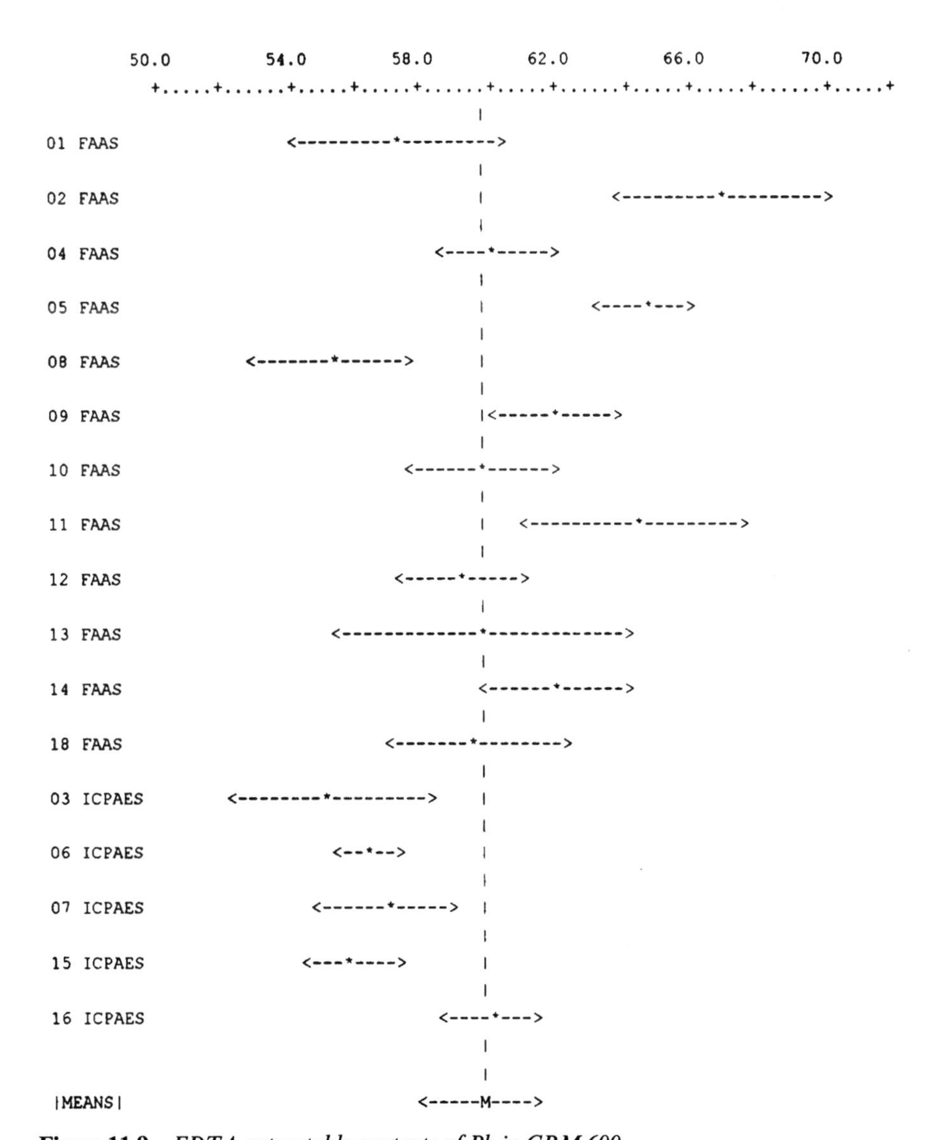

Figure 11.9 *EDTA extractable contents of Pb in CRM 600*

Certified Values

The certified values (unweighted mean of *p* accepted sets of results) and their uncertainties (half-width of the 95% confidence intervals) are given in Tables 11.27 and 11.28 as mass fractions of the respective extracts, EDTA and DTPA (based on dry mass), in mg kg^{-1}. Examples of bar graphs are shown in Figures 11.9 and 11.10. Additional information on the use of the material is given in the certification report [217].

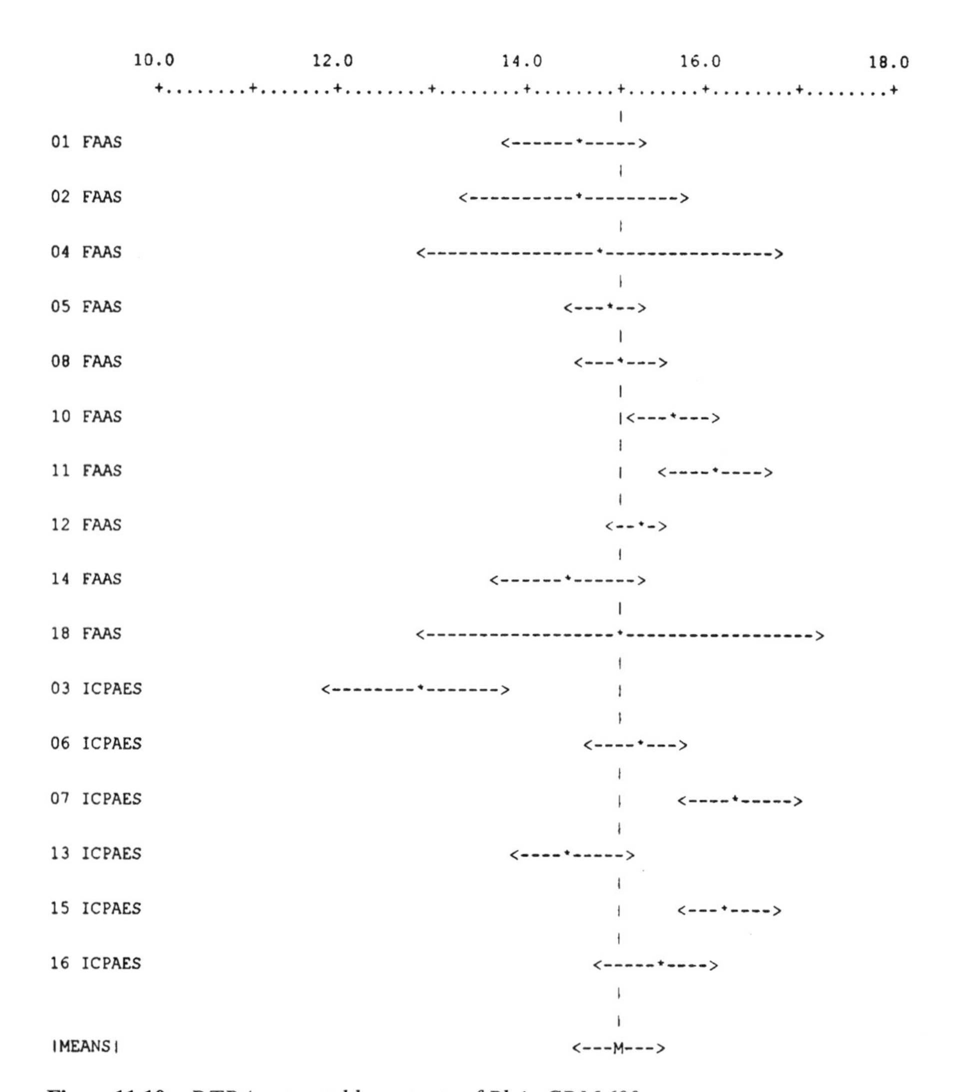

Figure 11.10 *DTPA extractable contents of Pb in CRM 600*

11.7 Annexes: Sequential and Single Extraction Schemes

Sequential Extraction Procedure

Apparatus

All laboratory-ware should be of borosilicate glass, polyethylene, polypropylene or PTFE, except for the centrifuge tubes, which should be of borosilicate glass or PTFE. Clean vessels in contact with samples or reagents with HNO_3

(4 mol L^{-1}) overnight and rinse with distilled water. Determine the blank as follows: to one vessel from each batch, taken through the cleaning procedure, add 40 mL of acetic acid (solution A, see below). Analyse this blank solution along with the sample solutions from step 1 described below. Use a mechanical shaker, preferably of the horizontal rotary or end-over-end type, at a speed of 30 rpm and record the speed. Carry out the centrifugation at 1500 g.

Reagents

Water. Glass-distilled water is normally suitable; simple deionized water may contain organically complexed metals and should not be used. Analyse a sample of distilled water with each batch of step 1 extracts.

Solution A (0.11 mol L^{-1} acetic acid). Add, in a fume-cupboard, 25 ± 0.2 mL of redistilled glacial acetic acid (or, for example, Analar or Suprapur grade acetic acid without distillation) to about 0.5 L of distilled water in a 1 L polyethylene bottle and make up to 1 L with distilled water. Make up 250 mL of this solution (0.43 mol L^{-1} acetic acid) with distilled water to 1 L to obtain an acetic acid solution of 0.11 mol L^{-1}. Analyse a sample of each batch of solution A.

Solution B (0.1 mol L^{-1} hydroxylamine hydrochloride or hydroxyammonium chloride). Dissolve 6.95 g of hydroxylamine hydrochloride in 900 mL of distilled water. Acidify with HNO_3 to pH 2 and make up to 1 L with distilled water. Prepare this solution on the same day as the extraction is carried out. Analyse a sample of each batch of solution B.

Solution C (300 mg g^{-1} hydrogen peroxide solution, *i.e.* 8.8 mol L^{-1}). Use the H_2O_2 as supplied by the manufacturer, *i.e.* acid-stabilized to pH 2–3. Analyse a sample of solution C.

Solution D (1 mol L^{-1} ammonium acetate). Dissolve 77.08 g of ammonium acetate in 900 mL of distilled water, adjust to pH 2 with HNO_3 and make up to 1 L with distilled water. Analyse a sample of each batch of solution D.

Method

Determine the extractable contents of the following trace metals: Cd, Cr, Cu, Ni, Pb and Zn, using the procedure below. Carry out all extractions on the sediment as received in the glass bottle. Before sub-sampling the sediment, with a suitable plastic (see apparatus above) spatula, shake the contents of the sample bottle with the PTFE ball supplied in the bottle for 3 minutes.

Dry a separate 1 g sample of the sediment in a layer of about 1 mm depth in an oven at 105 °C for 2 h and weigh. From this a correction "to dry mass" is obtained and applied to all analytical values reported (quantity per g dry sediment). Perform the extractions by shaking in a mechanical shaker at 20 ± 2 °C. Measure the temperature of the room at the start and at the end of the extraction procedure. The sediment should be continually in suspension during the extraction. If this is not verified, the shaking speed should be adapted in order to ensure a continuous suspension of the mixture.

Perform the sequential extraction procedure according to the steps described below:

Step 1: Add 40 mL of solution A to 1 g of sediment (as received) in a 100 mL centrifuge tube and extract by shaking for 16 hours at ambient temperature (overnight). No delay should occur between the addition of the extractant solution and the beginning of the shaking. Separate the extract from the solid residue by centrifugation and decantation of the supernatant liquid into a high-pressure polyethylene container. Stopper the container and analyse the extract immediately or store it at 4 °C prior to analysis. Wash the residue by adding 20 mL of distilled water, shaking for 15 min and centrifuging. Decant the supernatant and discard, taking care not to discard any of the solid residue. Break the "cake" obtained upon centrifugation by using a vibrating rod prior to the next step.

Step 2: Add 40 mL of solution B to the residue from step 1 in the centrifuge tube and extract by shaking for 16 h at ambient temperature (overnight). No delay should occur between the addition of the extractant solution and the beginning of the shaking. Separate the extract from the solid residue by centrifugation and decantation as in step 1. Retain the extract in a stoppered polyethylene tube, as before, for analysis. Wash the residue by adding 20 mL of distilled water, shaking for 15 min, and centrifuging. Decant the supernatant liquid and discard, taking care to avoid discarding any of the solid residue. Retain the residue for step 3. Break the "cake" obtained upon centrifugation with a vibrating rod prior to the next step.

Step 3: Add carefully, in small aliquots to avoid losses due to violent reaction, 10 mL of solution C to the residue in the centrifuge tube. Cover the vessel with a watch glass and digest at room temperature for 1 h with occasional manual shaking. Continue the digestion for 1 h at 85 °C and reduce the volume to a few mL by further heating of the uncovered vessel in a steam bath or equivalent. Add a further aliquot of 10 mL of solution C. Heat the covered vessel again to 85 °C and digest for 1 h. Remove the cover and reduce the volume of the liquid to a few mL. Add 50 mL of extracting solution D to the cool moist residue and shake for 16 h at ambient temperature (overnight). No delay should occur between the addition of the extractant solution and the beginning of the shaking. Separate the extract by centrifugation and decant into a high-pressure polyethylene tube. Stopper and retain as before for analysis.

Important

- The calibrant solutions should be made up with the appropriate extracting solutions
- With each batch of extractions a blank sample (*i.e.* a vessel with no sediment) should be carried out through the complete procedure
- The sediment should be continually in suspension during the extraction. If this is not verified, the shaking speed should be adapted in order to ensure a continuous suspension of the mixture

Note: After centrifugation, a filtration step (0.45 μm) is recommended for ICP determinations. Where ETAAS is the final method of element determination, the method of standard additions is strongly recommmended for calibration.

EDTA Extraction Protocol

The extraction should be performed in 250 mL pre-cleaned borosilicate glass, polypropylene or PTFE bottles using an end-over-end shaker. All laboratory glassware should be cleaned with HCl, rinsed with distilled water, cleaned with 0.05 mol L^{-1} EDTA and rinsed again with distilled water.

Extractants should be prepared according to the following procedure. 0.05 mol L^{-1} EDTA should be prepared as an ammonium salt solution by adding in a fume cupboard 146.12 ± 0.05 g of EDTA free acid to 800 ± 20 mL distilled water and by partially dissolving by stirring in 130 ± 5 mL of saturated ammonia solution (prepared by bubbling ammonia gas into distilled water). The addition of ammonia should be continued until all the EDTA has dissolved. The obtained solution should be filtered through a filter paper of porosity 1.4–2.0 μm (capable of retaining particles of 8.0 μm size) into a 10 L polyethylene container and diluted with water to 9.0 ± 0.5 L. The pH should be adjusted to 7.00 ± 0.05 by the addition of a few drops of either ammonia or hydrochloric acid as appropriate. The solution should therefore be diluted with distilled water to 10.0 ± 0.1 L, well mixed and stored in a stoppered polyethylene container.

Extraction should be batch-wise, followed by centrifugation according to the following procedure. A 5 g soil sample should be transferred to an extraction bottle in which 50 mL of 0.05 mol L^{-1} EDTA is added. The mixture obtained should be shaken on an end-over-end shaker operating at 30 rpm for 1 h in a room at 20 °C.

The temperature of the room should be measured at the beginning and at the end of the extraction, as well as the temperature of the extracting solution in the bottle at the end of the shaking period. The extracts should be immediately filtered through a filter paper (porosity 0.4–1.1 μm capable of retaining particles of 2.7 μm size) previously rinsed with 0.05 mol L^{-1} EDTA followed by distilled water. The filtrates should be collected in polyethylene bottles. Blank extractions (*i.e.* without soil) should be carried out for each set of analyses using the same reagents as described above.

The sample for analysis should be taken as it is. Before a bottle is opened it should be manually shaken for 5 min to rehomogenize the content. The results should be corrected for dry mass: this correction must be performed on a separate portion of 1 g taken at the same time from the same bottle by drying in an oven at 105 ± 2 °C for 2–3 h until constant mass is attained (successive weighings should not differ by more than 1 mg).

Acetic Acid Extraction Protocol

The extraction should be performed in 250 mL pre-cleaned borosilicate glass, polypropylene or PTFE bottles using an end-over-end shaker. All laboratory glassware should be cleaned with HCl, rinsed with distilled water, cleaned with 0.05 mol L^{-1} EDTA and rinsed again with distilled water.

Extractants should be prepared according to the following procedure. 0.43 mol L^{-1} acetic acid should be prepared by adding in a fume cupboard 250 ± 2 mL of redistilled glacial acetic acid to about 5 L of distilled water in a 10 L polyethylene container. The solution should be diluted with distilled water to 10 L volume, mixed well and stored in a stoppered polyethylene container.

Extraction should be batch-wise, followed by centrifugation according to the following procedure. A 5 g soil sample should be transferred to an extraction bottle in which 200 mL of 0.43 mol L^{-1} acetic acid is added. The mixture should be mixed by shaking in an end-over-end shaker as described above for 16 h (*e.g.* overnight) in a room at 20 °C. The temperature of the room should be measured at the beginning and at the end of the extraction as well as the temperature of the extracting solution in the bottle at the end of the shaking period. The extracts should be immediately filtered through a filter paper (porosity 0.4–1.1 μm capable of retaining particles of 2.7 μm size) previously rinsed with 0.05 mol L^{-1} EDTA followed by distilled water. The filtrates should be collected in polyethylene bottles. Blank extractions (*i.e.* without soil) should be carried out for each set of analyses using the same reagents as described above.

The sample for analysis should be taken as it is. Before a bottle is opened it should be manually shaken for 5 min to rehomogenize the content. The results should be corrected for dry mass: this correction must be performed on a separate portion of 1 g taken at the same time from the same bottle by drying in an oven at 105 ± 2 °C for 2–3 h until constant mass is attained (successive weighings should not differ by more than 1 mg).

NH_4NO_3, $CaCl_2$ and $NaNO_3$ Extraction Protocols Used in the Certification (Indicative Values)

In addition to EDTA and acetic acid, weak extraction procedures were applied by some laboratories in the course of the certification, *e.g.* 0.1 mol L^{-1} NH_4NO_3, 0.01 mol L^{-1} $CaCl_2$ and 0.1 mol L^{-1} $NaNO_3$. These procedures are outlined below. It is emphasized that they have not been validated by an interlaboratory trial and that the values given in this report are indicative only.

Extraction with 0.1 mol L^{-1} NH_4NO_3

Materials and chemicals:

1) Cleaning: all glassware and containers are cleaned with 4 mol L^{-1} HNO_3 and rinsed with distilled water
2) Apparatus: end-over-end shaker, centrifuge, acid-washed filter paper, 0.45 μm membrane filter, polyethylene or PTFE extraction bottles (100–150 mL) preconditioned with 4 mol L^{-1} HNO_3 and 50 mL polyethylene bottles preconditioned with HNO_3

3) Reagents: concentrated HNO_3 (Suprapur), 1.40–1.43 g mL^{-1}, 1% (v/v) HNO_3 and 0.1 mol L^{-1} NH_4NO_3 (dissolve 80.04 g NH_4NO_3 in doubly distilled water)

Procedure: manually shake the sample bottle for 5 min to homogenize the contents and take sample for analysis directly from the bottle. Operations should be carried out at 20 ± 2 °C. Weigh out accurately 20 g soil and extract in extraction bottle on shaker operating at 50–60 rpm for 2 h; filter supernatant solution through filter paper into 50 mL bottle, discarding the first 5 mL of filtrate; stabilize the solution by adding 1 mL HNO_3. If solids remain, centrifuge or filter through membrane filter. Analyse solution immediately.

Carry out two blank extractions (without soil) with each set of analyses using the above procedure. Correct the results to dry mass basis by drying a separate 1 g portion (taken at the same time as the analysis sample) at 105 ± 2 °C for 2–3 h to constant (±1 mg) mass.

Extraction with 0.01 mol L^{-1} $CaCl_2$

Materials and chemicals:

1) Cleaning: all glassware and containers are cleaned with 4 mol L^{-1} HNO_3 and rinsed with distilled water
2) Apparatus: centrifuge, polytethylene or PTFE bottles (250 mL), end-over-end shaker
3) Reagents: 0.01 mol L^{-1} $CaCl_2$ (dissolve 1.47 g $CaCl_2.2H_2O$ in doubly distilled water or equivalent). Verify that the Ca concentration is 40 ± 1 ng L^{-1} by EDTA titration

Procedure: manually shake the sample bottle for 5 min to homogenize contents and take sample for analysis directly from the bottle. Operations should be carried out at 20 ± 2 °C. Weigh 10.00 g soil into a 250 mL polyethylene bottle; add 100.0 mL of 0.01 mol L^{-1} $CaCl_2$ solution and extract on shaker for 3 h at 30 rpm; decant about 60 mL into a centrifuge tube and centrifuge for 10 min at 3000 *g*; measure and report the room temperature before and after the extraction and also the temperature of the extracting solution at the end of the extraction; measure pH in the extract before centrifugation; analyse immediately.

Carry out two blank extractions (without soil) with each set of analyses using the above procedure. Filtration is not recommended because of contamination risk. Dilutions, if required, are made with acidified (HNO_3) $CaCl_2$ solution. Correct the results to dry mass basis by drying a separate 1 g portion (taken at the same time as the analysis sample) at 105 ± 2 °C for 2–3 h to constant (±1 mg) mass.

Extraction with 0.1 mol L^{-1} $NaNO_3$

Materials and chemicals:

1) Cleaning: all glassware and containers are cleaned with 1 mol L^{-1} HNO_3
2) Apparatus: end-over-end shaker, centrifuge, 0.45 μm membrane filter, polyethylene extraction bottles (200 mL), polyethylene bottles (50 mL)
3) Reagents: concentrated Suprapur HNO_3 (65%), 1 mol L^{-1} HNO_3 (prepared from analytical grade by dilution of 70 mL conc. HNO_3 with Milli-Q water or equivalent), 0.1 mol L^{-1} $NaNO_3$ (dissolve 8.5 g $NaNO_3$ in 1 L Milli-Q water)

Procedure: manually shake the sample bottle for 1 min to homogenize contents and take sample for analysis directly from the bottle. Weigh accurately 40 g soil and extract in extraction bottle with 100 mL of 0.1 mol L^{-1} $NaNO_3$ solution at 20 ± 2 °C for 2 h at 120 rpm. Centrifuge for 10 min at 4000 *g*; remove the supernatant by syringe; fit the membrane filter and filter into 50 mL bottle; add 2 mL conc. HNO_3 to 50 mL volumetric flask and make up to volume with filtered extract (prevention of microbial growth); analyse immediately (note that solutions are stable for 1 week at 20 ± 5 °C).

Correct the results to dry mass basis by drying a separate 1 g portion (taken at the same time as the analysis sample) at 105 ± 2 °C for 2–3 h to constant (±1 mg) mass. Correct also the results for the dilution of the extract with HNO_3 (48 mL to 50 mL).

DTPA Extraction Protocol

The extraction should be performed in 250 mL pre-cleaned borosilicate glass, polypropylene or PTFE bottles using an end-over-end shaker. All laboratory glassware should be cleaned with HCl, rinsed with distilled water, cleaned with 0.05 mol L^{-1} EDTA and rinsed again with distilled water.

Extractants should be prepared according to the following procedure. The DTPA extracting solution should be prepared containing 0.005 mol L^{-1} diethylenetriaminepentaacetic acid (DTPA) [$C_{14}H_{23}N_3O_{10}$], 0.01 mol L^{-1} triethanolamine (TEA) [$(HOCH_2CH_2)_3N$] and adjusted to pH 7.3. To prepare 10 L of this solution, dissolve 149.2 g reagent grade TEA, 19.67 g DTPA and 14.7 g calcium chloride in approximately 200 mL distilled water. Allow sufficient time for the DTPA to dissolve, and dilute to approximately 9 L. Adjust the pH to 7.3 ± 0.5 with HCl while stirring and dilute to 10 L. This solution is stable for several months.

Extraction should be batch-wise, followed by centrifugation or filtration according to the following procedure. A 10 g soil sample should be transferred to an extraction bottle in which 20 mL of 0.005 mol L^{-1} DTPA solution is added. The mixture should be mixed by shaking in an end-over-end shaker as described above for 2 h in a room at 20 ± 2 °C. The temperature of the room should be measured at the beginning and at the end of the extraction as well as the temperature of the extracting solution in the bottle at the end of the shaking period. The extracts should be separated immediately. To separate, decant a portion of the extract in a centrifuge tube and centrifuge for 10 min at about 3000 *g*. The supernatant liquid must be stored in a polyethylene container at 4 °C until analysis.

Alternatively, the separation can be performed by filtration through a filter paper (porosity 0.4–1.1 μm capable of retaining particles of 2.7 μm size) previously rinsed with 0.05 mol L^{-1} EDTA followed by distilled water. The filtrates should be collected in polyethylene bottles and stored at 4 °C until analysis. Blank extractions (*i.e.* without soil) should be carried out for each set of analyses using the same reagents as described above.

The sample for analysis should be taken as it is. Before a bottle is opened it should be manually shaken for 5 min to rehomogenize the content. The results should be corrected for dry mass: this correction must be performed on a separate portion of 1 g taken at the same time from the same bottle by drying in an oven at 105 ± 2 °C for 2–3 h until constant mass is attained (successive weighings should not differ by more than 1 mg).

CHAPTER 12

European Network on Speciation

12.1 Introduction

The importance of speciation analysis is beyond any doubt, as has been illustrated in the various chapters of this book. The determination of chemical species is necessary for understanding the bio-geochemical cycle of contaminants in the terrestrial and aquatic ecosystems and for detecting possible harmful substances which might be toxic to biota and humans. Besides the importance of this tool for risk assessment studies, speciation is also highly relevant for testing the quality of products, *e.g.* the amount of essential elements in food products, or impurities in pharmaceutical products or chemical substances, *etc.* and is, therefore, of potential interest to industry.

12.2 The Paradox of Speciation

The evolution of the awareness for speciation is quite paradoxical. Fatalities were necessary to alert the public about toxic "forms" of elements, *e.g.* the high toxicity of methylmercury identified in Minamata (Japan) in the 1950s or organotin impurities in medicaments ("Stalinon" problem) in the 1960s both caused human deaths. Studies of trace element partitioning in sediments and soils were initiated in the 1970s, *e.g.* to evaluate the mobility and/or bioavailability of heavy metals, but speciation was still an "academic exercise" rather than a regulatory tool. At the beginning of the 1980s, the high mortality of oysters in Arcachon Bay (France) due to tributyltin leached from antifouling paints, and subsequent economic problems in this area, justified strong research efforts which were subsequently successful and opened the way for a larger field of investigations. The 1980s saw the start of increasing awareness from the legislation point of view since national regulations were implemented *e.g.* the ban on TBT-based antifouling paints or the systematic monitoring of methylmercury in fish by control laboratories, and the mentions of "elements and their compounds" appeared in EC Directives related to environmental protection (see Chapter 2). This decade was also marked by a flourishing development of new instruments and methods for the determination of a wide variety of chemical species. One could have thought that the 1990s would have been a key decade for speciation with an expanding market for new instruments and a large range of regulations. This trend was, however, not so marked. Huge efforts were actually made to improve the state-of-the-art of speciation analysis within Europe, of which this book gives a comprehensive summary. The projects funded by the European Commission enabled the cross-checking of techniques developed by expert EU laboratories and demonstration of the comparability of data, which was finalized in many instances by the production of reference materials certified for their chemical species contents. Six to seven years were necessary to obtain a good picture of the state-of-the-art for the speciation of some elements, *e.g.* As, Hg, Pb and Sn, and to certifiy suitable reference materials for quality control purposes. While these efforts were highly necessary and permitted the identification of robust methodologies and, adversely, improper techniques, a gap existed in the transfer of knowledge from expert laboratories to routine laboratories. This was recognized by some manufacturers at the beginning of the 1990s (*e.g.* development of the HG-GC-QFAAS method by Perkin-Elmer), but efforts were limited and still did not result in the creation of new markets. Moreover, information on the importance of pursuing research and development efforts in the area of speciation was not successfully passed to decision makers, *e.g.* regulators and industrialists.

12.3 The Trend

The end of the 1990s will represent a cornerstone for speciation. Some projects have recently started, aimed at developing simple methodologies readily marketable and usable by routine control laboratories (provided along with Standard Operating Procedures). Certified reference materials are now increasingly available for the quality control of speciation analysis and all the elements are met for speciation to be considered as the same level as trace organic analysis.

There is still, however, a need to break the circle of what could be called the "chapel syndroma", *i.e.* "vertical" chapels of experts working in their specific fields without collaboration with experts from different fields (creating 'vertical' walls between various disciplines) and "horizontal" chapels hampering the communication of the results to decision makers and end-users (*e.g.* consisting of legislators and industrialists) and routine control laboratories. The best strategy for a way forward is to share views and expertise among experts from different disciplines with complementary experience, to communicate with decision makers and end-users, and to discuss efforts necessary for the transfer of knowledge to routine laboratories. This has been well understood by the European scientific community and a network entitled "speciation '21" funded by the SM&T programme and conducted over two years (1998–2000) will hopefully create the decisive input for placing speciation at the level where it should be.

12.4 The Network "Speciation '21"

Aim of the Network

The network aims to tackle problems related to the lack of communication between scientists, industry representatives and legislators for the possible improvement of written standards and EC regulations. As mentioned previously, legislation at present mainly concerns total element content which is in many cases useless for an accurate risk evaluation (*e.g.* for environmental contamination, food quality or health risks). It is hence timely to share expertise and views among different communities in order to improve the situation. This network has been selected for funding after the approval of a first step which was granted in 1996. The implementation phase was positively evaluated in January 1997.

The main objective of the network is to bring together scientists with a background in analytical chemistry interested in speciation method development with potential users from industry and representatives from legislative bodies. The network will organize a series of meetings to debate all the important questions for organizing the information on environmental, food and occupational health aspects of speciation. The findings and conclusions will be summarized in a series of general papers that will be published in the open literature, recapitulating the essential information gathered to date, outline the state of the art for each topic and recommend legislative actions; the network also issues a Newsletter and a WWW page.

The programme covers an inaugural workshop, expert meetings (after 6 months and 18 months), and two workshops (after 12 months and at the end of the project). The inaugural meeting has been successfully held in Überlingen (Germany) on 4–5 December 1997. The Web page is being developed and will include the minutes of the meetings, workshop proceedings *etc.*

Participants in the Network

The project is coordinated by the University of Gent (Belgium). The partners in the project are: Chalmers University of Technology, Göteborg (Sweden); CSL Food Science Laboratory, Norwich (United Kingdom); Degussa AG, Zentrale Forschungseinrichtungen, Hanau (Germany); ENEA, Divisione de Chimica Ambientale, Roma (Italy); EUROMETAUX, Brussels (Belgium); GKSS-Forschungszentrum, Geesthacht (Germany); Imperial Chemicals Industries, London (United Kingdom); National Food Agency, Søborg (Denmark); Perkin-Elmer, Überlingen (Germany); Prolabo S.A., Fontenay-sous-Bois (France); Rhône-Poulenc Industrialization, Saint-Fons (France); Sveriges Provnings- och Forskingsinstitut, Boras (Sweden); Technische Universität Wien, Inst. für Analytische Chemie, Wien (Austria); Universidad Complutense, Departamento de Química Analítica, Madrid (Spain); Universitaire Instelling Antwerpen, MiTAC, Antwerp (Belgium); Université de Pau, Lab. Chimie Bio-inorganique & Environnement, Pau (France); University of

Plymouth, Dept. of Environmental Sciences, Plymouth (United Kingdom); Vrije Universiteit, Inst. voor Milieuvraagstukken, Amsterdam (The Netherlands)

Work Programme

The network will be organized through a series of workshops and expert meetings on different themes which will address the needs of industry, standardization and Community policy in the field of speciation related to product characterisitics, occupational safety, waste management, *etc.* The organization of expert meetings will be based on a series of items to be discussed in round-tables with a maximum of 15 participants per meeting. Each round-table will be chaired by a selected expert and minutes will be taken by a rapporteur. A summary of all round-table discussions will be systematically prepared by the organizer. Participants (not belonging to the partnership described above) will be invited to expert meetings.

Programme and Dissemination Strategy

The first workshop was organized by Perkin-Elmer GmbH in Überlingen (Germany) on 4–5 December 1997, and was aimed at launching the network and discussing the programmes of the different expert meetings. Beside this launch meeting, the dissemination strategy consists of a "Speciation Newsletter" to be published twice a year and widely distributed (free of charge) to all interested persons; in addition, the World Wide Web page is being developed by the University of Plymouth and contains updated information on the progress of the network.

Expert Meetings on Environmental Issues

The discussions were limited to organotin and chromium speciation, in particular the improvement of organotin determinations in the environment (including legislative aspects and analytical developments) and improved strategies for remediating chromium contaminated soils (including discussions of remediation strategies, environmental risk assessment and related analytical methodology). A first semestrial meeting was organized by Complutense University in Segovia (Spain). A second semestrial meeting will be organized by ENEA (Italy).

Expert Meetings on Food Issues

Discussions will focus on elements/compounds which are critical with respect to legislation, industry, nutrition, medicine and analytical chemistry and are important for the characterization of food quality for the consumer. The chemical species of concern will be those of As and Hg (*e.g.* in fish and shellfish), Se (*e.g.* from cereals, fish, nuts), Cu (*e.g.* from cereals, nuts, dairy products), and Fe (*e.g.* from meat, fruits, vegetables, cereals). A first semestrial meeting will be organized by the CSL Food Science Laboratory (UK). A second semestrial meeting will be organized by the Danish National Food Agency.

Expert Meetings on Occupational Health/Hygiene Issues

This part of the network encompasses (1) biological monitoring of trace elements by measuring biomarkers of exposure, (2) identification of chemical species of inhalable particles that will be the basis of toxicity studies, and (3) study of trace element kinetics in the body due to industrial exposure. A first semestrial meeting will be organized by the University of Gent (Belgium). A second semestrial meeting will be organized by the GKSS-Forschungszentrum (Germany).

Mid-term Workshop

A Workshop will be organized at the end of the first year (1998) to present the results of the first series of expert meetings and seek the opinion of invited participants. While participation to the expert meetings is restricted to partners and selected experts (up to a maximum of 15), the workshop will be opened to a wider participation (with a maximum of 40 participants). This workshop will be organized in close collaboration with the Free University of Amsterdam (The Netherlands).

Review of the State of the Art and Transfer of Knowledge

A detailed review of the state of the art will be carried out by the University of Gent (occupational health issues), the University of Pau (environment issues) and the CSL Food Science Laboratory (food issues) in close collaboration with the SM&T programme. This review, along with the summaries of the round-table discussions, will serve the basis of a book to be edited by the three partners and the SM&T programme and published by a selected publisher. The review will include aspects of Community policy, industrial competiveness and analytical methods currently available for speciation (with validation and performance characteristics), and will identify future requirements and research needs.

The work package will also address transfer of knowledge through the establishment of a register of methods with comments on cost, ease of use, fitness for purpose, validation, *etc.*, and review the findings of the expert meetings with respect to speciation needs in terms of legislation and benefits to industry. These documents will be made accessible through the network WWW page. Finally, the Speciation Newsletter will disseminate the main outcome of the network throughout the European Union.

Final Workshop and Publication

The final Workshop will be organized by the University of Pau (France). The participation will be opened to all the partners and a maximum of 30 additional participants. The workshop will present the results of the expert meetings, and discuss the identified research needs and recommendations to legislators and industry with the expert (as a means of validating these recommendations).

Additional Information

Information on the SM&T programme can be obtained through the SM&T help-desk (e-mail: smt-helpdesk@dg12.cec.be) or through the Internet (http://www.cordis.lu/smt/home/html). The Certified Reference Materials produced by the programme can be purchased at the Institute for Reference Materials and Measurements (IRMM), Retieseweg, B-2440 Geel, Belgium (Fax: +32 14 590406), where a catalogue can be obtained free of charge; further information on existing BCR CRMs is available through the Internet under http://www.irmm.jrc.be/mrm.html.

Information on the "Speciation '21" network will be available on-line through the Internet before the middle of 1998 (please contact Dr Les Pitts at the University of Plymouth by e-mail at lpitts@plymouth.ac.uk).

References

1 E. Prichard, in: *Quality in the Analytical Chemistry Laboratory*, E.J. Newman (ed.), Wiley, Chichester (1995)
2 Ph. Quevauviller, E.A. Maier and B. Griepink (eds.), *Quality Assurance for Environmental Analysis*, Elsevier, Amsterdam (1995)
3 H. Günzler (ed.), *Accreditation and Quality Assurance in Analytical Chemistry*, Springer, Berlin (1996)
4 *Measurements and Testing Newsletter*, European Commission, Brussels, **1**, 1 (1993)
5 *Measurements and Testing Newsletter*, European Commission, Brussels, **3**, 2 (1995)
6 *Measurements and Testing Newsletter*, European Commission, Brussels, **4**, 1 (1996)
7 E.A. Maier, Ph. Quevauviller and B. Griepink, *Anal. Chim. Acta*, **283**, 590 (1993)
8 B. Griepink, *Fresenius' J. Anal. Chem.*, **334**, 606 (1989)
9 Ph. Quevauviller, *Mikrochim. Acta*, **123**, 3 (1996)
10 Ph. Quevauviller, W. Cofino, R. Cornelis, P. Fernandez, R. Morabito and H. van der Sloot, *Int. J. Environ. Anal. Chem.*, **67**, 173 (1998)
11 Ph. Quevauviller, E.A. Maier and B. Griepink, in: *Element Speciation in Bioinorganic Chemistry*, S. Caroli (ed.), Wiley, Chichester (1996)
12 Ph. Quevauviller, *Analyst*, **120**, 597 (1995)
13 E.A. Maier, *Trends Anal. Chem.*, **10**, 340 (1991)
14 B. Griepink and M. Stoeppler, in: *Hazardous Metals in the Environment*, M. Stoeppler (ed.), Elsevier, Amsterdam (1992)
15 D.E. Wells, J. De Boer, L.G.M.Th. Tuinstra, L. Reutergardh and B. Griepink, *Fresenius' J. Anal. Chem.*, **332**, 591 (1988)
16 P.J. Craig (ed.), *Organometallic Compounds in the Environment*, Longman, Harlow (1986)
17 W.M.R. Dirkx, R. Lobinski and F.C. Adams, in: *Quality Assurance for Environmental Analysis*, Ph. Quevauviller, E.A. Maier and B. Griepink (eds.), Elsevier, Amsterdam, pp. 360–411 (1995)
18 R. Morabito, S. Chiavarini and C. Cremisini, in: *Quality Assurance for Environmental Analysis*, Ph. Quevauviller, E.A. Maier and B. Griepink (eds.), Elsevier, Amsterdam, pp. 437–465 (1995)
19 A. Amran, F. Lagarde, M.J.F. Leroy, A. Lamotte, C. Demesmay, M. Ollé, M. Albert, G. Rauret and J.F. López-Sánchez, in: *Quality Assurance for Environmental Analysis*, Ph. Quevauviller, E.A. Maier and B. Griepink (eds.), Elsevier, Amsterdam, pp. 288–307 (1995)
20 I. Drabaek and Å Iverfeldt, in: *Quality Assurance for Environmental Analysis*, Ph. Quevauviller, E.A. Maier and B. Griepink (eds.), Elsevier, Amsterdam, pp. 308–319 (1995)
21 R. Lobinski, W.M.R. Dirkx, J. Szpunar-Lobinska and F.C. Adams, in: *Quality*

Assurance for Environmental Analysis, Ph. Quevauviller, E.A. Maier and B. Griepink (eds.), Elsevier, Amsterdam, pp. 320–352 (1995)
22 C. Cámara, M.G. Cobo, M.A. Palacios, R. Muñoz and O.F.X. Donard, in: *Quality Assurance for Environmental Analysis*, Ph. Quevauviller, E.A. Maier and B. Griepink (eds.), Elsevier, Amsterdam, pp. 237–263 (1995)
23 S. Hetland, I. Martinsen, B. Radzuk and Y. Thomassen, *Anal. Sci.*, **7**, 1029 (1991)
24 A.M. Ure, Ph. Quevauviller, H. Muntau and B. Griepink, *Int. J. Environ. Anal. Chem.*, **51**, 135 (1993)
25 Ph. Quevauviller, H.A. van der Sloot, A. Ure, H. Muntau, A. Gomez and G. Rauret, *Sci. Total Environ.*, **178**, 133 (1996)
26 Ph. Quevauviller, *Fresenius' J. Anal. Chem.*, **354**, 515 (1996)
27 Ph. Quevauviller, O.F.X. Donard, E.A. Maier and B. Griepink, *Mikrochim. Acta*, **109**, 169 (1992)
28 Ph. Quevauviller, *Fresenius' J. Anal. Chem.*, **351**, 345 (1995)
29 Ph. Quevauviller, *Appl. Organomet. Chem.*, **8**, 715 (1994)
30 Ph. Quevauviller and E.A. Maier, *EUR Report*, 16000 EN, EC, Brussels (1994)
31 Ph. Quevauviller, E.A. Maier and B. Griepink, in: *Element Speciation in Bioinorganic Chemistry*, S. Caroli (ed.), Wiley, Chichester, pp. 195–222 (1996)
32 A. Astruc, R. Lavigne, V. Desauziers, R. Pinel and M. Astruc, *Appl. Organomet. Chem.*, **3**, 267 (1989)
33 E. Bailey and A.G.F. Brooks, *Mikrochim. Acta*, **109**, 121 (1992)
34 J. Dachs, R. Alzaga, J.M. Bayona and Ph. Quevauviller, *Anal. Chim. Acta*, **286**, 319 (1994)
35 J.M. Bayona, in: *Quality Assurance for Environmental Analysis*, Ph. Quevauviller, E.A. Maier and B. Griepink (eds.), Elsevier, Amsterdam, pp. 466–489 (1995)
36 S. Zhang, Y.K. Chau, W.C. Li and A.S.Y. Chau, *Appl. Organomet. Chem.*, **5**, 431 (1991)
37 R. Morabito, M. Abalos, J.M. Bayona, M.B. de la Calle and Ph. Quevauviller, *Abstr. 26th Int. Symp. Environ. Anal. Chem.*, Vienna, 9–12 April 1996
38 R. Ritsema, F.M. Martin and Ph. Quevauviller, in: *Quality Assurance for Environmental Analysis*, Ph. Quevauviller, E.A. Maier and B. Griepink (eds.), Elsevier, Amsterdam, pp. 490–505 (1995)
39 Ph. Quevauviller, F. Martin, C. Belin and O.F.X. Donard, *Appl. Organomet. Chem.*, **7**, 149 (1993)
40 G. Rauret, R. Rubio, A. Padró and J. Albertí, *Proc. 12th Int. Symp. Microchem. Techniques*, Córdoba, 6–11 September 1992
41 N. Violante, F. Petrucci, F. La Torre and S. Caroli, *Spectroscopy*, **7**(7), 36 (1992)
42 J.R. Ashby and P.J. Craig, *Sci. Total Environ.*, **78**, 219 (1989)
43 Ph. Quevauviller, M. Astruc, L. Ebdon, H. Muntau, W. Cofino, R. Morabito and B. Griepink, *Mikrochim. Acta*, **123**, 163 (1996)
44 O.F.X. Donard, *Thesis*, Univ. Bordeaux (1987)
45 R. Harrison and S. Rapsomanikis (eds.), *Environmental Analysis Using Chromatography Interfaced with Atomic Spectroscopy*, Ellis Horwood, Chichester, 1988, p. 189
46 Ph. Quevauviller, R. Ritsema, R. Morabito, W.M.R. Dirkx, S. Chiavarini, J.M. Bayona and O.F.X. Donard, *Appl. Organomet. Chem.*, **8**, 541 (1994)
47 D.E. Wells, J. De Boer, L.G.M.Th. Tuinstra, L. Reutergardh and B. Griepink, *EUR Report*, 12496 EN, EC, Brussels (1992)
48 T.A. Rymen, B. Griepink and S. Fachetti, *Chemosphere*, **20**, 1291 (1990)

49 M. Ochsenkühn-Petropoulou, G. Poulea and G. Parikassis, *Mikrochim. Acta*, **109**, 93 (1992)

50 J.A. Stäb, B. van Hattum, P. de Voogt and U.Th. Brinkman, *Mikrochim. Acta*, **109**, 101 (1992)

51 Ph. Quevauviller, M.B. de la Calle-Guntiñas, E.A. Maier and C. Cámara, *Mikrochim. Acta*, **118**, 131 (1995)

52 R.J. Maguire, J.H. Carey and E.J. Hale, *J. Agric. Food Chem.*, **31**, 1060 (1983)

53 Ph. Quevauviller, O.F.X. Donard and A. Bruchet, *Appl. Organomet. Chem.*, **5**, 125 (1991)

54 F.A. Cotton and G. Wilkinson, *Advanced Inorganic Chemistry*, Wiley, New York (1988)

55 P.M. Stang and P. Seligman, *Proc. Oceans '86 Organotin Symposium*, Washington, September 23–25, vol. 4, p. 1256 (1986)

56 T.H. Hartley, *Computerized Quality Control: Programs for the Analytical Laboratory*, Ellis Horwood, Chichester, 2nd edn., (1990)

57 W.A. Shewhart, *Economic Control of Quality of Manufactured Products*, Van Nostrand, New York (1931)

58 *Certification of Reference Materials: General and Statistical Principles*, ISO-Guide 33 (1985)

59 E.A. Maier, in: *Quality Management in Chemical Laboratories*, Cofino W.P. and Griepink B. (eds.), Elsevier, (1988)

60 A. Ure, Ph. Quevauviller, H. Muntau and B. Griepink, *EUR Report*, 14763 EN, EC, Brussels (1993)

61 H.A. van der Sloot, D. Hoede, G.J. de Groot, G.J.L. van der Wegen and Ph. Quevauviller, *EUR Report*, 16133 EN, EC, Brussels (1995)

62 W. Horwitz, Project 27/87, *Nomenclature for Interlaboratory Studies*, IUPAC (1991)

63 D.W. Allen, J.S. Brookst, J. Unwin and D. McGuinness, *Appl. Organomet. Chem.*, **1**, 311 (1987)

64 E.A. Maier, B. Griepink, C.G. van der Paauw and A.M. Rietveld, *EUR Report*, 14063 EN, EC, Brussels (1992)

65 E.A. Maier, L.G.M.Th. Tuinstra, A.H. Roos and B. Griepink, *EUR Report*, submitted (1998)

66 B. Griepink, *Fresenius' J. Anal. Chem.*, **338**, 486 (1990)

67 B. Griepink, Ph. Quevauviller, E.A. Maier and S. Vandendriessche, *Fresenius' J. Anal.. Chem.*, **346**, 530 (1993)

68 J. Pauwels, C. Hofmann and C. Vandecastele, *Fresenius' J. Anal. Chem.*, **348**, 418 (1994)

69 Ph. Quevauviller, E.A. Maier, K. Vercoutere, H. Muntau and B. Griepink, *Fresenius' J. Anal. Chem.*, **343**, 335 (1994)

70 Ph. Quevauviller, M. Astruc, L. Ebdon, V. Desauziers, P.M. Sarradin, A. Astruc, G.N. Kramer and B. Griepink, *Appl. Organomet. Chem.*, **8**, 629 (1994)

71 E.A. Maier, in: *Techniques for Environmental Analysis*, D. Barceló (ed.), Elsevier, Amsterdam (1993)

72 H.F.R. Reijnders, Ph. Quevauviller, D. van Renterghem, B. Griepink and H. van der Jagt, *Fresenius' J. Anal. Chem.*, **348**, 439 (1994)

73 Certification of Reference Materials—General and Statistical Principles, *ISO Guide 35*, International Organization for Standardization, Geneva (1985)

74 P. de Bièvre, *Fresenius' J. Anal. Chem.*, **337**, 766 (1990)

75 J.-C. Wolff, P.D.P. Taylor and P. de Bièvre, *Anal. Chem.*, **68**, 3231 (1996)

76 Ph. Quevauviller, G.N. Kramer and B. Griepink, *Mar. Pollut. Bull.*, **24**, 601 (1992)
77 A.A. Brown, L. Ebdon and S.J. Hill, *Anal. Chim. Acta*, **286**, 391 (1994)
78 D.E. Wells, W.P. Cofino, Ph. Quevauviller and B. Griepink, *Mar. Pollut. Bull.*, **26**, 368 (1993)
79 W.P. Cofino and D.E. Wells, *Mar. Pollut. Bull.*, **29**, 149 (1994)
80 W. Smith and A. Smith (eds.), *Minamata*, Rinehart and Winston, New York (1975)
81 Ph. Quevauviller, I. Drabæk, H. Muntau and B. Griepink, *Appl. Organomet. Chem.*, **7**, 413 (1993)
82 W.J. Youden, *Anal. Chem.*, **32**, 23A (1960)
83 J.H. Petersen and I. Drabaek, *Mikrochim. Acta*, **109**, 125 (1992)
84 DOLT-1, Dogfish Liver. *Description Sheet*, National Research Council of Canada, Ottawa
85 DORM-1, Dogfish Muscle. *Description Sheet*, National Research Council of Canada, Ottawa
86 Ph. Quevauviller, I. Drabæk, H. Muntau and B. Griepink, *EUR Report*, 15902 EN, EC, Brussels (1994)
87 Ph. Quevauviller, I. Drabaek, H. Muntau, M. Bianchi, A. Bortoli and B. Griepink, *Trends Anal. Chem.*, **15**, 160 (1996)
88 P.J. Craig and P.D. Bartlett, *Nature*, **275**, 635 (1978)
89 F. Baldi, F. Parati and M. Filippelli, *Water Air Soil Pollut.*, in press
90 F. Baldi, M. Pepi and M. Filippelli, *Appl. Environ. Microbiol.*, **59**, 2479 (1993)
91 J.R. Postgate, *The Sulphate-Reducing Bacteria*, 2nd edn., Cambridge University Press, Cambridge (1984)
92 Ph. Quevauviller, G.U. Fortunati, M. Filippelli, F. Baldi, M. Bianchi and H. Muntau, *Appl. Organomet. Chem.*, **10**, 537 (1996)
93 M. Horvat, V. Manduc, L. Liang, N.S. Bloom, S. Padberg, Y.-L. Lee, H. Hintelmann and J. Benoit, *Appl. Organomet. Chem.*, **8**, 533 (1994)
94 H. Hintelmann and R. Falter, *Abstr. Fourth Int. Conf. Mercury Global Pollutant*, Hamburg, Sept. 1996, p. 284 (1996)
95 Ph. Quevauviller, *Fresenius' J. Anal. Chem.*, **358**, 419 (1997)
96 R.J. Maguire, *Appl. Organomet. Chem.*, **1**, 475 (1987)
97 Ph. Quevauviller and O.F.X. Donard, in: *Element Speciation in Bioinorganic Chemistry*, S. Caroli (ed.), Wiley, New York, pp. 331–357 (1996)
98 Ph. Quevauviller, B. Griepink, E.A. Maier, H. Meinema and H. Muntau, presented at *Euroanalysis VII Int. Conf.*, Vienna, 26–31 August 1990
99 J. Dachs, R. Alzaga, J.M. Bayona and Ph. Quevauviller, *Anal. Chim. Acta*, **286**, 371 (1994)
100 C. Rivas, S. Hill, Ph. Quevauviller and L. Ebdon, *Anal. Chem.*, in press
101 M.B. De la Calle, R. Scerbo, S. Chiavarini, Ph. Quevauviller and R. Morabito, *Appl. Organomet. Chem.*, **11**, 693 (1997)
102 M. Abalos, J.M. Bayona and Ph. Quevauviller, *Appl. Organomet. Chem.*, in press
103 Ph. Quevauviller, M. Astruc, L. Ebdon, G.N. Kramer, B. Griepink, *Appl. Organomet. Chem.*, **8**, 639 (1994)
104 F.M. Martin, C.M. Tseng, O.F.X. Donard, C. Belin and Ph. Quevauviller, *Anal. Chim. Acta*, **286**, 343 (1994)
105 L. Ebdon, *BCR Internal Report* (1991)
106 Ph. Quevauviller and O.F.X. Donard, *Fresenius' J. Anal. Chem.*, **339**, 6 (1991)
107 S. Zhang, Y.K. Chau, W.C. Li and A.S.Y. Chau, *Appl. Organomet. Chem.*, **5**, 431 (1991)

108 Ph. Quevauviller, M. Astruc, L. Ebdon, G.N. Kramer and B. Griepink, *EUR Report*, 15337 EN, EC, Brussels (1993)

109 A. Lamberty, R. Morabito and Ph. Quevauviller, *EUR Report*, EC, Brussels, (1998)

110 A.M. Caricchia, S. Chiavarini, C. Cremisini, R. Morabito and R. Scerbo, *Anal. Sci.*, **7**, 1193 (1991)

111 S. Chiavarini, C. Cremisini and R. Morabito, in: *Element Speciation in Bioinorganic Chemistry*, S. Caroli (ed.), John Wiley & Sons, Chichester (1996)

112 A.M. Caricchia, S. Chiavarini, C. Cremisini, R. Morabito and R. Scerbo, *Anal. Chim. Acta*, **286**, 329 (1994)

113 Ph. Quevauviller, R. Morabito, W. Cofino, H. Muntau and M.J. Campbell, *EUR Report*, 17921 EN, EC, Brussels (1997)

114 R. Morabito, P. Soldati, M.B. de la Calle and Ph. Quevauviller, *Appl. Organomet. Chem.*, in press

115 C.N. Hewitt and R.M. Harrison, in: *Organometallic Compounds in the Environment*, P.J. Craig (ed.), Longman, London, p. 160 (1986)

116 C.N. Hewitt and R.M. Harrison, *Environ. Sci. Technol.*, **21**, 260 (1986)

117 R.M. Harrison, M. Radojevic and S.J. Wilson, *Sci. Total Environ.*, **50**, 129 (1986)

118 R.J.C. Van Cleuvenbergen, W.M.R. Dirkx, Ph. Quevauviller and F.C. Adams, *Int. J. Environ. Anal. Chem.*, **47**, 21 (1992)

119 Ph. Quevauviller, Y. Wang, A.B. Turnbull, W.M.R. Dirkx, R.M. Harrison and F.C. Adams, *Appl. Organomet. Chem.*, **9**, 89 (1995)

120 Ph. Quevauviller, Y. Wang and R.M. Harrison, *Appl. Organomet. Chem.*, **8**, 703 (1994)

121 S.J. Hill, A. Brown, C. Rivas, S. Sparkes and L. Ebdon, in: *Quality Assurance for Environmental Analysis*, Ph. Quevauviller, E.A. Maier and B. Griepink (eds.), Elsevier, Amsterdam, pp. 412–434 (1995)

122 J.S. Blais and W.D. Marshall, *J. Anal. At. Spectrom.*, **4**, 641 (1989)

123 Y. Wang, R.M. Harrison and Ph. Quevauviller, *Appl. Organomet. Chem.*, **10**, 69 (1996)

124 Ph. Quevauviller, L. Ebdon, R. Harrison, Y. Wang and M.J. Cambpell, *EUR Report*, 18025 EN, EC, Brussels (1997)

125 Ph. Quevauviller, L. Ebdon, R. Harrison and Y. Wang, *Analyst*, in press

126 Ph. Quevauviller, L. Ebdon, R. Harrison, Y. Wang and M.J. Campbell, *EUR Report*, 18046 EN, EC, Brussels (1998)

127 Ph. Quevauviller, L. Ebdon, R. Harrison and Y. Wang, *Appl. Organomet. Chem.*, in press

128 A.C. Chapman, *Analyst*, **51**, 548 (1926)

129 K. Shiomi, Y. Kakchaski, H. Yamanaka and T. Kikuchi, *Appl. Organomet. Chem.*, **1**, 177 (1987)

130 W.R. Cullen and M. Dodd, *Appl. Organomet. Chem.*, **3**, 79 (1989)

131 Y. Shibata and M. Morita, *Anal. Chem.*, **61**, 2116 (1989)

132 E.H. Larsen, G. Pritzl, S.H. Hansen, *J. Anal. At. Spectrom.*, **8**, 1075 (1993)

133 F. Lagarde, M. Amran, M. Leroy, C. Demesmay, M. Ollé, A. Lamotte, H. Muntau, B. Griepink and E. Maier, *Fresenius' J. Anal. Chem.*, submitted (1998)

134 E.A. Maier, C. Demesmay, M. Ollé, A. Lamotte, F. Lagarde, R. Heimburger, M.J.F. Leroy, Z. Asfari and H. Muntau, *EUR Report*, EUR 17889 (1997)

135 P. Thomas and K. Sniatecki, *Fresenius' J. Anal. Chem.*, **351**, 410 (1995)

136 A. El Moll, R. Heimburger, F. Lagarde and M.J.F. Leroy, *Fresenius' J. Anal. Chem.*, **354**, 550 (1996)

137 G.M. Monplaisir, J.S. Blais, M. Quinteiro and W.D. Marshall, *J. Agric. Food Chem.*, **39**, 1448 (1991)
138 J. Alberti, R. Rubio and G. Rauret, *Fresenius' J. Anal. Chem.*, **351**, 415 (1995)
139 F. Lagarde, Z. Asfari, M. Leroy, C. Demesmay, M. Ollé, A. Lamotte, P. Leperchec, B. Griepink and E. Maier, *Fresenius' J. Anal. Chem.*, submitted (1998)
140 P. Morin, M.B. Amran, S. Favier, R. Heimburger and M. Leroy, *Fresenius' J. Anal. Chem.*, **342**, 357 (1992)
141 R. Muñoz Olivas, O.F.X. Donard, C. Cámara and Ph. Quevauviller, *Anal. Chim. Acta*, **286**, 343 (1994)
142 M.G. Cobo, M.A. Palacios, C. Cámara, F. Reis and Ph. Quevauviller, *Anal. Chim. Acta*, **286**, 371 (1994)
143 M.G. Cobo-Fernández, M.A. Palacios, C. Cámara and Ph. Quevauviller, *Quím. Anal.*, **14**, 169 (1995)
144 Ph. Quevauviller, C. Cámara, M.A. Palacios and M.G. Cobo, *EUR Report*, 18044 EN, EC, Brussels (1998)
145 C. Cámara, M.G. Cobo, M.A. Palacios, R. Muñoz and Ph. Quevauviller, *Analyst*, in press
146 S. Dyg, R. Cornelis, B. Griepink and P. Verbeeck, in: *Metal Speciation in the Environment*, NATO ASI Series G: Ecological Sciences, vol 23, F. Adams and J. Broeckaert (eds.), Springer, Berlin, pp. 361–376 (1990)
147 S. Dyg, Th. Anglov and J.M. Christensen, *Anal. Chim. Acta*, **286**, 273 (1994)
148 K. Vercoutere, R. Cornelis, S. Dyg, L. Mees, J. Molin Christensen, K. Byrialsen, B. Aaen and Ph. Quevauviller, *Mikrochim. Acta*, **123**, 109 (1996)
149 K. Vercoutere, R. Cornelis and Ph. Quevauviller, *EUR Report*, 17605 EN, EC, Brussels (1997)
150 K. Vercoutere, R. Cornelis, J.M. Christensen, K. Byrialsen and Ph. Quevauviller, *EUR Report*, 18026 EN, EC, Brussels (1997)
151 K. Vercoutere and R. Cornelis, in: *Quality Assurance for Environmental Analysis*, Ph. Quevauviller, E.A. Maier and B. Griepink (eds.), Elsevier, pp. 195–213 (1995)
152 R. Nusko and K.G. Heumann, *Anal. Chim. Acta*, **286**, 283 (1994)
153 NIOSH, Publ. No 88–110a; PB 88–231774, Cincinnati, OH (1988)
154 H. Beere and P. Jones, *Anal. Chim. Acta*, **293**, 237 (1994)
155 M. Lederer, *The Periodic Table for Chromatographers*, Wiley, Chichester (1992)
156 B. Fairman and A. Sanz Medel, in: *Quality Assurance for Environmental Analysis*, Ph. Quevauviller, E.A. Maier and B. Griepink (eds.), Elsevier, Amsterdam, pp. 216–231 (1995)
157 A.C. Alfrey, G.R. Le Gendre and W.D. Kaehny, *N. Engl. J. Med.*, **294**, 184 (1976)
158 C.N. Martyn, *Spec. Publ. R. Soc. Chem.*, **73**, 37 (1989)
159 C.T. Driscoll, J.P. Baker, J.J. Bisogni and C.L. Schofield, *Nature*, **284**, 161 (1980)
160 B.O. Rosseland, O.K. Skogheim, F. Krogland and E. Hoell, *Water, Air Soil Pollut.*, **30**, 751 (1986)
161 K. Sadler and A.W.H. Turnpenny, *Water, Air, Soil Pollut.*, **30**, 593 (1986)
162 T. Okura, K. Goto and T. Yotuyanagi, *Anal. Chem.*, **34**, 581 (1962)
163 H.M. May, P.A. Helmke and M.L. Jackson, *Chem. Geol.*, **24**, 259 (1979)
164 R.B. Barnes, *Chem. Geol.*, **15**, 177 (1975)
165 C.T. Driscoll, *Int. J. Environ. Anal. Chem.*, **16**, 267 (1984)
166 H.M. Seip, L. Müller and A. Naas, *Water, Air, Soil Pollut.*, **23**, 81 (1984)
167 X. Goenaga and D.J.A. Williams, *Environ. Pollut.*, **14**, 131 (1988)
168 B. Fairman and A. Sanz-Medel, *Int. J. Environ. Anal. Chem.*, **50**, 161 (1993)

169 B.D. LaZerte, C. Chun, D. Evans and F. Tomassini, *Environ. Sci. Technol.*, **22**, 1106 (1988)
170 E.J.S. Rogeberg and A. Henriksen, *Vatten*, **41**, 48 (1985)
171 J.I. García Alonso, A. Lopez García, A. Sanz-Medel, E. Blanco Gonzales, L. Ebdon and P. Jones, *Anal. Chim. Acta*, **225**, 339 (1989)
172 P.M. Bertsch and M.A. Anderson, *Anal. Chem.*, **61**, 535 (1989)
173 L.L. Hendrickson, M.A. Turner and R.B. Corey, *Anal. Chem.*, **54**, 1633 (1982)
174 P.G.C. Campbell, M. Bisson, R. Bougie, A. Tessier and J.P. Villeneure, *Anal. Chem.*, **55**, 2246 (1983)
175 C.L. Chakrabarti, Y. Lu, J. Cheng, M.H. Back and W.H. Schroeder, *Anal. Chim. Acta*, **267**, 47 (1993)
176 E. Courtyn, C. Vandecasteele and R. Dams, *Sci. Total Environ.*, **90**, 191 (1991)
177 B.D. LaZerte, *J. Can. Fish Aquat. Sci.*, **41**, 766 (1984)
178 P. Jones, *Int. J. Environ. Anal. Chem.*, **44**, 1 (1991)
179 P. Jones, L. Ebdon and T. Williams, *Analyst*, **113**, 641 (1988)
180 P.M. Bertsch, R.I. Barnhisel, G.W. Thomas, W.J. Layton and S.L. Smith, *Anal. Chem.*, **58**, 2583 (1986)
181 F. Thomas, A. Masion, J.Y. Boltero, J. Rouiller, F. Gerévrier and D. Boudot, *Environ. Sci. Technol.*, **25**, 1553 (1991)
182 L.O. Öhman and S. Sjöberg, *J. Chem. Soc., Dalton Trans.*, 2513 (1983)
183 T.J. Sullivan, H.M. Seip and I.P. Muniz, *Int. J. Environ. Anal. Chem.*, **26**, 61 (1986)
184 B. Fairman, A. Sanz-Medel, M. Gallego, M.J. Quintela, P. Jones and R. Benson, *Anal. Chim. Acta*, **286**, 401 (1994)
185 D.H. Oughton, B. Salbu, H.E. Bjørnstad and J.P. Day, *Analyst*, **117**, 619 (1992)
186 Ph. Quevauviller, K. Vercoutere and B. Griepink, *Mikrochim. Acta*, **108**, 195 (1992)
187 A.G. Cox, I.G. Cook and C.W. McLeod, *Analyst*, **110**, 331 (1985)
188 A. Tessier, P.G.C. Campbell and M. Bisson, *Anal. Chem.*, **51**, 844 (1979)
189 M. Meguellati, D. Robbe, P. Marchandise and M. Astruc, *Proc. Int. Conf. Heavy Metals in the Environment, Heidelberg*, CEP-Consultants, Edinburgh, vol. 1, p. 1091 (1987)
190 W. Salomons and U. Förstner, *Environ. Technol. Lett.*, **1**, 506 (1980)
191 E.A. Thomas, S.N. Luoma, D.J. Cain and C. Johansson, *Water, Air, and Soil Pollut.*, **14**, 215 (1980)
192 M. Kersten and U. Förstner, *Mar. Chem.*, **22**, 299 (1987)
193 A. Ure, Ph. Quevauviller, H. Muntau and B. Griepink, *Int. J. Environ. Anal. Chem.*, **51**, 135 (1993)
194 Ph. Quevauviller, A. Ure, H. Muntau and B. Griepink, *Int. J. Environ. Anal. Chem.*, **51**, 129 (1993)
195 Ph. Quevauviller, G. Rauret, R. Rubio, J.-F. López-Sánchez, A. Ure, J. Bacon and H. Muntau, *Fresenius' J. Anal. Chem.*, **357**, 611 (1997)
196 Ph. Quevauviller, M. Lachica, E. Barahona, G. Rauret, A. Ure, A. Gomez, H. Muntau, *Sci. Total Environ.*, **178**, 127 (1996)
197 Ph. Quevauviller, G. Rauret, H. Muntau, R. Rubio, J.-F. López-Sánchez , H. Fiedler and B. Griepink, *Fresenius' J. Anal. Chem.*, **349**, 808 (1994)
198 Ph. Quevauviller, M. Lachica, E. Barahona, A. Gomez, G. Rauret G. A. Ure and H. Muntau, *Fresenius' J. Anal. Chem.*, **360**, 505 (1998)
199 Ph. Quevauviller, G. Rauret, J.-F. López-Sánchez, R. Rubio, A. Ure and H. Muntau, *Sci. Total Environ.*, **205**, 223 (1997)

200 W. Salomons and S. Scheltens, *Changes in Speciation of Metals in Sediments Over a 12 Year Period*, Report T44, Inst. for Soil Fertility, Haren, The Netherlands (1987)
201 U. Förstner, *in: Chemical Methods for Assessing Bioavailable Metals in Sludges*, R. Leschber, R.A. Davis and P. L'Hermite (eds.), Elsevier, London (1985)
202 W. Salomons and U. Förstner, *Metals in the Hydrocycle*, Springer, New York (1984)
203 M. Meguellati, D. Robbe, P. Marchandise and M. Astruc, *Proc. Int. Conf. Heavy Metals in the Environment, Heidelberg*, pp. 1090–1093 (1983)
204 H.D. Fiedler, J.-F. López-Sánchez, R. Rubio, G. Rauret, Ph. Quevauviller, A.M. Ure and H. Muntau, *Analyst*, **119**, 1109 (1994)
205 A. Sahuquillo, R. Rubio and G. Rauret, in: *Quality Assurance for Environmental Analysis within the BCR-Programme*, Ph. Quevauviller, E.A Maier and B. Griepink (eds.), Elsevier, Amsterdam (1995)
206 Ph. Quevauviller, G. Rauret, J.-F. López-Sánchez, R. Rubio, A. Ure and H. Muntau, *EUR Report*, 17554 EN, EC, Brussels (1997)
207 Ph. Quevauviller, G. Rauret, A. Ure, J. Bacon and H. Muntau, *EUR Report*, 17127 EN, EC, Brussels (1997)
208 J.R. Sanders, T.M. Adams and B.T. Christensen, *J. Sci. Food Agric.*, **37** 1155 (1986)
209 P.M. Clayton and K.G. Tiller, *CSIRO Div. Soils Tech. Papers*, **41**, 1 (1987)
210 A.U. Haq and M.H. Miller, *Agron. J.*, **64**, 779 (1972)
211 E. Lakanen and R. Ervio, *Acta Agr. Fenn.*, **123**, 223 (1971)
212 M. Sillanpaa, *FAO Soil Bull.*, **48**, 444 (1982)
213 M.L. Raisanen, *Analyst*, **117**, 623 (1992)
214 P.N. Soltanpour and A.P. Schwab, *Commun. Soil Sci. Plant Anal.*, **8**, 195 (1977)
215 W.L. Lindsay and W.A. Norwell, *Soil Sci. Soc. Am. J.*, **42**, 421 (1978)
216 G.A. O'Connor, *J. Environ. Qual.*, **17**, 715 (1988)
217 Ph. Quevauviller, M. Lachica, E. Barahona, G. Rauret, A. Gomez and H. Muntau, *EUR Report*, 17555 EN, EC, Brussels (1997)

Appendix

Assessment Forms

INFORMATION SHEET

CERTIFICATION OF [CHEMICAL FORM]
IN [MATRIX]

LABORATORY: ....

....

RESPONSIBLE: ....

PERSON ATTENDING: ....
THE MEETING

ELEMENTS/COMPOUNDS DETERMINED: ....

....

METHOD(S) USED (IN BRIEF):

Sample intake	Pre-treatment, extraction, derivatization	Separation	Detection

PROJECT

NAME

Interlaboratory study (or certification)

Name of the participant: ....

Laboratory: ....

PLEASE COMPLETE *ALL* DATA REQUESTED

Water content (%)

	1	2	3
Sample mass (g)			
% humidity			

Sample intake (g)

1	2	3	4	5

(if more than five replicates are taken, please complete on an additional form)

CALIBRATION

Reagents used:

- Chemical form: ....
- Water of crystallization/verification of stoichiometry, storage conditions (describe method):

  ....

  ....
- Dried:
- Purity:
- Solvent: .... purity: ....

In case of a ready made calibrant solution:

- Chemical form: .... in .... solution
- Concentration of final solution: ....
- Verified using: ....

Description of the method of calibration

* Calibration graph:

- Number of points measured: 1 ☐ 2 ☐ 3 ☐ 4 ☐ more ☐
- Concentration(s) of measured calibrant solutions:

  ....
- Matrix matching

  reagents used and their purity, concentration in final solution:

  ....

  ....

* Standard additions:

- Number of additions:
- Chemical form of element added: ....
- Amount of element added for each addition:

  ....

– Additions done before digestion ☐

after digestion ☐

* Other calibration methods:

– Specify: ....

....

– NAA: – Sandard(s) co-irradiated ☐

separately irradiated ☐

⟶ flux monitoring ☐

specify: ....

– Treatment after irradiation (dissolution, spiking . . .):

....

....

– Geometry for measurement (specify):

....

....

Linear range: from .... to ....

Sensitivity: ....

General

Indicate the measures taken to avoid contamination (*e.g.* clean room class . . . , clean benches . . .)

....

SAMPLE TREATMENT

Sample homogenization (if prescribed)

- Grinding and sieving ☐
  - Grinder (specify) ....
  - Particle size (sieve apertures) .... µm
- Mixing ☐
  - Manual shaking of the bottle ☐
  - Mechanical ☐
  - Other (specify) .... ☐
- Storage of rehomogenized sample
  - Under Ar ☐
  - In a dried desiccator ☐
  - Silica gel ☐
  - P_2O_5 ☐
  - Other (specify) .... ☐
  - Light (specify) .... ☐
  - Temperature .... °C

Chemical pretreatment

- pH adjustment
  - Addition of hydroxide: .... or acid: .... to pH =
  - Buffer composition: .... pH =
- Saponification ☐
- Others (specify): .... ☐
- Digestion/extraction ☐

  (if yes, proceed with p.6)

EXTRACTION PROCEDURE

Sample size .... mg

Extractant

- Acid (specify): .... ☐

  Quantity .... mg
- Solvent (specify) .... ☐

  Quantity .... mg
- Mixture (specify) .... ☐

  Quantity .... mg

Mixing during extraction

- Manual shaking ☐
- Ultrasonic ☐
- Mechanical shaking ☐
- Soxhlet ☐

Drying of the extract

- Silica gel ☐
- Forisil ☐
- Anhydrous Na_2SO_4 ☐
- Others (specify): .... ☐

Clean-up of the final extract ☐

Specify: ....

Pre-concentration of the compound(s)

- Evaporation ☐

  (specify if solvent added)

– Open flask evaporation ☐
– Adsorption/elution ☐
  * Final volume: .... mL
    Dryness ☐

Internal standard for extraction efficiency ☐

Compound ....

Quantity ....

Final content in the sample ....

Recovery experiments

Same matrix: .... ☐
Similar matrix (specify): .... ☐
Synthetic solutions (specify): .... ☐

Recovery experiment	% recovery of total amount
1	
2	
3	
4	
5	
mean	
S.D.	

Storage of the extract prior to analysis

– Glass or quartz ☐
– Polymer (specify) ☐
– Temperature (specify): .... °C
– Light conditions (specify): ....

MATRIX DIGESTION

General

Measures taken to avoid contamination (clean room class . . . , clean benches *etc.*):

....

Chemical blank: ....

Acid digestion

- Container used (open vessel, reflux app.): .... ☐
- Pressurized ☐
- Microwave ☐

  Programme applied (power): ....
- Procedure

	amount of acids (conc.)	purity (vol) =	mixing ratio (*)	temperature	duration
step 1	....	....	....	....	....
step 2	....	....	....	....	....
step 3	....	....	....	....	....
step 4	....	....	....	....	....
step 5	....	....	....	....	....

(*) Please indicate whether the temperature is measured in the heating block or in the digest

solution:..

Alkaline fusion

- Procedure

	reagent purity	temperature	duration
step 1	....	....	....
step 2	....	....	....
step 3	....	....	....

- Dissolution of the melt
  - Solvent: ....
  - Purity: ....
  - Procedure: ....

  ....

  ....

Programmed dry ashing

- Container used (type, size): ....
- Procedure

	temperature	duration
step 1	....	....
step 2	....	....
step 3	....	....
step 4	....	....

- Ashing aids: ....
- Dissolution of the ash
  - Solvent: ....
  - Purity: ....
  - Procedure: ....

  ....

  ....

Low temperature ashing oxygen plasma

- Container used: ....
- Power applied: ....
- Duration: ....
  - Solvent: ....
  - Purity: ....
  - Procedure: ....

Combustion

- Gas: .... flow: ....
- Other reagents used: .... purity: ....

Methods without digestion

- Addition of substance, *e.g.* for sample dilution: ....
- Pressed pellets ☐
- Other treatment (specify): ....

....

....

Treatment before final determination

- Sample dilution
  - Mass basis ☐
  - Volume basis ☐

  ⟶ check of glassware
- Sample preconcentration/purification
  - Extraction (specifications: see separate form) ☐
  - Ion exchange resins ☐

    Type: ....

    Column: ....

    Eluent(s): ....

  Describe method: ....

  ....

  ....
  - Complexation ☐

    Ligand: ....

  Describe method: ....

  ....

  ....

– Precipitation/filtration ☐

Describe method: ....

....

....

– Yield determination of those methods ☐

Other (specify):

....

....

Addition of an internal standard ☐

– Type: ....

Added before digestion ☐

Added after digestion ☐

Addition of non-radioactive carriers ☐

(NAA only)

Addition of spike (IDMS) ☐

Added before digestion ☐

Added after digestion ☐

Natural abundance of isotopes verified ☐

Addition of other reagents ☐

e.g. for ion strength, buffers:

....

....

BLANK DETERMINATIONS

- Cleaning of vessels:

  Describe method: ....

  ....

  ....

- Number of blank determinations (at least one per occasion of analysis):

- Reagents used: ....

  Procedure: ....

  ....

  ....

DERIVATIZATION

Derivatization procedure

- Hydride generation ($NaBH_4$) ☐

  Reagent concentration: .... purity: ....

- $SnCl_2$ ☐

  Reagent concentration: .... purity: ....

- Ethylation ($NaBEt_4$) ☐

Reagent concentration: .... purity: ....

- Grignard reagent (specify): ☐

....

Reagent concentration: .... purity: ....

- Media: addition of – CH_3COOH ☐

  – HCl ☐

  – HNO_3 ☐

- Carrier gas: .... flow: .... mL/min
- On-line treatment/injection ☐
- Separate derivatization ☐
- Concentration in final fraction: .... ng/mM
- Addition of complexing agent: ☐

  Specify:

- Addition of other agent (*e.g.* anti-foaming): ☐

  Specify: ....

- Verification of the derivatization yield: ☐

  Describe method: ....

  ....

  ....

SEPARATION

GAS CHROMATOGRAPHY

Column characteristics

–	Length:	....	m	
–	Glass column			☐
–	Fused silica column			☐
–	Internal diameter:	....	mm	
–	Stationary phase	....		
–	Chemically bound			☐
–	Load	....	%	
–	Film thickness	....	µm	
–	Support type	....		
–	Particle size	....	mesh	

Carrier gas type

– Flow .... mL/min

Make-up gas type

– Flow .... mL/min

Injector temperature .... °C

Detector temperature .... °C

Column temperature .... °C

GAS CHROMATOGRAPHY (continued)

Injections

- On-column injection ☐
- Splitless injection ☐
- Split closing: .... min
- Split injection ☐
- Split ratio ....
- Injected volume .... μL

Quantitation

- Peak height ☐
- Peak area ☐
- Manually ☐
- Electronically ☐
- Integration over whole chromatogram ☐
- Internal standard: ....
- External standard: ....
- Baseline correction Electronically ☐

  Manually ☐
- Calibration Linear range:

....

Bracketing standards

....

Identification

Specify and describe in detail on the attached sheets:

....

SEPARATION

HIGH PERFORMANCE LIQUID CHROMATOGRAPHY

Precolumn treatment ☐

– Column – sorbent (specify): ....

Analytical column

– Reversed phase ☐

– Adsorption ☐

– Type: ....

– Length: .... cm

– Internal diameter: .... mm

– Packing:

Type: ....

Particle size .... µm

Preconditioning: ....

Mobile phase

– Solvent (specify): ....

– Flow rate: .... mL/min

Programme

– Gradient details (isocratic, *etc.*):

– Maximum elution time: min

HIGH PERFORMANCE LIQUID CHROMATOGRAPHY (continued)

Quantitation

- Peak height ☐
- Peak area ☐
- Manually ☐ Electronically ☐
- Integration over whole chromatogram ☐
- Internal standard: ....
- External standard: ....
- Baseline correction Electronically ☐
  Manually ☐
- Calibration Linear range: ....
  Bracketing standards: ....

Identification

Specify and describe in detail on the attached sheets:

....

INDUCTIVELY COUPLED PLASMA ATOMIC EMISSION SPECTROMETRY (ICP-AES)

Plasma and sample introduction

– Nebulizer: Type pneumatic ultrasonic □
  Composition: glass □
  quartz □
  PTFE □
– Spray chamber: Type ....
  Composition: glass □
  quartz □
  PTFE □
– Sample consumption rate: .... mL/min
– Hydride generation: □
  (specifications: see separate form)
– Other pretreatment: specify .... □
– Torch: Type ....
  Observation height:
– Purge: vacuum □ argon □ N_2 □

Spectrometry

– Spectrometer: Simultaneous □
  Sequential □
– Line measured: .... nm resolution: ....
– Integration time: ....
– Background subtraction: □
  specify: ....
– Scanning possible: □
– Add if possible spectrum

INDUCTIVELY COUPLED PLASMA / MASS SPECTROMETRY + MASS SPECTROMETRY (ICP-MS and MS)

Plasma and sample introduction

– Nebulizer: Type .... ☐

Composition: glass ☐

quartz ☐

PTFE ☐

– Spray chamber: Type ....

Composition: glass ☐

quartz ☐

PTFE ☐

– Sample consumption rate: .... mL/min

– Sample pre-treatment (specify): ....

Mass spectrometry

– *m/e* measured: ☐

– Scanning: ....

Total measuring time: ....

Mass range: ....

– Peak hopping: .... ☐

Measuring time/channel: ....

Channels/peak: ....

– Interferences possible (specify): ....

Correction: .... ☐

– Blank value: ....

– Add if possible spectrum

ATOMIC ABSORPTION SPECTROMETRY (AAS)

Atomization	Flame	☐	Electrothermal	☐
	Cold vapour	☐	Other	☐

– Flame: gas mixture

– Hydride generation ☐ ☐

(specifications: see separate form)

– Other pretreatment: .... ☐

– Electrothermal:	Graphite	☐	Pyrocoated	☐
	Tube	☐	L'vov platform	☐

Other (specify): ....

Matrix modifier: ....

Injection volume: ....

Programme:	Drying	....	Temp	....	Time	....
	Ashing	....	Temp	....	Time	....
	Atomization	....	Temp	....	Time	....

Other steps (specify) ....

Background correction:

Deuterium	☐
Zeeman	☐
Smith–Hieftje	☐
Tungsten iodide	☐

Absorbance

– Peak height ☐

– Peak area ☐

– Line measured .... nm

– Add if possible time/absorbance/graph function

ION CHROMATOGRAPHY

IC Conditions

- Instrument/model: ....
- Detector type (specify): ....

  ....
- Conductivity detector: ....
- UV detection (give wavelengths) ....
- Others ....

Pre-column

- Stationary phase: ....

  ....
- Length: .... cm
- Internal diameter: .... mm
- Particle size (mesh or μm) ....

Analytical column

- Stationary phase: ....

  ....
- Length: .... cm
- Internal diameter: .... mm
- Particle size (mesh or μm) ....
- Mobile phase: ....
- Flow rate: .... mL/min

Suppressor column

- Stationary phase: ....

  ....
- Length: .... cm
- Internal diameter: .... mm
- Particle size (mesh or μm) ....
- Regeneration time and medium:

SPECTROMETRY

Give a description of the method used

....

....

....

....

....

....

....

....

....

....

....

....

....

....

....

....

....

....

....

ACTIVATION ANALYSIS (INAA, RNAA)

Irradiation

- Thermal neutrons ☐
- Epithermal neutrons ☐
- Fast neutrons ☐
- Photons ☐
- Charged particles ☐

Nuclear reaction(s): ....

....

Flux: ....

Flux gradient correction ☐

Irradiation time: ....

- Sample: Size ....

  Dimension ....

- Calibrants: Co-irradiated ☐

  Separately irradiated ☐

  Flux monitor: ....

  Composition: ....

- Shielding: ☐
- Radiochemical separation (specify) ☐

....

....

....

Counting

- Decay time: ....
- Counting time: ....
- X-ray counting: ☐
- Gamma counting: ☐

  Geometry ....

  Peak deconvolution ....

Element/matrix

RESULTS

No.	Final digest volume (mL)	Mean total signal	Mean net signal (total – blank)	Conc. in solution (μg mL^{-1})	Blank (mg/kg)	Reproducibility (%)	content in sample (μg kg^{-1})
1							
2							
3							
4							
5							

Subject Index

AAS, 11, 14, 15, 16, 17, 22, 44, 46, 77, 85, 87, 89, 104, 113, 114, 115, 120, 122, 143, 147
Accreditation, 26
Accuracy, 6, 12, 17, 21, 22, 23, 37, 111, 135, 145, 147
Acetic acid, 47, 72, 75, 77, 79, 80, 81, 88,106, 107, 109, 181, 182 197, 198, 200, 201, 202, 203, 204, 205, 207, 209, 219, 221, 222
Acetone, 45
Acid reactive aluminium, 173, 174
Adsorption, 18, 120, 143, 157, 162, 246
AFS, 46, 142
Air-drying, 62, 64, 85, 88, 89, 125, 185, 199
Al species, 172, 173, 175, 176, 177, 179, 180
Alkylated organotin derivatives, 76
Alkyllead compounds, 104, 105, 112, 114
Alumina column, 79, 80, 81, 111, 142
Aluminium, 172, 173, 174, 175, 176, 177, 178, 179, 180
Aluminium complexes, 177
Amberlite, 158, 159, 160, 161, 163, 168, 173
Ammonium acetate, 50, 82, 111, 159, 182, 184, 196, 197, 198, 208, 219
Ammonium nitrate, 205, 210, 222, 223
Anion exchange, 131, 132, 133, 143, 145, 158, 159, 160, 163
Antifouling, 69, 226
Aqua regia, 62, 65
Arsenic, 130, 131, 135, 136, 137, 138, 139
Arsenobetaine, 13, 15, 16, 130, 131, 132, 133, 134, 135, 136, 137, 138, 139
Arsenocholine, 13, 130, 133, 134, 135, 138
Arsenosugars, 130, 135
Assigned values, 37, 38, 87, 149, 154
As(III), 8, 130, 134, 138
As(V), 130, 134, 135, 138

Back-extraction, 45, 47, 48, 50, 57, 62, 63, 73, 74, 75, 109, 158, 160
Bacteria, 61, 62
Bar-graphs, 39, 59, 60, 64, 68, 84, 86, 88, 93, 99, 100, 101, 102, 114, 115, 117, 118, 119, 125, 128, 129, 139, 150, 151, 152, 153, 165, 166, 167, 170, 196, 197, 198, 205, 207, 208, 209, 212, 217, 218
BCR, 2, 4, 17, 18, 20, 31, 33, 34, 36, 37, 38, 41, 56, 88, 98, 104, 135, 136, 158, 165, 177, 184, 185
Benzene, 46, 97
Bioavailability, 181, 226
Borosilicate, 55
Bottling, 56, 65, 85, 88, 89, 90, 96, 116, 126, 136, 147, 148, 191, 199, 200, 213, 214
Buffer solution, 156, 158, 159, 160, 168
Butylation, 13, 47, 50, 108, 111
Butyltins, 69, 70, 71, 72, 74, 75, 77, 81, 88, 89, 95, 97, 98, 99

Calcium chloride, 184, 205, 209, 222, 223
Calibrants, 13, 14, 17, 18, 19, 21, 35, 36, 46, 47, 48, 49, 50, 53, 54, 55, 59, 66, 67, 70, 71, 72, 73, 74, 75, 76, 77, 78, 79, 80, 81, 97, 101, 105, 107, 108, 109, 110, 111, 112, 114, 115, 118, 123, 124, 130, 131, 133, 134, 137, 138, 141, 142, 143, 146, 147, 157, 158, 159, 186, 187, 188, 220, 242
Calibration, 17, 18, 20, 22, 25, 27, 28, 36, 38, 44, 46, 47, 48, 49, 50, 55, 57, 72, 73, 74, 75, 77, 78, 79, 80, 81, 82, 90, 106,

Calibration (*cont.*)
107, 108, 109, 110, 111, 115, 118, 124, 132, 135, 138, 141, 142, 143, 145, 148, 158, 159, 163, 188, 196, 197, 200, 210, 216, 242, 243
Calibration graph, 45, 47, 48, 49, 50, 73, 75, 79, 80, 81, 82, 106, 107, 110, 111, 141, 142, 242
Capillary GC, 15, 45, 46, 47, 48, 50, 53, 55, 57, 65, 67, 73, 74, 77, 78, 79, 80, 106, 107, 108, 110, 111, 142, 148
Capillary zone electrophoresis, 15, 16, 134, 137
Cartridge, 132, 133, 135
Cation exchange, 133, 173
CEN, 3
Centrifugation, 47, 65, 72, 75, 77, 78, 81, 82, 96, 163, 188, 203, 210, 218, 219, 220, 224, 225
Certification, 4, 5, 20, 24, 26, 27, 28, 35, 36, 37, 38, 41, 42, 43, 44, 55, 56, 59, 63, 64, 67, 70, 71, 75, 76, 82, 83, 84, 85, 87, 93, 94, 95, 99, 105, 106, 112, 120, 123, 124, 130, 131, 133, 136, 140, 141, 143, 146, 148, 149, 154, 156, 157, 160, 161, 162, 165, 166, 171, 179, 180, 182, 183, 184, 185, 190, 194, 197, 207, 212, 240, 241
Certified reference materials, 2, 10, 12, 21, 22, 23, 27, 28, 30, 31, 33, 34, 35, 37, 47, 48, 49, 50, 56, 57, 58, 59, 62, 64, 65, 66, 67, 74, 77, 82, 83, 86, 87, 89, 90, 93, 94, 95, 97, 98, 100, 101, 102, 103, 108, 121, 126, 139, 146, 147, 163, 164, 166, 168, 169, 170, 171, 191, 195, 201, 203, 204, 205, 206, 210, 213, 214, 227
Certified values, 38, 56, 61, 68, 93, 94, 103, 128, 139, 165, 169, 170, 171, 195, 204, 205, 216, 217
Chemical additives, 30
Chemical forms, 181, 240, 242
Chemical species, 4, 6, 8, 10, 13, 14, 18, 24, 61, 140, 226, 227
Chemiluminescence, 157, 164
Chelation, 142
Chromium, 155, 156, 157, 158, 161, 162, 163, 164, 165, 167, 168, 169, 170, 189
Clean-up, 12, 13, 14, 25, 38, 41, 47, 48, 51, 64, 74, 75, 78, 79, 80, 81, 97, 107, 110, 111, 131, 132, 133, 136, 245
Collection, 29, 56, 62, 64, 85, 89, 95, 125, 185, 190
Comparability, 7, 181, 227
Complexation, 13, 50, 63, 94, 108, 109, 110, 111, 161, 173, 174, 249
Conditioning, 41
Confidence interval, 20, 60
Contamination, 20, 34, 36, 96, 98, 104, 114, 135, 143, 167, 181, 188, 189, 191, 200, 213, 243
Control charts, 21
Council decisions, 9, 10, 69
Cr(III), 155, 156, 157, 158, 160, 161, 162, 163, 164, 165, 166, 167, 168, 169
Cr(VI), 8, 12, 155, 157, 158, 160, 161, 162, 163, 164, 165, 167, 168, 169, 170, 171
Cryogenic trapping, 15, 46, 62, 72, 77, 85, 88, 106, 109, 110, 132
CSV, 148
CVAAS, 48, 49, 51, 53, 57, 62, 63, 65, 67, 149
CVAFS, 47, 48, 49, 50, 63
Cysteine, 45, 47, 48, 54, 56

(Na)DDTC, 47, 63, 74, 94, 107, 108, 110, 111, 112, 121, 126, 158
Decomposition, 115, 116, 122, 123
Definitive methods, 35, 37
Degradation, 19, 61, 87, 88, 89, 91, 95, 97, 99, 101, 104, 112, 135, 138
Derivatization, 11, 13, 18, 22, 25, 47, 48, 49, 50, 63, 66, 67, 69, 72, 73, 74, 75, 76, 77, 78, 79, 80, 81, 82, 83, 87, 89, 95, 99, 101, 104, 106, 107, 108, 109, 110, 111, 115, 124, 131, 252
Detection, 7, 11, 16, 23, 24, 25, 38, 41, 46, 47, 48, 49, 50, 51, 67, 69, 72, 73, 74, 75, 77, 78, 79, 80, 81, 82, 83, 86, 87, 89, 104, 106, 107, 108, 109, 110, 111, 131, 132, 133, 136, 141, 142, 143, 157, 158, 159, 173, 174, 179, 193, 204, 212
Detection limits, 41, 161
Dialysis, 172, 173, 177
Dichloromethane, 75, 80, 82, 96
Diethyl ether, 74, 79, 80, 81, 107, 108, 110, 135
Digestion, 25, 28, 48, 57, 59, 62, 63, 77, 78, 80, 81, 112, 114, 137, 243, 244, 247, 250
Dimethylarsinic acid, 130, 132, 134, 135, 137, 138, 139

Dibutyltin, 12, 72, 73, 74, 75, 76, 77, 78, 79, 80, 81, 82, 87, 89, 90, 91, 92, 93, 94, 95, 96, 97, 98, 99, 100, 102, 103
Diphenyltin, 80, 82, 83, 96, 97, 98, 99, 103
Directives, 1, 2, 4, 8, 10, 69, 155, 226
Distillation, 47, 48, 49, 50, 59, 63, 64, 66, 67, 68, 76
DMSO, 51, 53
DPASV, 114, 117, 118, 124
DPCSV, 157
Drinking water, 155
Driscoll methods, 173, 174, 175, 177, 178, 179
Drying, 71, 76, 89, 99, 108, 191, 221, 245
DTPA, 181, 182, 208, 211, 212, 213, 214, 215, 216, 217, 218, 224

ECD, 11, 16, 17, 32, 41, 45, 47, 48, 51, 57
Electrode, 75, 114
Electrodeposition, 158, 160
Ethylmercury, 49
EDTA, 106, 107, 108, 109, 110, 112, 117, 119, 126, 181, 182, 184, 196, 197, 198, 200, 201, 202, 203, 204, 205, 207, 208, 210, 211, 212, 213, 214, 215, 216, 217, 221, 222, 224, 225
EDXRF, 138
Enzymatic digestion, 77, 82, 99, 138
(Z)ETAAS, 16, 46, 106, 112, 113, 114, 117, 127, 142, 158, 159, 161, 163, 168, 169, 173, 185, 186, 187, 188, 189, 190, 191, 201, 209, 210, 213, 221, 259
Ethanol, 46
Ethylation, 13, 14, 73, 77, 78, 79, 80, 81, 86, 104, 106, 109, 110, 142
Evaporation (rotary), 74, 75, 80, 82, 111, 126, 132, 133, 135
Exchangeable, 175
Extract, 25, 41, 42, 45, 46, 51, 52, 53, 54, 55, 65, 71, 74, 76, 77, 78, 80, 82, 114, 121, 126, 127, 130, 131, 132, 133, 134, 135, 136, 137, 204, 208, 211, 213, 220, 223, 224, 225, 245
Extractability, 31, 98, 193
Extractable trace elements, 7, 17, 181, 183, 184, 185, 186, 189, 192, 193, 194, 195, 196, 197, 200, 201, 204, 205, 209, 210, 211, 214, 215, 217, 219
Extraction, 7, 11, 12, 22, 25, 26, 28, 33, 38, 41, 45, 46, 47, 48, 49, 50, 54, 57, 61, 62, 64, 65, 67, 69, 72, 73, 74, 75, 76, 77, 78, 79, 80, 81, 82, 83, 85, 86, 88, 91, 94, 95, 99, 101, 104, 106, 107, 108, 109, 110, 111, 117, 118, 119, 120, 122, 126, 127, 131, 132, 133, 135, 136, 138, 158, 160, 161, 163, 173, 181, 185, 186, 187, 189, 190, 191, 193, 194, 196, 197, 200, 201, 210, 211, 212, 213, 216, 219, 220, 221, 222, 223, 224, 225, 244, 245, 246

FAAS, 159, 184, 185, 187, 188, 189, 190, 191, 203, 206, 209, 210, 213, 259
FID, 16, 32
Filter, 155, 158, 159, 160, 161, 162, 166, 167, 168, 169, 179
Filtration, 120, 123, 126, 132, 133, 138, 173, 179, 188, 203, 210, 211, 220, 223, 224
Fish, 42, 44, 52, 54, 55, 56, 130, 131, 134, 135, 136, 137, 178
Flow injection analysis, 174
Fluorescence, 176
Fluoride, 175, 176, 179
Fluoride electrode, 175, 176
Fluorimetric detection, 174
FPD, 11, 16, 32, 73, 74, 78, 79, 80, 83, 87, 99, 101, 148
Fluorimetry, 82, 83
Fractionation, 173, 174, 175, 178
Freeze-drying, 30, 56, 96, 116, 120, 126, 136, 162
Freezing, 56, 95, 96
FTIR, 46, 62

GC, 11, 13, 14, 15, 16, 18, 22, 32, 41, 45, 46, 47, 49, 50, 62, 63, 72, 77, 83, 85, 87, 97, 99, 101, 104, 106, 107, 109, 113, 114, 117, 119, 120, 121, 132, 133, 137, 142, 148, 227, 253
Glass, 18, 34, 53, 85, 88, 90, 116, 121, 125, 126, 132, 133, 136, 162, 199, 213, 218, 221
GLC, 15, 47
Grignard reactions, 13, 14, 18, 47, 50, 73, 74, 76, 97, 104, 108, 110, 111, 114, 121, 124, 126, 252
Grinding, 56, 85, 88, 89, 90, 96, 116, 125, 136, 185, 187, 244
Groundwater, 181

HAAS, 134
Harmonization, 1, 2, 3, 182
Heating, 72, 85, 88, 89, 146
Hexane, 67, 73, 74, 78, 79, 80,81, 97, 106, 107, 108, 110, 112, 113, 114, 123, 126, 127, 160
HICP, 134
Homogeneity, 28, 31, 32, 34, 42, 57, 58, 62, 65, 66, 70, 85, 90, 96, 98, 103, 105, 121, 122, 126, 127, 136, 137, 143, 147, 148, 162, 163, 168, 169, 182, 185, 191, 192, 196, 198, 199, 200, 213, 214
Homogenization, 30, 31, 56, 65, 88, 89, 125, 126, 136, 146, 147, 185, 191, 195, 196, 198, 199, 207, 213, 244
HPLC, 11, 14, 15, 16, 22, 36, 49, 50, 63, 67, 75, 82, 83, 87, 99, 101, 104, 111, 112, 115, 117, 124, 133, 136, 137, 141, 142, 143, 146, 147, 159, 172, 176, 255, 256
Hydride generation, 13, 18, 65, 72, 73, 77, 78, 83, 85, 86, 87, 88, 94, 95, 106, 115, 127, 131, 132, 137, 141, 142, 143, 147, 252
Hydrochloric acid, 44, 45, 46, 47, 48, 53, 54, 59, 73, 74, 75, 78, 81, 86, 96, 106, 107, 110, 120, 141, 142, 144, 146, 162, 222, 224
Hydrogen peroxide, 112, 113, 114, 159, 188, 219
Hydrolysis, 156

ICP-AES, 15, 16, 32, 82, 99, 131, 132, 134, 138, 184, 185, 186, 187, 188, 189, 190, 200, 209, 210, 257
ICPMS, 15, 16, 22, 32, 36, 49, 50, 75, 77, 78, 82, 83, 87, 99, 101, 108, 111, 112, 131, 133, 137, 142, 143, 145, 146, 148, 158, 159, 185, 186, 187, 188, 189, 190, 258
IDMS, 35, 36, 157, 158, 159
Improvement, 2, 94, 157, 158, 159
Improvement schemes, 20, 24, 25, 36, 130
Indicative values, 87, 94, 99, 194, 205, 209, 210, 222
Instability, 34, 55, 61, 86, 98, 116, 122, 137, 143, 144, 148, 150, 154, 201, 214
Interferences, 13, 20, 23, 56, 63, 64, 66, 86, 87, 115, 117, 134, 145, 186, 187, 188, 189
Interlaboratory studies, 4, 5, 11, 19, 20, 23, 24, 26, 32, 35, 36, 37, 38, 41, 42, 43, 44, 51, 52, 54, 61, 62, 69, 70, 71, 83, 84, 94, 104, 105, 106, 112, 114, 115, 116, 117, 118, 130, 131, 133, 134, 140, 141, 143, 145, 146, 149, 154, 156, 160, 161, 162, 177, 178, 179, 180, 181, 182, 183, 185, 186, 187, 188, 190, 198, 206, 207, 210, 241
Internal standard, 17, 45, 47, 48, 72, 73, 74, 78, 79, 80, 81, 96, 108, 109, 110, 111, 124, 246, 250
Ion chromatography, 112, 113, 114, 157, 158, 159, 164, 260
Ion exchange, 44, 51, 135, 168, 172, 174, 175
Ion selective electrode, 173, 175
Irradiation (gamma), 30, 61, 62, 63, 87, 88, 137, 243
ISO, 22, 23, 35, 37
Isotope, 108, 111
Isotope dilution, 83, 99, 111, 112

Karl Fischer titration, 85, 90

Labile aluminium, 174, 175, 177
LC, 15, 16, 131, 132, 133, 134
Leaching, 26, 34, 69, 75, 81, 158, 159, 160, 161, 162, 168, 169
Lead, 104, 105, 106, 111, 112, 113, 114, 115, 117, 119, 120, 121
Leakage, 149
Ligands, 176, 180
Losses, 20, 36, 55, 91, 114, 120, 132, 135, 144, 211
Lyophilization, 156, 157, 162, 164

Matrix matching, 141, 142, 146, 163, 242
Measurements and testing, 3, 18, 79, 80, 95, 104, 106, 107, 109, 110, 111, 124, 172, 181
Mechanical shaking, 72, 73, 78, 80, 82, 109, 132, 133, 244, 245
Mercury, 41, 44, 46, 47, 49, 50, 51, 52, 53, 55, 56, 57, 58, 61, 62, 63, 65, 66, 67, 68, 136
Methanol, 50, 73, 75, 77, 78, 79, 80, 81, 82, 83, 97, 99, 109, 110, 131, 132, 133, 135, 138
Methylation, 19

Methylmercury, 6, 8, 12, 14, 15, 41, 44, 45, 46, 47, 48, 49, 50, 51, 52, 53, 54, 55, 56, 57, 58, 59, 60, 61, 62, 63, 64, 65, 66, 67, 68, 226
MIBK, 158, 159, 160, 161, 163, 168, 173, 174
Microbial, 19, 19, 30, 88, 137
Microcolumn preconcentration, 142, 145, 147, 158
Microwave, 48, 57, 79, 81, 99, 133, 137, 141, 143, 247
Mineralization, 57, 65
MIP-AES, 13, 46, 47, 50, 64, 67, 80, 81, 87, 99, 107, 108, 110, 119, 142, 148
Mobility, 181, 216, 226
Moisture, 56, 72, 76, 85, 89, 90, 96, 109, 110, 111, 213
Monomeric aluminium, 173, 174, 175, 176, 177
Monomethylarsonic acid, 130, 134, 135, 138
Monobutyltin, 12, 19, 72, 73, 75, 76, 77, 78, 79, 80, 81, 82, 87, 89, 91, 94, 95, 96, 97, 98, 99, 100, 102, 103
Monophenyltin, 79, 80, 81, 82, 83, 96, 97, 98, 99, 103
MS, 11, 16, 32, 74, 75, 81, 87, 97, 108, 110, 111, 120, 133, 137
Mussel, 14, 52, 55, 56, 69, 70, 75, 77, 80, 83, 95, 96, 97, 98, 99, 101, 102, 103, 130, 131, 135, 136

NAA, 15, 32, 33, 36, 52, 55, 143, 169, 262
Nitric acid, 48, 57, 65, 112, 113, 143, 147, 158, 159, 200, 218, 219, 222, 223, 224
NMR, 76, 123, 133, 137, 173, 176, 177
Nonane, 74, 79
Norms, 1, 2, 155

Operationally-defined, 7, 28, 173, 178, 179, 181
Organic compounds, 86
Organometallic compounds, 4, 7, 19, 37
Outlying values, 39, 63, 145, 148
Oxidation, 134, 148, 161
Oxidation states, 4, 5, 7, 36

Packed GC, 15, 48, 51, 55, 107, 110
Pasteurization, 30
Pentane, 73, 74, 108, 111, 123
Pentylation, 13, 18, 73, 74, 75, 80, 81, 82, 97, 108, 111
Performance, 1, 2, 20, 22, 23, 24, 25, 26, 27, 28, 33, 36, 39, 41, 51, 69, 97, 157
Phenyltins, 69, 75, 81, 95, 97, 98, 99
Polarography, 75, 91
Polyethylene, 57, 64, 85, 88, 90, 96, 121, 143, 144, 146, 147, 151, 197, 198, 199, 218, 221, 222, 223, 225
Polypropylene, 150, 151, 153, 154, 218
Post-column reaction, 157, 159, 176
Precision, 55, 101, 111, 135
Preparation, 29, 42, 56, 61, 70, 85, 87, 89, 95, 112, 114, 120, 123, 125, 133, 146, 162, 166, 182, 184, 190, 197, 212
Proficiency testing, 2, 37
Propylation, 107, 108, 109, 110, 114, 126
PTFE, 18, 53, 55, 131, 132, 133, 143, 166, 168, 218, 221, 222, 223, 224
Purification, 76, 135, 136, 138
Purity, 17, 36, 53, 76, 105, 112, 118, 123, 124, 133, 137, 138, 242

QFAAS, 16, 48, 63, 72, 74, 77, 89, 99, 106, 107, 109, 110, 131, 132, 137, 141, 227
Quality, 2, 3, 4, 6, 8, 14, 35, 36, 37, 56, 124, 155, 226
Quality assurance, 2, 11, 16, 20, 21, 23, 26
Quality control, 4, 6, 16, 18, 21, 23, 26, 56, 69, 75, 95, 97, 99, 105, 115, 135, 143, 145, 154, 178, 179
Quartz ampoules, 156

Radioactive tracer, 12
Rainwater, 104, 105, 106, 108, 112, 115, 116, 117, 120, 121, 122, 123, 125
Reconstitution, 162, 163, 164
Recoveries, 11, 12, 25, 45, 47, 48, 49, 50, 59, 63, 64, 67, 72, 73, 76, 77, 78, 79, 81, 82, 85, 86, 91, 94, 95, 97, 99, 101, 102, 108, 109, 110, 111, 118, 119, 120, 122, 126, 127, 135, 136, 246
Reduction, 62, 65, 141, 143, 146, 164
Reference materials, 4, 8, 20, 21, 22, 23, 24, 26, 27, 28, 29, 30, 31, 32, 35, 36, 37, 42, 55, 56, 57, 62, 63, 70, 79, 85, 86, 91, 95, 98, 104, 105, 109, 112, 116, 120, 121, 122, 124, 125, 131, 135, 136, 149, 150, 154, 155, 156, 160, 162, 179, 180, 181, 182, 183, 184, 185, 190, 197, 206, 212

Reference method, 178
Reporting forms, 30
Representativeness, 28, 29, 87, 116
Reproducibility, 6, 34, 53, 115, 138
Resin, 175

Shaking, 47, 72, 74, 77, 78, 80, 82, 108, 111, 114, 119, 121, 126, 159, 186, 187, 188, 189, 190, 196, 203, 204, 219, 220, 221, 225
Sieving, 64, 65, 85, 89, 90, 96, 116, 125, 136, 185, 191, 198, 199, 207, 213, 244
Single extraction, 7, 181, 182, 195, 196, 206, 218
Sediment, 14, 17, 20, 29, 30, 34, 42, 47, 61, 62, 63, 64, 65, 69, 70, 71, 72, 73, 74, 75, 83, 84, 85, 86, 87, 89, 91, 92, 94, 95, 181, 182, 183, 184, 185, 187, 188, 190, 191, 195, 219, 220, 226
Selectivity, 14, 168
Selenate, 140, 141, 142, 143, 144, 145, 146, 147, 148, 149, 151, 152, 153
Selenite, 140, 141, 143, 144, 145, 146, 147, 148, 149, 150, 151, 152, 153
Selenium, 140, 141, 142, 143, 144, 146, 147, 149, 150, 151, 153, 154
Sensitivity, 27, 111, 147, 176, 208, 243
Separation, 11, 14, 23, 25, 35, 38, 41, 44, 46, 47, 48, 49, 50, 51, 54, 62, 63, 67, 69, 72, 73, 74, 75, 77, 78, 79, 80, 81, 82, 87, 104, 107, 108, 109, 110, 111, 117, 131, 132, 133, 134, 135, 136, 141, 158, 162, 163, 164, 168, 173, 174, 179, 255
Sequential extraction, 7, 181, 182, 184, 185, 195, 218, 220
Silica gel, 75, 97, 107
Sodium hydroxide, 46, 50, 73, 77, 78, 107, 131, 132, 159, 161
Sodium nitrate, 205, 210, 222, 224
Sodium sulfate, anhydrous, 53, 74, 81, 96, 108, 114, 121, 126, 127, 245
Sodium tetraethylborate, 47, 48, 49, 67, 73, 78, 79, 80, 81, 106, 107, 110, 119, 142
Sodium tetrahydroborate, 46, 48, 49, 50, 62, 65, 72, 73, 77, 78, 88, 106, 109, 131, 132, 141, 142, 143
Soil, 17, 20, 29, 30, 34, 134, 172, 181, 182, 183, 184, 195, 196, 197, 198, 199, 201, 202, 203, 204, 205, 206, 211, 213, 216, 221, 223, 226
Solutions (aqueous), 51, 52, 53, 69, 70, 71, 83, 105, 112, 114, 115, 116, 118, 119, 130, 134, 140, 141, 143, 144, 145, 146, 147, 154, 161, 162
Spectrometric detection, 157, 158, 159, 169
Spiking, 11, 12, 25, 28, 47, 48, 49, 50, 53, 55, 59, 63, 69, 71, 72, 76, 77, 78, 79, 80, 81, 82, 83, 84, 86, 87, 91, 97, 101, 107, 108, 109, 110, 111, 119, 120, 121, 122, 126, 128, 158, 160, 250
Stability, 18, 19, 28, 29, 32, 33, 34, 42, 51, 53, 55, 57, 58, 59, 61, 62, 66, 67, 70, 85, 86, 88, 89, 91, 92, 95, 96, 98, 99, 100, 103, 104, 105, 112, 114, 115, 116, 122, 123, 127, 128, 137, 140, 143, 144, 145, 148, 149, 150, 151, 154, 156, 163, 164, 169, 179, 180, 182, 183, 184, 185, 192, 193, 194, 201, 202, 203, 211, 214, 215
Stabilization, 30, 63, 87, 88, 145, 147, 154, 160
Stainless-steel, 34, 56, 166, 191
Standard additions, 11, 17, 18, 45, 46, 47, 48, 49, 50, 57, 67, 72, 73, 75, 77, 78, 79, 80, 81, 82, 90, 95, 106, 107, 108, 109, 110, 111, 115, 118, 189, 210, 212, 242
Standard methods, 2, 4, 7, 179, 208
Standardized, 12, 22, 28, 38, 39, 67, 178, 187
Statistical control, 21, 26, 30, 37
Statistical evaluation, 39, 40, 61, 94, 124
Sterilization, 85, 89, 90, 207
Stoichiometry, 27, 28, 36, 76, 105, 112, 242
Storage, 33, 34, 53, 56, 57, 65, 66, 71, 85, 86, 89, 90, 91, 95, 96, 98, 112, 116, 122, 125, 127, 132, 133, 134, 140, 144, 145, 147, 150, 151, 162, 163, 164, 169, 179, 180, 193, 201, 211, 213, 214, 242, 244
Sulfuric acid, 46, 47, 48, 49, 50, 62, 65, 74, 108, 121, 126, 131, 132, 143, 144, 169
Supercritical fluid extraction, 12, 50, 64, 67, 73, 79, 83, 101, 110
Systematic errors, 20, 23, 24, 25, 26, 39, 68, 77, 84, 99, 115, 212

Technical evaluation, 38, 39, 40, 59, 61, 66, 71, 86, 91, 94, 99, 118, 124, 127, 138, 148, 160, 164, 169, 186, 193, 201, 211, 216

Thiosulfate, 46, 48, 63, 65
Tin, 69, 71, 89, 91, 97, 98
TMAH, 77, 80, 81
Toluene, 45, 46, 47, 48, 50, 51, 53, 54, 55, 57, 62, 63, 65, 73, 75, 123
Toxicity, 172, 178, 179, 180, 181, 226
Trace elements, 4, 13, 37, 86, 136, 181, 184, 185, 191, 199, 200, 201, 205, 207, 209, 213, 215, 216, 226
Traceability, 18, 22, 24, 97
Transport, 84, 85, 89, 90, 91, 122, 197
Tributyltin, 6, 7, 8, 10, 12, 16, 18, 19, 34, 69, 70, 71, 73, 75, 76, 77, 78, 79, 80, 81, 82, 83, 84, 85, 86, 87, 88, 89, 90, 91, 92, 93, 94, 95, 96, 97, 98, 99, 100, 101, 103, 226
Triethyllead, 104, 112, 113, 114, 115, 117
Triethyltin, 72
Trimethyllead, 104, 106, 107, 108, 109, 110, 111, 112, 113, 114, 115, 116, 117, 118, 119, 120, 121, 122, 123, 124, 125, 126, 127, 128, 129
Triphenyltin, 10, 13, 15, 69, 78, 79, 80, 81, 82, 83, 96, 97, 98, 99, 101, 103
Tripropyltin, 73, 78, 81, 96
Treatment, 30, 58, 62, 244
Tropolone, 73, 74, 75, 78, 80, 81, 82, 96

Ultrafiltration, 175
Ultrasonic, 74, 77, 78, 109, 126, 127, 131, 132, 133, 138, 158, 159, 245
Uncertainty, 20, 31, 33, 35, 36, 57, 62, 68, 91, 94, 98, 99, 103, 121, 122, 137, 149, 163, 165, 168, 171, 192, 193, 195, 214
Urban dust, 104, 105, 106, 109, 110, 111, 112, 116, 119, 120, 125, 126, 127, 128, 129
UV digestion, 157
UV irradiation, 13, 44, 50, 64, 67, 112, 131, 132

Validation, 12, 20, 177, 180
Volatilization, 134, 135, 136
Voltammetry, 16, 36

Water, 20, 29, 33, 155, 160, 172, 178, 179, 180
Welding dust, 155, 158, 161, 166, 167, 168, 169
Westöö extraction, 45

XRF, 15, 22, 32, 136, 185, 191, 199, 213

Yields, 13, 14, 36, 76, 80, 86, 95, 112, 138
Youden plot, 39, 51, 53, 54

Z-scores, 39